Ulrich Schubert, Nicola Hüsing

Synthesis of
Inorganic Materials

WILEY-VCH

Further Titles of Interest:

H.W. Roesky, K. Möckel
Chemical Curiosities
Spectacular Experiments and Inspired Quotes

ISBN 3-527-29414-7

R. Faust, G. Knaus, U. Siemeling
H.-J. Quadbeck-Seeger (Ed.)
World Records in Chemistry

ISBN 3-527-29574-7

C. Elschenbroich, A. Salzer
Organometallics
Second Edition

ISBN 3-527-28164-9

U. Diederichsen, T.K. Lindhorst, B. Westermann,
L.A. Wessjohann
Bioorganic Chemistry

ISBN 3-527-29665-4

Ulrich Schubert, Nicola Hüsing

Synthesis of Inorganic Materials

WILEY-VCH

Weinheim · New York · Chichester · Brisbane · Singapore · Toronto

Prof. Dr. U. Schubert
Dr. N. Hüsing
Institute of Inorganic Chemistry
Vienna University of Technology
Getreidemarkt 9/153
A-1060 Wien
Austria

Library of Congress Card No. applied for

A catalogue record for this book is available from the British Library

Die Deutsche Bibliothek – CIP-Cataloguing-in-Publication Data:
A catalogue record for this publication is available from Die Deutsche Bibliothek

© WILEY-VCH Verlag GmbH, D-69469 Weinheim (Federal Republic of Germany), 2000

ISBN 3-527-29550-X

Printed on acid-free paper.

Composition: Kühn & Weyh, D-79111 Freiburg
Printing: Strauss Offsetdruck, D-69509 Mörlenbach
Bookbinding: Osswald & Co., D-67433 Neustadt (Weinstraße)
Printed in the Federal Republic of Germany.

Foreword

Within a textbook, there are several possible ways in which materials may be treated, according to:

- their chemical composition (organic polymers, metals, oxides, nitrides, carbides, etc.);
- their physical state (ceramics, glass, composites, polymers, etc.);
- their properties and applications (electronic materials, magnetic materials, optical materials, etc.);
- technological aspects (powder preparation, sintering methods, preparation of films or coatings, etc.); and
- the methods for their preparation (solid state reactions, polycondensations, gas phase reactions, etc.).

In most materials science and solid state chemistry textbooks – even those that are highly recommended ones – there is a regrettable lack of chemical information. Materials science is often reduced to physical and technological aspects, and chemistry is only introduced to discuss bonding and to describe structures. Processes for the preparation and modification of materials – the most important contribution of chemistry to material science – are mostly treated just in passing – or perhaps not at all!

With this textbook we are attempting to fill this gap. The book is not intended as a substitute for existing, physically or technologically oriented textbooks on materials science, but rather to complement chemical aspects. The nucleus of this book was a lecture course on "Inorganic materials from molecular precursors" given by the authors at the Vienna University of Technology. The selection of suitable precursors and the development of correct conditions to obtain a product with the desired composition and properties and a suitable (micro-) structure is what chemists can contribute to materials science.

Nevertheless, this textbook is intended for use not only by chemistry students (for whom, we have tried to keep the number of physical formulae at a minimum), but also by physics and materials science students (for whom, we have tried to keep the required chemical prerequisites at a minimum). The glossary at the end of the book may help to bridge the gap between chemical/physical/materials science fundamentals.

In an up-to-date textbook, SI units should be used exclusively. However, different scientific communities still have their own habits concerning physical units (e.g., ceramists prefer Pascal, while CVD people prefer bar or atmosphere as the pressure unit). We therefore decided to leave some transformations of physical units to the reader – with the help of a table at the inner cover of the book.

Since the main focus of this book is on syntheses, we did not treat all major inorganic materials comprehensively, neither did we discuss naturally occurring materials such as lime, asbestos, gypsum, etc. Instead, some materials were selected as examples to discuss the ways in which (natural or artificial) chemical compounds are transformed into materials. For the same reason, materials properties and technological aspects are discussed only exemplarily.

The book is organized according to preparation processes. Since many materials can be prepared using several methods, this organization inevitably has the consequence that some materials are treated in more than one chapter. For example, perovskites can be prepared by solid state reactions, by sol–gel processing, by hydrothermal processes, or by CVD, and is therefore discussed in different chapters.

One difficulty we were facing was to avoid writing a textbook on preparative inorganic chemistry, i.e., to discuss the preparation of inorganic *materials* instead of inorganic *compounds*. In fact, almost any chemical compound is a potential "material", and therefore the distinction is not always obvious. Not being too conservative, we considered materials as compounds that are used technically, or which have the potential for being used. In order to stay as close as possible to the real world of material science, we have tried to introduce a relevant technically applied material in most sub-chapters. The properties and uses of this material are discussed exemplarily.

At the end of each chapter, a selection of more recent books and review articles is provided, which may help the reader to study those processes under discussion in greater detail.

Acknowledgements

We thank Wiley-VCH for encouraging us to start this project, in particular Dr. Gudrun Walter, who did not give up hope that this book would eventually be published.

Harald Schauer made invaluable contributions to this book with his careful and patient drawing of the majority of figures.

We thank the following companies and institutions who have provided us with information on, and photographs of, their products that allowed us to illustrate the technical relevance of the methods discussed:

- *Agfa-Gevaert*, Leverkusen (Dr. Wolfgang Schmidt), on the preparation of silver halides for photographic emulsions (Figure 4-15),
- *Austria Microsysteme Intl.*, Unterpremstätten (Dr. Holger Wille), on CVD processes for microelectronics (Figure 3-15),
- *Degussa*, Hanau, on the Aerosil process (Figures 3-26, 3-27, and 3-28),
- *ESK*, Kempten (Dr. Karl A. Schwetz), on the preparation of silicon carbide by carbothermal reduction. In particular, we wish to point out that the cover of this book was composed from photographs provided by ESK.
- *Fraunhofer-Institut für Silicatforschung*, Würzburg (Dieter Sporn and Dr. Klaus Rose) on sol–gel materials (Figures 4-43 to 4-45, 4-65, and 4-67),
- *Hoechst AG*, Frankfurt, on aerogels (Figure 6-15),
- *Institut für Anorganische Technologie, Vienna University of Technology* (Prof. Dr. Benno Lux and Dr. Roland Haubner) on diamond films (Figure 3-20),
- *Philips Forschungslaboratorien*, Aachen (Dr. Ulrich Nieman), on halogen lamps (Figure 3-3),
- *Wacker-Chemie*, Burghausen (Dr. Johann Weis), on silicones (Figures 5-4 and 5-5).

The following publishers transferred copyrights for the reproduction of figures from their publications:

Academic Press: C. J. Brinker, G. W. Scherer, *Sol–Gel Science: The Physics and Chemistry of Sol-Gel-Processing*: Figures 4-17, 4-42, 4-47 and 4-52.
W. W. Porterfield, *Inorganic Chemistry – A Unified Approach*: Figure 7-11.
American Chemical Society, Chemistry of Materials: Figure 2-22 (*10*, **1998**, 2898), Figure 2-26 (*8*, **1996**, 1752), Figure 4-18 (*4*, **1992**, 549);
Advances in Chemistry, Series: Figure 4-29 (*245*, **1995**, 535).

Chapman & Hall: A. W. Weimer (Ed.), *Carbide, Nitride and Boride Materials - Synthesis and Processing*: Figures 2-9, 2-18 and 3-5.

Elsevier: Aerosol Science and Technology: Figure 3-31 and 3-32 (*19*, **1993**, 434), Figure 7-9 (*19*, **1993**, 416);
Endeavour: Figure 4-38 and 4-39 (**1969**, 114);
Microelectronic Engineering: Figure 4-43 (*29*, **1995**, 161);
Thermochimica Acta: Figure 3-4 (*299*, **1997**, 50).

Gordon and Breach Publishers: B. Marciniec, J. Chojnowski (Eds.), *Progress in Organosilicon Chemistry*: Figure 5-6.

Harry Deutsch Verlag: K. T. Wilke, *Kristallzüchtung*: Figure 4-37.

Materials Research Society (MRS Bulletin): Figure 3-6 (*18*(6), 1993, 25), Figures 6-1 and 6-11 (*14*(4) (1994) 24), Figure 6-4 (*14*(4) (1994) 15) .

Nature Publishing Group (Nature): Figure 5-20 (*402*, **1999**, 278).

Pergamon Press: C. Suryanarayana, *Non-Equilibrium Processing of Materials*: Figure 4-11, 4-12 and 4-13.
Current Opinion in Solid State & Materials Science: Figure 3-12 (*3*, **1998**, 147), Figure 6-29 (*1*, **1996**, 798).
Progress in Materials Science: Figure 2-10 (*37*, **1993**, 123).
Progress in Solid State Chemistry: Figure 2-19 (*19*, **1989**,.28);

Kluwer: Journal of Materials Science: Figure 6-6 (*18*, **1983**, 1899).

The Royal Society of Chemistry: J. E. Shelby, *Introduction to Glass Science and Technology*: Figure 4-1.

Springer: Marine Biology: Figure 4-21 (*19*, **1973**, 323).

van Nostrand Reinhold: R. Szostak, *Molecular Sieves – Principles of Synthesis and Identification*: Catalysis Series, **1989**, Figures 6-21, 6-23, 6-24, 6-25 and 6-26.

VEB-Verlag Leipzig: A. Petzold, *Anorganisch-nichtmetallische Werkstoffe*: Figure 4-6.

Vulkan-Verlag: Silicone: Chemie und Technologie: Figure 5-8.

W. B. Saunders Company: Orthopedic Clinics of North America: Figure 4-28 (*20*, 1989, 1).

Wiley/Wiley-VCH: Advanced Materials: Figures 4-33 and 4-34 (*7*, **1995**, 609), Figure 6-17 (*5*, **1993**, 127);
Angewandte Chemie: Figure 2-27 (*107*, **1995**, 2515), Figures 3-29 and 3-30 (*101*, **1989**, 815), Figure 5-12 (*101*, **1989**, 1773);
D. M. Adams, *Inorganic Solids:* Figure 2-21;
D. W. Bruce, D. O'Hare (Eds.), *Inorganic Materials:* Figure 4-24;
Chemie in unserer Zeit: Figure 4-22 (*33*, **1999**, 11), Figure 6-22 (*42*, **1995**, 1);
Encyclopedia of Chemical Technology, Kirk-Othmer: Figure 4-40;
R. K. Iler, *The Chemistry of Silica*: Figure 4-51;
L. V. Interrante, M. Hampden-Smith (Eds.), *Chemistry of Advanced Materials – an Overview*: Figures 3-14 and 3-21, Table 4-2;

T. Kodas, M. Hampden-Smith (Eds.), *The Chemistry of Metal CVD*: Figures
3-16 and 3-18;
S. Mann (Ed), *Biomimetic Materials Chemistry*: Figure 4-25, Figure 4-26,
Figure 4-27 and 4-32;
Nachrichten aus Chemie, Technik und Laboratorium: Figure 4-30 (*47*, **1999**,
1405);
Ullmann's Encyclopedia of Technical Chemistry: Figure 4-41;
M. T. Weller, *Anorganische Materialien*: Figure 2-2;
J. Zarzycki (Ed.), *Materials Science and Technology*: Figures 4-7 and 4-10;
Zeitschrift für Anorganische und Allgemeine Chemie: Figure 2-20 (*277*,
1954, 156).

The following figures were taken from internet pages:

Figure 2-3: www.spectraweb.ch/~hwiehl/supra.html
Figure 2-8: www.ifm.liu.se/Matephys/new_page/research/sic/Chapter3.html
Figure 2-11:
www.benchmarkceramics.com/Technologies/CCS/techinfo.html
Figure 2-12:
www.railway-technology.com/contractors/track/elektro/elektro1.html
Figure 4-19: www.physics.ucsb.edu/~bettye/research/biomin/biomin.html
Figure 4-20: www.indiana.edu/~diatom/secret.html
Figure 4-23: www.soton.ac.uk/~serg/biotech/mtb-main.htm
Figure 4-31: mmg.sechrest.com/mmg/thr/index.html
Figure 4-36: www.tewna.com/crystal/mcf/e_filter01.html and
www.vanlong.com/Prodindex/osc/oscmain.html
Figure 4-53: www.metu.edu.tr/home/www70/who/ctas/lab/cordi.htm
Figure 6-5: indigo4.gi.rwth-aachen.de/flyer/feinguss/omg_e.htm
Figure 6-16: stardust.jpl.nasa.gov/spacecraft/capsule.html
Figure 6-18:
ruby.chemie.uni-freiburg.de/Vorlesung/Gif_bilder/Silicate/zsm5_a.gif
Figure 7-2:
www.phys.ttu.edu/~tlmde/thesis/CARBON_NANOTUBES.html

Table of Contents

Abbreviations

AACVD	aerosol-assisted chemical vapor deposition
a.c.	alternating current
Ac	acetyl
acac	acetylacetonate = 2,4-pentanedionate
AFM	atomic force microscope (atomic force microscopy)
AIBN	azobis(isobutyronitrile)
ALE	atomic layer epitaxy
aq	aqueous
Ar	aryl
a.u.	arbitrary units
b.p.	boiling point
Bu	butyl
C	critical point
cat	catalyst
CBE	chemical beam epitaxy
CD	compact disc
CDJP	controlled double jet precipitation
cmc	critical micelle concentration
CMC	ceramic matrix composite
CMR	colossal magnetoresistance
COD	cyclcooctadiene
Cp	cyclopentadienyl
CTAB	cetyltrimethylammonium bromide
CVD	chemical vapor deposition
CVI	chemical vapor infiltration
D	dimensional
d.c.	direct current
DMF	dimethylformamide
DMSO	dimethylsulfoxide
DRAM	dynamic random access memory
diglyme	ethyleneglycol dimethylether
diphos	1,2-bis(diphenylphosphino)ethane
dpm	dipivaloylmethanate (= thd or tmhd)
dppp	1,3-bis(diphenylphosphino)propane
EDTA	(ethylenedinitrilo)tetraacetic acid
Eq	equation
Et	ethyl
G	free energy
g	gaseous
GMR	giant magnetoresistance

HA	hydroxylapatite
Hex	hexyl
hfac	1,1,1,5,5,5-hexafluoroacetylacetonate (= 1,1,1,5,5,5-hexafluoro-2,4-pentanedionate)
HIP	hot isostatic pressing
HTV	high-temperature vulcanizing
IEP	isoelectric point
IR	infrared
L	ligand; or Lewis base
l	liquid
LB	Langmuir–Blodgett (technique)
LC	liquid crystal; or liquid crystalline
LCVD	laser-assisted or laser-induced CVD
LED	light-emitting diode
Ln	lanthanoid
LPCVD	low-pressure CVD
LPS	liquid phase sintering
LR	liquid rubber
M	molar; or metal
MBE	molecular beam epitaxy
MCM	Mobil composition of matter
Me	methyl
MLE	molecular layer epitaxy
MMC	metal matrix composite
MO...	metal organic ...
NLO	non-linear optic
OAc	acetate
OM...	organometallic ...
p	para; or pressure
p_c	critical pressure
PACVD	plasma-assisted CVD
PE	polyethylene
PECVD	plasma-enhanced CVD
Ph	phenyl
phen	phenanthroline
PMMA	poly(methylmethacrylate)
PMC	polymer matrix composite
POSS	polyhedral oligomeric silsesquioxane
Pr	propyl
PTFE	poly(ethyleneterephthalate)
PZC	point of zero charge

PZT	lead zirconate titanate
PVD	physical vapor deposition
py	pyridine
R	organic group
RAM	random access memory
r.f.	radio frequency
RHEED	reflection high-energy electron diffraction
ROP	ring-opening polymerization
RPCVD	remote-plasma chemical vapor deposition
RTV	room-temperature vulcanizing
s	solid
SAM	self-assembled monolayers
SAW	surface acoustic wave
SBU	secondary building unit
SCF	supercritical fluid
sec	secondary
SHS	self-propagating high-temperature synthesis
SIMIT	size-induced metal-insulator transition
SSM	solid-state metathesis
STM	scanning tunneling microscope
t_{gel}	gel time
T_{ad}	adiabatic temperature
T_c	critical temperature
T_g	glass transition (glass transformation) temperature
T_m	melting temperature
TBA	*tert.* butylarsine
TEM	transmission electron microscopy
TEOS	tetraethoxysilane (tetraethylorthosilicate)
tert	tertiary
thd	tetramethylheptanedionate (= tmhd or dpm)
THF	tetrahydrofuran
tmhd	tetramethylheptanedionate (= thd or dpm)
TMOS	tetramethoxysilane (tetramethylorthosilicate)
Tr	triple point
TTT curve	time–temperature-transformation curve
UHV	ultra-high vacuum
UV	ultraviolet
Vi	vinyl
VPE	vapor phase epitaxy
VTMS	vinyltrimethylsilane
wt	weight
XRD	X-ray diffraction
YBCO	yttrium barium copper oxide (high-temperature superconductor)

1 Introduction

Although efforts to make new materials, or to improve existing materials, are as old as mankind, only in our age has "materials science" matured into a unique area. In the earliest ages of civilization, materials were made using a "learning by doing" approach. For example, the first piece of iron – which does not occur in elemental form in nature – was most probably made by chance, when an iron ore and a carbon source were accidentally heated together. From today's point of view, the properties of this "new material" would be regarded as minimal, but at the time of its discovery, this was a quantum leap for mankind. Over the millennia, the iron and steel-making process has been advanced to a highly sophisticated state, as has the quality of the materials produced. The manufacture of glass and ceramics has developed in a similar manner.

The improved properties and better performance of the traditional materials were achieved by advances in materials technology. However, this was only made possible through a better understanding of the underlying basic chemical and physical principles. The route from the first primitive man-made piece of glass to the highest-quality glass currently used in the lenses of large telescopes, for example, was paved by a detailed understanding of the glass chemistry, the structure of glass, crystallization phenomena, and the physics of melts.

During the twentieth century, materials science has led to major changes in the way that we live. For example, only fifteen years ago the manuscript of this book would have been written on a (probably portable) typewriter after making time-consuming searches in several scientific libraries, the figures would have been drawn with ink, and the book would have been eventually produced after time-consuming cut-and-paste corrections, typesetting by the printer and photographic reproduction of the figures. Instead, writing of the book and drawing of the figures was done on a (rather voluminous) personal computer in the early stages and completed on two small (again portable) laptops, with much of the literature research having been done using electronic databases and by the internet. Editing and revision of the manuscript was again carried out on computers, and eventually the "camera-ready" manuscript was delivered electronically to the publishing company. This major change in how we worked was only made possible by the dramatic advances in information technology and data storage capacities that have been developed in a rather short period. As in many other areas, the rate of progress was, and is, determined by the speed at which materials are newly developed or improved, and by advances in their processing. Further development in information technology will to a large degree depend on how materials science allows further miniaturization, and on the development of materials capable of optical or optoelectronic data transfer.

In our generation, many new materials have been discovered with properties that were beyond imagination, such as the high-temperature superconductors, "intelligent" and "adaptive" materials, or nanostructured materials. Furthermore, the demand for longer product life-times, higher quality and efficiency, etc. has also placed new challenges on the synthesis and processing methods of known materials. Major changes in materials production technologies were induced by the impact of environmental issues ("sustainable development", ▶ glossary). An increasingly important goal for materials developments is the minimization of energy and raw materials consumption. For example, materials for moving components in motors or gas turbines with better thermomechanical properties or a lower mass improve the efficiency of the motor or turbine and, as a consequence, save fuel and reduce the emission of waste gases.

Many of the scientific and technological findings that have been discovered in a variety of laboratories, together with the worldwide increase in knowledge, form the basis for new technologies in the 21st century. Several trends are apparent:

- Materials become increasingly specialized and multi-functional. In contrast to many materials that have a broad range of applications, many modern materials are tailor-made for only one special application.
- It is realized that the transition between molecules and solids results in materials with unique physical or chemical properties. The so-called nanomaterials (see Chapter 7) with a characteristic dimension in the lower nanometer range have the potential of revolutionizing materials design for many applications.
- The boundary between "natural" and "artificial" is beginning to disappear. Material science is beginning to learn from nature ("biomimetic" approach; see Section 4.3.3).
- Theoretical methods have matured in a way that they are able to predict materials properties. A visionary example is carbon nitride (C_3N_4), which is reputed to be harder than diamond and to have interesting applications as a high-temperature semiconductor. However, this prediction has still to be verified by the synthesis of this compound.

Although modern materials science is interdisciplinary, and more than a blend of some established communities, such as chemistry, physics or engineering, the experience from these disciplines is essential to carry the development of some material from the raw materials or precursors to the final application.

The triangle *Synthesis and Processing – Composition and Structure – Properties and Performance* represents the essential relations in material science. The properties of a material of a given composition depend to a very high degree on the way that it was made or processed, this being a consequence of different structures (on any length scale). Vice versa, certain applications require certain structural features and certain chemical compositions of the employed materials, which again requires the deliberate design or modification of the synthesis and processing procedures.

For example, SiO_2 can be crystallized as quartz (for oscillator crystals, for example) by hydrothermal treatment (see Section 4.4). SiO_2 as an insulating layer in a microelectronic device would be made by chemical vapor deposition (see Chapter 3.2). Silica with a high surface area, used as adsorbent or for thermally insulating materials, for example, is produced either by the aerosol process, where agglomerated spherical, amorphous particles are obtained (see Section 3.3) or as aerogels with a highly porous network structure via sol–gel processing (see Section 6.3). Sol–gel processing also allows the preparation of amorphous SiO_2 powders or dense films. Although the composition of the obtained material is SiO_2 in each case, completely different routes of preparation are required.

Materials are traditionally grouped in the following classes:

- Natural materials (wood, bone, leather, cellulose, minerals, etc.)
- Metals
- Semiconductors
- Non-metallic inorganic materials (oxide ceramics, non-oxide ceramics, glass)
- Organic polymers
- Composite materials (polymer matrix, ceramic matrix, metal matrix)

This classification is based on the fundamental properties and somehow also reflects the different structural and bonding features. In this book, we wish to emphasize the chemical aspects of materials science. Chemistry is the science of transforming substances into other substances on a molecular level, and of investigating the concomitant changes in composition, structure and properties. We therefore do not follow the traditional classification of materials, but instead structure the book according to methods of how substances can be transformed.

When does an inorganic compound qualify as a "material", in other words, how does this book differ from one on preparative inorganic chemistry? We do not attempt to define the term "material", as many chemical compounds are potential materials, and thus the distinction is not always obvious. Not being too conservative, we considered materials to be compounds that are utilized for some technical application, or which have at least the potential for being used in such a manner. We will, as examples, select several types of materials in order to discuss the ways in which (natural or artificial) chemical compounds are transformed into materials, and show both the options and problems that originate from the various preparation methods, independently of a particular chemical composition.

2 Solid-State Reactions

In this chapter, reactions will be discussed in which at least one of the reactants is in the solid state. A large variety of inorganic solids has been prepared by reacting a solid with another solid, a liquid (melt) or a gas, usually at high temperatures. We will first deal with reactions between two (or more) solid compounds (Section 2.1) followed by a section covering reactions between a solid and a gas (Section 2.2). In Section 2.3, thermal decomposition reactions will be treated in which also a solid and a gas phase are involved. However, the gas is a product rather than a reactant. No single section is devoted to solid/liquid reactions, because in many reactions which start with solid compounds, a liquid phase (melt) is formed at the reaction temperature, i.e., many "solid/solid" reactions are actually "solid/liquid" reactions. It is sometimes difficult to determine what physical phases are involved in a given reaction. The final section (2.4) describes intercalation reactions, in which guest species (atoms, molecules or ions) are inserted into a crystalline host lattice. Intercalation can occur with compounds of any physical state.

2.1 Reactions Between Solid Compounds

When solid compounds are employed to react with each other at high temperatures, this does not necessarily imply that all components are still in the solid state at the temperatures required for the reaction to occur. A liquid phase (melt) or even gaseous intermediates may be involved to provide mass transport. We will first discuss some general principles of solid-state reactions, which are often called the "ceramic method" (Section 2.1.1) and then turn to two important industrial processes, carbothermal reduction (Section 2.1.2), and combustion synthesis (Section 2.1.3). Fundamental processes during sintering, by which powders are converted to dense solid bodies, are dealt with in Section 2.1.4.

2.1.1 Ceramic Method

The oldest and still most common method of preparing multicomponent solid materials is by direct reaction of solid components at high temperatures. Since solids do not react with each other at room temperature – even if thermodynamics favors product formation – high temperatures are necessary to achieve appreciable reaction rates. The advantage of solid-state reactions is the ready availability of the precursors and the low cost for powder production on the industrial scale. We will not cover all the classes of compounds that can be prepared by solid–solid reactions. Rather, we will discuss some fundamental issues of this method after an introductory example.

An example

The high-temperature superconductor (▶ glossary) $YBa_2Cu_3O_{7-x}$ (YBCO) can be prepared by heating an intimate mixture of yttrium oxide (Y_2O_3), barium peroxide (BaO_2), and cupric oxide (CuO). The starting compounds should be fine-grained in order to maximize surface areas and hence reaction rates (see below). The powder is pressed into a pellet to ensure an intimate contact between the grains (▶ glossary). The pellets are placed in an alumina boat and heated in a furnace to 930 °C over a period of 8–12 h, held at this temperature for 12–16 h, allowed to cool to 500 °C and held there for 12–16 h. After heating to 930 °C, the composition is about $YBa_2Cu_3O_{6.5}$. By annealing the material to 500 °C, it reacts further with oxygen from the air, and the final composition is about $YBa_2Cu_3O_{6.9}$. The heating–cooling sequence is shown graphically in Figure 2-1. The oxygen content x in $YBa_2Cu_3O_{7-x}$ depends very much on the oxygen partial pressure during annealing.

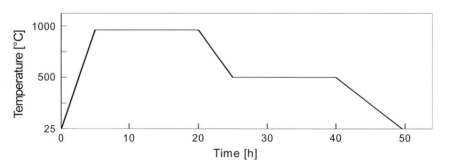

Figure 2-1. Heating protocol for the preparation of $YBa_2Cu_3O_{6.9}$ (see text).

Other precursor compounds may be employed for the synthesis of YBCO. For example, $BaCO_3$ can be used as the source of barium. *In-situ* decomposition of barium carbonate at high temperature gives a fine-grained BaO with high surface area and reactivity. Apart from carbonates, nitrates and other oxy salts that decompose at high temperatures are often employed in solid-state reactions. The heating program depends very much on the form and reactivity of the reactants. If one or more of the reactants is an oxy salt, the mixture is heated first at an appropriate temperature for a few hours to allow decomposition to occur in a controlled manner. If this stage is omitted and the mixture is heated directly at a higher temperature, decomposition may occur very vigorously.

For solid-state reactions in general, some caution is necessary in choosing a suitable container material which is chemically inert to the reactants at the high temperatures. Platinum, silica, stabilized zirconia and alumina containers are generally used for the synthesis of metal oxides, while graphite containers are employed for sulfides and other chalcogenides or pnictides. If one of the constituents is volatile or sensitive to the atmosphere, the reaction is carried out in sealed evacuated capsules.

$YBa_2Cu_3O_{7-x}$ is superconducting at about $-181\,°C$, which is well above the temperature of liquid nitrogen (b.p. $-196\,°C$) for cooling of the magnet. The unit cell of $YBa_2Cu_3O_7$ is shown in Figure 2-2. It is derived from the perovskite structure (▶ glossary).

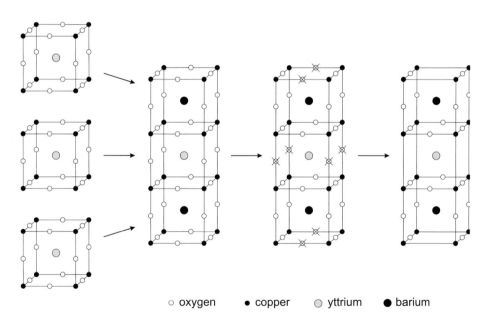

○ oxygen   ● copper   ◌ yttrium   ● barium

Figure 2-2. Deduction of the unit cell of $YBa_2Cu_3O_7$ (right) from the perovskite structure. Three perovskite unit cells (left) are stacked above each other. Then the oxygen atoms marked by an x are removed.

Perovskites are a class of solid compounds of the general formula $M^{2+}M^{4+}O_3$ with interesting electrical properties (piezoelectricity (▶ glossary), ferroelectricity (▶ glossary), high-temperature superconductivity, ▶ glossary) (Figure 2-3). The M^{2+} and oxide ions occupy the anion sites of the rock salt structure, while the M^{4+} ions are in one-quarter of the octahedral sites (those only surrounded by oxide ions). $YBa_2Cu_3O_7$ is an oxygen-deficient variation of this structure type. Its unit cell (Figure 2-2) consists of three stacked perovskite (▶ glossary) unit cells with Ba^{2+} and Y^{3+} sharing the M^{4+} positions and $^2/_9$ of the oxide ions removed for charge balancing (part of the copper is in the +II and part in the +III oxidation state). The critical temperature for superconductivity depends on the oxygen content x of $YBa_2Cu_3O_{7-x}$. It is highest for x = 0; superconductivity breaks down for x <0.6.

Figure 2-3. In 1933, Walter Meissner and Robert Ochsenfeld discovered that a superconducting material will repel a magnetic field (the "Meissner–Ochsenfeld effect"). A body made of a superconducting material thus floats above the surface of a magnet.

General aspects of solid-state reactions

In order to understand the difference between reactions in solution and in the solid state, and the problems associated with solid-state reactions, let us consider the thermal reaction of two crystals of the compounds A and B which are in intimate contact across one face (Figure 2-4). When no melt is formed during the reaction, the reaction has to occur initially at the points of contact between A and B, and later by diffusion of the constituents through the product phase.

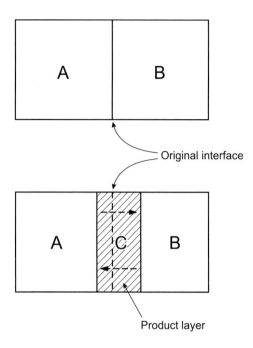

Figure 2-4. Reaction of two crystals (A and B) sharing one face. After initial formation of a product layer C, ions from A and B have to counter-diffuse through the product layer to form new product at the interfaces A/C and B/C.

The first stage of the reaction is the formation of nuclei of the product phase C at the interface between A and B. This may be difficult, if a high degree of structural reorganization is necessary to form the product. After nucleation (▶ glossary) of product C has occurred, a product layer is formed. At this stage, there are two reaction interfaces: one between A and C, and another between C and B. In order for further reaction to occur, counter-diffusion of ions from A and B must occur through the existing product layer C to the new reaction interfaces.

As the reaction progresses, the product layer becomes thicker. This results in increasingly longer diffusion paths and slower reaction rates, because the product layer between the reacting particles acts as a barrier. In the simple case where the rate of the reaction is controlled by lattice diffusion through a planar layer, the rate law has a parabolic form:

$$\frac{\mathrm{d}x}{\mathrm{d}t} = k \cdot x^{-1} \tag{2-1}$$

where x is the amount of reaction (here equal to the thickness of the growing product layer), t is time, and k is the rate constant.

Ions are normally regarded as being trapped on their appropriate lattice sites, and it is difficult for them to move to adjacent sites. Only at very high temperatures do the ions have sufficient energy to diffuse through the crystal lattice. As a rule of thumb, two-thirds of the melting temperature of one component are sufficient to activate diffusion sufficiently and hence to enable the solid-state reaction.

The formation of the perovskite (▶ glossary) barium titanate ($BaTiO_3$) by solid-state reaction of $BaCO_3$ and TiO_2 may serve as an example to illustrate this point, but also to show that such considerations are sometimes oversimplifying the facts. $BaTiO_3$ is an important material for the fabrication of thermistors, capacitors, optoelectronic devices, and DRAMs. BaO (formed by decomposition of $BaCO_3$) has the rock salt structure (cubic close packing of the oxide ions; Ba^{2+} ions in octahedral sites), while TiO_2 (rutile structure) has a hexagonal close packing of the oxide ions and Ti^{4+} ions in half of the octahedral sites.

The formation of $BaTiO_3$ takes place in at least three stages:

1. First BaO reacts with the outer surface regions of TiO_2 grains (▶ glossary) to form nuclei and a surface layer. This requires reorganization of the oxide lattice at the $TiO_2/BaTiO_3$ interface.
2. Further reaction of BaO and the previously formed $BaTiO_3$ leads to the formation of the intermediate Ba-rich phase Ba_2TiO_4. The formation of this phase is necessary for the migration of the Ba^{2+} ions.
3. Ba^{2+} ions from the Ba-rich phase Ba_2TiO_4 migrate into the remaining TiO_2 to form $BaTiO_3$.

From the above discussion it is clear that reaction between two solids may not occur even if thermodynamic considerations favor product formation. There are three important factors that influence the *rate* of reaction between solids:

1. The area of contact between the reacting solids and hence their surface areas.
2. The rate of nucleation (▶ glossary) of the product phase.
3. Rates of diffusion of ions through the various phases, and especially through the product phase.

Apart from the problems arising from nucleation and diffusion, the ceramic method suffers from several additional disadvantages:

- Undesirable phases may be formed, such as $BaTi_2O_5$ during the synthesis of $BaTiO_3$.

- The homogeneous distribution of dopants, important for many ceramic materials, is sometimes difficult to achieve.
- There are only limited possibilities for an *in-situ* monitoring of the progress of the reaction. Instead, physical measurements (such as X-ray diffraction) are periodically carried out. Because of this difficulty, mixtures of reactants and products are frequently obtained. Separation of the desired product from these mixtures is generally difficult, if not impossible.
- In many systems the reaction temperature cannot be raised as high as necessary for reasonable reaction rates, because one or more components of the reacting mixture may volatilize.

In order to overcome some of the problems associated with the ceramic method, particularly to reduce the reaction times, it is necessary to optimize the critical parameters.

Surface area of solids. The surface area of a given amount of solid depends on the particle size. This is shown by a simple calculation. A cubic crystal with a volume of 1 cm^3 has six faces each with an area of 1 cm^2 and, therefore, a total surface area of 6 cm^2. When this crystal is now cut ten times parallel to each face (Figure 2-5), 10^3 cubic crystallites are obtained with a dimension of $0.1 \times 0.1 \times 0.1$ cm each. The 10^3 smaller cubes have the same mass and volume as the large cube, but their total surface area now is ten times larger ($10^3 \times 6 \times 0.01$ cm^2).

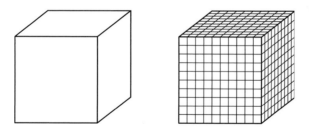

Figure 2-5. When a cubic crystal of $1 \times 1 \times 1$ cm (total surface area 6 cm^2) is cut ten times parallel to each face, 10^3 cubic crystallites of $0.1 \times 0.1 \times 0.1$ cm are obtained. The total volume is the same but the total surface area increases to 60 cm^2.

Grinding of the 1 cm^3 crystal or ball-milling for some time typically results in particles with an average size of about 10 μm (10^{-3}cm). Hence, the total surface area of the powder would be 6×10^3 cm^2 (0.6 m^2) if all grains are cubic crystallites. Note that a 10 μm particle size still represents diffusion distances of about 10^4 unit cells! The advantage of further decreasing the crystallite size in the nanometer range is obvious from these considerations (see Chapter 7).

Although the surface area of solids largely controls the area of contact between reacting grains (▶ glossary) in a mixture, it does not appear directly in the equation for the rate of reaction, such as Eq. 2-1. However, it is included indirectly since there is an inverse correlation between the thickness of the product layer, x, and the area of contact. For example, when two cubic particles of 10 μm size react with each other, the product layer at 50 % conversion is 10 μm thick. When the dimension of the crystals is decreased to 1 μm, the total surface area for a given mass of reactants is increased by a factor of 10 (see above), but the thickness of the product layer at 50 % conversion is only 1 μm. According to Eq. 2-1, this will result in a faster reaction.

In practice, it is rather unlikely that all the surfaces of the reacting solids will be in intimate contact, and usually the contact area is considerably less than the total surface area. The area of contact may be increased somewhat by pressing the reacting powder into a pellet. However, even at relatively high pressures, the crystal contacts are not maximized. A further increase in contact area and reduction in pellet porosity may be achieved by compressing the pellet at high temperatures, i.e., by hot pressing. However, the densification process is usually slow and may require several hours.

Various modifications have been employed to increase diffusion rates (decrease of diffusion path lengths) in solid–solid reactions:

- Small particle sizes, achieved through grinding, ball milling (Chapter 7), spray drying (Section 3.3), freeze drying, etc. It is possible to reduce the particle size to several tens of nanometers. Solid-state reactions are often greatly facilitated by cooling and grinding the sample periodically. This is because sintering and grain (▶ glossary) growth of both reactant and product phases may occur during heating, causing a reduction in the surface area of the mixture. Grinding maintains a high surface area and brings fresh surfaces into contact.
- Intimate mixture of the reactants by coprecipitation, sol–gel processing (Section 4.5), etc.
- Reduction of the diffusion distances by incorporating the cations in the same solid precursor.
- Performing solid-state reactions in molten fluxes or high-temperature solvents (e.g., Bi, Sn, halide salts, or alkaline metals).

Nucleation (▶ glossary). The overall process in solid-state reactions may not only be controlled by diffusion of reactants, as discussed above, or by the rate of the reaction at the phase boundary, but also by nuclei growth. Nucleation-limited reactions are represented by the Avrami–Erofeyev equation (Eq. 2-2 and Figure 2-6),

$$x(t) = 1 - e^{kt^n} \tag{2-2}$$

where n is a real number, usually between 1 and 3. For n >1, the function has a sigmoid shape. Initially a large number of nuclei is formed, and then the reaction front expands with the growth of the nuclei. When the resulting product regions touch each other, the reaction rate starts to decrease. Note that for most solid-state reactions, it is usually incorrect and misleading to think in terms of reaction order, since the reactions do not involve molecules. Nevertheless, the data may still be represented empirically in this way.

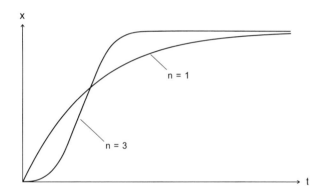

Figure 2-6. Avrami equation for two different values of n. The curve for n = 3 graphically shows the induction period at the beginning of the reaction.

The nucleation step is easier if there is a structural similarity between the product and at least one of the reactants, because this reduces the degree of structural reorganization necessary for nucleation to occur. For instance, in the reaction of MgO and Al_2O_3 to form spinel (▶ glossary), the spinel has a similar oxide ion arrangement to that in MgO (cubic close packing). Spinel nuclei may therefore form at the surface of the MgO crystals such that the oxide arrangement is essentially continuous across the MgO–spinel interface.

When nucleation is facilitated by a structural similarity, there is usually a clear orientational relationship between the structures of the reactant and product. Furthermore, the interatomic distances, should be similar. If the two structures have quite different interatomic distances then they cannot be matched over a large area of contact. For example, the oxygen–oxygen separation in MgO and BaO are quite different, although both compounds have the rock salt structure. A difference in interfacial lattice parameters of about 15 % between nucleus and substrate is the most that can be tolerated for oriented nucleation.

There are two types of oriented reactions: *epitactic reactions* and *topotactic reactions* (▶ glossary). In epitactic reactions, the structural relationship is restricted to the actual interface between the two crystals. For example, the two structures may have a common arrangement of oxide ions at the interface, but the structures at both sides of the interface may be different. Epitactic reactions therefore require only a two-dimensional structural similarity at the crystal interface. Topotactic reactions are more specific than epitactic ones because they require not only the structural similarity at the interface but also that this similarity continues into the bulk of both crystalline phases.

Topotactic and epitactic reactions are quite common since their nucleation step (▶ glossary) is usually easier than that for reactions in which there is no structural relationship between reactant and product.

The ease of nucleation of product phases also depends on the actual surface structure of the reacting phases. From a consideration of crystal structures it can be shown that in most crystals the structure cannot be the same over the entire crystal surface. For example, the (100) planes of BaO contain Ba^{2+} and O^{2-} ions while a (111) surface will be either a complete layer of Ba^{2+} ions or a complete layer of O^{2-} ions (Figure 2-7). Since different surfaces have different structures, their reactivity is likely to differ considerably. It is difficult, however, to give general rules about which surfaces are the most reactive.

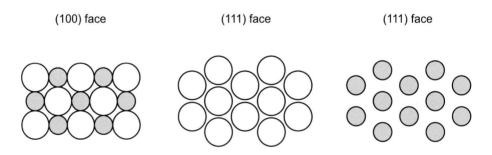

(100) face   (111) face   (111) face

Figure 2-7. Different planes of the BaO (rock salt) structure. The open circles represent the oxygen atoms, the smaller dark circles the metal atoms.

Diffusion of ions. Equation 2-1 relates the rate of solid-state reactions to the diffusion of ions through the bulk of the crystals and, especially, through the product phase. Diffusion path lengths are influenced, inter alia, by the particle size of the reactants, the degree of homogenization achieved during mixing, and the intimacy of contact between the grains (▶ glossary). Diffusion of ions is also enhanced greatly by the presence of crystal defects, especially vacancies and interstitials, but also occurs via structural defects such as dislocations and grain boundaries. In general, the rate of diffusion and, therefore, the reactivity of solids depends greatly on the types of crystal defects.

Diffusion path lengths are greatly reduced when the cations are brought in close contact. Instead of heating powder mixtures, solid compounds are used that already contain the different cations in an ideally atomic dispersion. These methods are nearly exclusively employed for the preparation of oxide materials. There are two modifications of this approach:

1. In the *coprecipitation* method, salts of different metals are precipitated together from a common solution. The precipitate then consists of an intimate mixture of two salts or a solid solution (▶ glossary).
2. In the *precursor* method, the cations are incorporated in the same solid precursor.

In both methods, the precipitate is thermally treated to give the desired oxide material. The decomposition temperatures are generally lower than the temperatures employed in the ceramic method due to the shorter diffusion distances of the ions. The precursor method can be subdivided in reactions in which the mixed-metal precursor thermally decomposes without change of the oxidation state of the involved elements, and those where the products are formed by redox reactions.

Coprecipitation. Salts of the required metals are dissolved in the same medium (usually water). They are coprecipitated either by concentrating the solution or by adding a precipitating reagent. Hydroxides, carbonates, oxalates, formates or citrates are often formed by the latter procedure. The obtained precipitate is heated to the required temperature in a desired atmosphere to produce the final product.

This may be illustrated by the synthesis of $ZnFe_2O_4$ spinel (▶ glossary). Oxalates of zinc and iron are dissolved in water in the ratio of 1:1. The solutions are then mixed and heated to evaporate the water. The precipitated fine powder is a solid solution (▶ glossary) that contains the cations mixed together, essentially on an atomic scale. The powder is filtered off and heated. Because of the high degree of homogenization, much lower reaction temperatures are sufficient for reaction to occur. The overall reaction may be written as in Eq. 2-3. The method has also been employed for the preparation of super-conducting (▶ glossary) $YBa_2Cu_3O_{7-x}$.

$$Fe_2(C_2O_4)_3 + Zn(C_2O_4) \longrightarrow ZnFe_2O_4 + 4\ CO + 4\ CO_2 \qquad (2\text{-}3)$$

Another example is the formation of $M_{1-x}M'_xO$ (M, M' = Ca, Mg, Mn, Fe, Co, Zn, Cd). Because their carbonates are isostructural (calcite structure), a large number of carbonate solid solutions $M_{1-x}M'_x(CO_3)$ containing two or more cations can be prepared. These solid solutions (▶ glossary) are ideal precursors for the synthesis of solid solutions of the corresponding oxides $M_{1-x}M'_xO$ (rock salt structure). The carbonates are thermally decomposed in vacuum or in flowing dry nitrogen by cleavage of CO_2. The facile formation of the oxides from the carbonates is due to the close relationship between the structures of calcite and rock salt. The $M_{1-x}M'_xO$ solid solutions can be used as precursors for preparing spinels (▶ glossary) and other complex oxides.

A number of ternary and quaternary metal oxides can be prepared by employing hydroxide, nitrate and cyanide solid solution (▶ glossary) precursors as well. For example, hydroxide solid solutions of the general formula $Ln_{1-x}M_x(OH)_3$ (Ln = La or Nd; M = Al, Cr, Fe, Co or Ni) and $La_{1-x-y}Ni_{x-}M_y(OH)_3$ (M = Co or Cu) are crystallizing in the $Ln(OH)_3$ structure. They are decomposed at relatively low temperatures (around 600 °C) to yield $LnNiO_3$, $LaNi_{1-x}Co_xO_3$, $LaNi_{1-x}Cu_xO_3$, etc.

The coprecipitation method is only successful if the metal salts have similar solubility and the precipitation rate is similar or if solid solutions are formed.

Precursor method. A variation of the coprecipitation method is that stoichiometric mixed-metal salts are precipitated. A selection is shown in Table 2-1. For example, ferrite spinels such as $NiFe_2O_4$ can be prepared by slowly heating the mixed-metal acetate $Ni_3Fe_6O_3(OH)(OAc)_{17} \cdot 2$ py to 200–300 °C to burn off the organic material, followed by heating in air at ~1000 °C for 2–3 days. The starting mixed-metal salt with the Ni:Fe ratio of exactly 1:2 is formed from a basic double acetate hydrate compound.

Another type of mixed-metal precursors are redox compounds. Compounds like $(NH_4)_2Cr_2O_7$ which contain both oxidizing $Cr_2O_7^{2-}$ and reducing NH_4^+ groups when ignited decompose autocatalytically to yield Cr_2O_3 (artificial volcano). The exothermicity of the reaction is due to the oxidation of NH_4^+ to N_2 and H_2O by the dichromate ion which itself is reduced to Cr(III). Fine, voluminous particles are obtained due to the gas-producing reaction. Similar precursors have been used to synthesize chromite spinels, MCr_2O_4 (M = Mg, Zn, Cu, Mn, Fe, Co, Ni) (see Table 2-1). For example, magnesium chromite, $MgCr_2O_4$,

is prepared by heating precipitated $(NH_4)_2Mg(CrO_4)_2 \cdot 6\ H_2O$ gradually to
1100–1200 °C (Eq. 2-4).

$$(NH_4)_2Mg(CrO_4)_2 \cdot 6\ H_2O \longrightarrow MnCr_2O_4 + N_2 + 10\ H_2O \qquad (2\text{-}4)$$

By careful control of the experimental conditions, these precursor methods
are capable of yielding phases of accurate stoichiometry. This is important since
several chromites and ferrites are valuable magnetic materials whose properties may be sensitive to purity and stoichiometry.

Table 2-1. Examples of mixed-metal precursors and the ceramic products obtained by
calcinations (▶ glossary).

Mixed-metal precursor	Ceramic product
$La[Co(CN)_6] \cdot 5\ H_2O$	$LaCoO_3$
$Ba[TiO(C_2O_4)_2]$	$BaTiO_3$
$M_3Fe_6O_3(OH)(OAc)_{17} \cdot 12$ py	MFe_2O_4 (M = Mg, Mn, Co, Ni) (ferrite spinels)
$(NH_4)_2M(CrO_4)_2 \cdot 6\ H_2O$	MCr_2O_4 (M = Mg, Ni) (chromites)
$(NH_4)_2M(CrO_4)_2 \cdot 2\ NH_3$	MCr_2O_4 (M = Cu, Zn) (chromites)
$MCr_2O_7 \cdot 4$ py	MCr_2O_4 (M = Mn, Co) (chromites)
$MFe_2(C_2O_4)_3 \cdot x\ N_2H_4$	MFe_2O_4 (M = Mg, Mn, Co, Ni, Zn) (ferrites)

Formation of metastable solids

The ceramic method yields crystalline, i.e., thermodynamically stable solids.
Many metastable solids (▶ glossary) are of great current interest. Such compounds cannot be prepared by the conventional high-temperature routes discussed in this chapter, and alternative strategies have been developed. Low-
temperature, chemistry-based approaches, sometimes called "soft chemistry"
(chimie douce), often allow a better control of the structure, stoichiometry, and
phase purity. Metastable solids can be obtained by three different approaches:

1. Synthesis under conditions where the solid is thermodynamically stable,
   followed by quenching to ambient conditions.
2. Prefabrication of a thermodynamically stable phase which is then transformed to a metastable phase by a low-temperature, soft-chemical
   operation.

3. Synthesis under non-equilibrium conditions; the products are kinetically controlled metastable compounds.

According to the organization of this book, the soft chemical routes are spread over several chapters, such as intercalation and ion-exchange reactions in Section 2.4, dehydration reactions in Section 2.3 or sol–gel techniques in Section 4.5.

2.1.2 Carbothermal Reduction

Carbothermal reduction is practiced commercially for the synthesis of many non-oxide ceramic powders such as carbides, nitrides or borides.

We will discuss this method for one of the most prominent carbothermal reduction processes, the so-called "Acheson process". Almost all silicon carbide, SiC ("carborundum"), produced world-wide is made by this method. SiC is mainly used in four areas of application:

1. As an abrasive. SiC powders are used for cutting and grinding precious and semi-precious stones and for fine grinding and lapping of metals and optical glasses. Bound with synthetic resins and ceramic binders, SiC grits are used in grinding wheels, whetstones, hones, abrasive cutting-off wheels, and monofiles for machining of all types of materials.
2. As a deoxidizer. SiC is used in cast iron and steel production as a deoxidant, and for carburization and siliconization.
3. As a refractory material (▶ glossary). SiC is applied as a structural refractory material with excellent thermal shock, oxidation and corrosion resistance in linings and skid rails for furnaces, and in kiln furniture.
4. As electric heating elements and electrical resistors. For example, SiC is used for heating elements operating in oxidizing atmospheres up to 1500 °C owing to its good electric conductivity combined with its excellent oxidation resistance.

The first commercial plant to produce SiC was built in 1896 in Niagara by Acheson to meet the demand for abrasives at this time. The basic design of the original electric furnace (Figure 2-8) has remained unchanged, despite larger sizes and better efficiency. Most Acheson furnaces are shaped like a trough. Graphite electrodes connected to a graphite core (that constitutes the initial electrical path through the reaction mixture) are laid in a mixture of carbon and sand. When an electric current is passed through the graphite core, the charge is heated from within by resistive heating. The reaction takes place at temperatures above 1577 °C. This results in the formation of a hollow cylinder of SiC, and the expulsion of carbon monoxide gas. The charge acts as a refractory container (▶ glossary) as well as a thermal insulator for the ingot being formed.

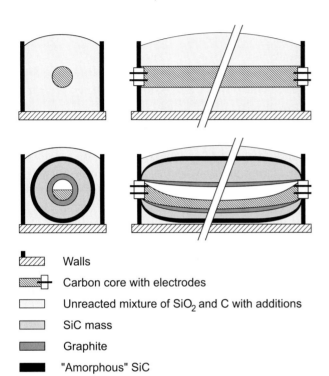

Walls

Carbon core with electrodes

Unreacted mixture of SiO_2 and C with additions

SiC mass

Graphite

"Amorphous" SiC

Figure 2-8. Section through an Acheson furnace before (above) and after the reaction (below).

Current conventional furnaces are about 12–18 m long, and the power intensities approach 260 kW·m^{-1}. The technological limit to furnace size and power intensity is the available electrical power supply technology. The manufacture of 1 kg SiC requires about 12 kW·h, and so most plants are located in areas where power is relatively inexpensive.

The overall reaction of the SiC formation is

$$SiO_2 + 3\,C \longrightarrow SiC + 2\,CO \tag{2-5}$$

The SiO_2 source can be sand, quartzite, or crystalline rock quartz. The most common carbon source is petroleum coke, the final residue from crude oil refining. Carbon black, graphite, charcoal, pyrolyzed organic polymers or other carbon sources can also be used. The particle size of both components is in the range of 5–10 mm; the size distribution should be narrow so that the final furnace mix is permeable to the CO gas that must leave the reaction zone. Porosity agents, such as sawdust, bagasse or rice hulls, are often incorporated into the furnace mix to make it additionally permeable to the escaping CO. High pressures of gas may locally be built up to form voids and channels to more porous parts of the mixture. The product from an Acheson furnace at the end of a run is a large hollow cylinder (Figure 2-8). The unreacted furnace mix is recycled.

Pure SiC is colorless and transparent. However, such pure SiC cannot be made by the Acheson process because nitrogen from the air is soluble in SiC, causing the crystals to take a green color. Black SiC is obtained in the presence of impurities, such as Al (maximum solubility 2 %) and B (maximum solubility 0.5 %) (see cover of the book). To make green SiC, the furnace mixture must be made from quartz and low metal content coke. The highest tonnage application for SiC (about 50 % of the produced SiC) is "metallurgical SiC", a silicon and carbon source for the iron casting industry. Purity is not important to its function. Purer SiC is needed for abrasive (about 40 %), refractory and other applications.

The SiC-forming reaction is much more complicated than given in Eq. 2-5. Originally, it was assumed that SiC solely results from solid-state reaction between SiO_2 and carbon. However, this is not reasonable for the relatively large particle sizes used in commercial SiC production. One of the current models is that SiC forms as a result of four sub-reactions (Eqs. 2-6 to 2-9), which provide mass transport via the vapor phase.

$$C(s) + SiO_2(s) \longrightarrow SiO(g) + CO(g) \tag{2-6}$$

$$SiO_2(s) + CO(g) \longrightarrow SiO(g) + CO_2(g) \tag{2-7}$$

$$C(s) + CO_2(g) \longrightarrow 2\,CO(g) \tag{2-8}$$

$$2\,C(s) + SiO(g) \longrightarrow SiC(s) + CO(g) \tag{2-9}$$

SiO_2 initially reacts at the contact points with coke particles in a solid-state reaction to liberate CO and gaseous SiO (Eq. 2-6). Reaction of the CO with further SiO_2 results in the formation of additional SiO and CO_2 (Eq. 2-7). Carbon dioxide is re-converted to CO by the Boudouard equilibrium (Eq. 2-8). SiC is formed by reaction of gaseous SiO directly at the surface of the carbon particles once the C/SiO_2 contact points are consumed (Eq. 2-9). Silicon is transported to the carbon particles in the form of SiO (Figure 2-9). As the product layer grows, the reaction surface decreases and the reaction gets slower.

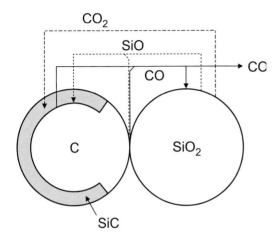

Figure 2-9. Material transport paths during the preparation of SiC by the Acheson process.

In general, the manufacture of non-oxide ceramic materials by carbothermal reduction can be carried out in a variety of ways, and various types of reactors have been developed (electric arc furnaces, moving bed furnaces, rotary tube reactors, fluidized bed reactors, etc.). A discussion of the advantages and disadvantages of the various reactor types would be beyond the scope of this book.

Carbides are made by the high-temperature reaction between carbon and metal oxides alone, as exemplified by the Acheson process discussed above. The synthesis of borides requires the presence of elemental boron that is usually formed *in situ* by reduction of B_2O_3. The energy required to manufacture the borides (with B_2O_3 as the boron source) is greater than for the corresponding carbides because carbon has to reduce both the metal oxide and B_2O_3. Nitrides are obtained in the presence of nitrogen sources, usually elemental nitrogen ("carbothermal nitridation"; see also Section 2.2). In all cases, the reactions are highly exothermic and thermodynamically favorable at very high temperatures. Since the reactions are reversible, it is advantageous to remove the by-product CO. Selected examples are shown in Table 2-2.

Although the final equilibrium products are determined solely from the temperature, pressure and chemical species present, the mechanism and rate of a given reaction depend on a number of additional variables, such as particle size, degree of mixing of the reactants, diffusion rates, gas concentration, porosity, and the presence of impurities, as already discussed in Section 2.1.1.

The fact that many of the carbothermal reductions are intrinsically fast is evidence that solid–solid reactions do not dominate the process. As already discussed for the Acheson process, gaseous species play a very important role. Some of the solid reactant metal oxides sublime, have a substantial vapor pressure, dissociate, or are reacted into volatile species within the temperature range for reaction.

In all carbothermal reductions, CO is a necessary by-product, and plays a substantial role in the thermodynamics and kinetics of the overall process. Reduction processes involving CO are believed to be the primary route for the reduction of certain oxides to produce gaseous suboxides. Carbothermal reductions involving SiO_2 almost always occur through gaseous SiO, as already discussed (Eqs. 2-6 and 2-7). Under reducing conditions, SiO is the dominant gaseous species at temperatures between about 1000 °C and 1700 °C. In the carbothermal nitridation of silicon, Si_3N_4 can form from SiO either by reaction with solid carbon (Eq. 2-10) or by reaction with CO in a gas phase reaction (Eq. 2-11).

Table 2-2. Examples for the carbothermal reduction, and minimum temperatures.

Reactions	Minimum temperatures (°C) at atmospheric pressure
Carbides	
$2 Al_2O_3 + 9 C \rightarrow Al_4C_3 + 6 CO$	1950
$2 B_2O_3 + 7 C \rightarrow B_4C + 6 CO$	1550
$SiO_2 + 3 C \rightarrow SiC + 2 CO$	1500
$TiO_2 + 3 C \rightarrow TiC + 2 CO$	1300
$WO_3 + 4 C \rightarrow WC + 3 CO$	700
$2 MoO_3 + 7 C \rightarrow Mo_2C + 6 CO$	500
Borides	
$Al_2O_3 + 12 B_2O_3 + 39 C \rightarrow 2 AlB_{12} + 39 CO$	1550
$V_2O_5 + B_2O_3 + 8 C \rightarrow 2 VB + 8 CO$	950
$V_2O_3 + 2 B_2O_3 + 9 C \rightarrow 2 VB_2 + 9 CO$	1300
$TiO_2 + B_2O_3 + 5 C \rightarrow TiB_2 + 5 CO$	1300
$2 TiO_2 + B_4C + 3 C \rightarrow 2 TiB_2 + 4 CO$	1000
Nitrides	
$Al_2O_3 + 3 C + N_2 \rightarrow 2 AlN + 3 CO$	1700
$B_2O_3 + 3 C + N_2 \rightarrow 2 BN + 3 CO$	1000
$3 SiO_2 + 6 C + 2 N_2 \rightarrow Si_3N_4 + 6 CO$	1550
$2 TiO_2 + 4 C + N_2 \rightarrow 2 TiN + 4 CO$	1200
$V_2O_5 + 5 C + N_2 \rightarrow 2 VN + 5 CO$	600

$$3 \, C + 3 \, SiO + 2N_2 \longrightarrow Si_3N_4 + 3 \, CO \qquad (2\text{-}10)$$

$$3 \, CO + 3 \, SiO + 2 \, N_2 \longrightarrow Si_3N_4 + 3 \, CO_2 \qquad (2\text{-}11)$$

Other examples for the formation of suboxides by CO are given in Eqs. 2-12 and 2-13.

$$B_2O_3 + CO \longrightarrow B_2O_2(g) + CO_2 \qquad (2\text{-}12)$$

$$Al_2O_3 + 2 \, CO \longrightarrow Al_2O(g) + 2 \, CO_2 \qquad (2\text{-}13)$$

For carbothermal reductions involving B_2O_3, there is an increasing vapor pressure of B_2O_3 at temperatures above the melting point of $450\,°C$. Above temperatures of approximately $1450\,°C$, B_2O_2 (and other B/O species) are becoming more important. For Al_2O_3 systems under reducing conditions, the major gaseous aluminum species above $1200\,°C$ are gaseous Al and Al_2O. Reaction of the suboxides (B_2O_2 or Al_2O) with solid carbon provides a mechanism for the formation of the carbides similar to Eq. 2-9.

Liquid phases (melts) may also be involved, particularly for reactions with B_2O_3. In the formation of B_4C, there is a change in mechanism at about $1707\,°C$. The liquid phase reduction of B_2O_3 (Eq. 2-14) dominates at lower temperatures, and the gas phase reaction at higher temperatures (Eq. 2-15).

$$7 \, C(s) + 2 \, B_2O_3(l) \longrightarrow B_4C(s) + 6 \, CO(g) \qquad (2\text{-}14)$$

$$5 \, C(s) + 2 \, B_2O_2(g) \longrightarrow B_4C(s) + 4 \, CO(g) \qquad (2\text{-}15)$$

2.1.3 Combustion Synthesis

The "combustion synthesis" approach uses highly exothermic reactions. Such reactions typically have high activation energies and generate substantial amounts of heat. Once the reactions are initiated by the rapid input of energy from an external source, sufficient heat is released to render the reactions self-sustaining. The reactants are thus heated rapidly (10^3 to 10^6 $K \cdot s^{-1}$) to very high temperatures. The reactions are so fast that they are pseudo-adiabatic, i.e., all the energy produced by the exothermic reaction is used to heat the sample.

Combustion syntheses can be conducted in the self-propagating mode and the simultaneous combustion mode, although in reality many combustion synthesis reactions lie in between:

- *Self-propagating mode*, also referred to as "self-propagating high-temperature synthesis" (SHS): The combustion reaction is initiated at one point and then propagates rapidly through the reaction mixture in the form of a combustion wave (Figure 2-10, route A and Figure 2-11). In this mode, the heat of the reaction furnishes more than 90 % of the energy required for the synthesis.

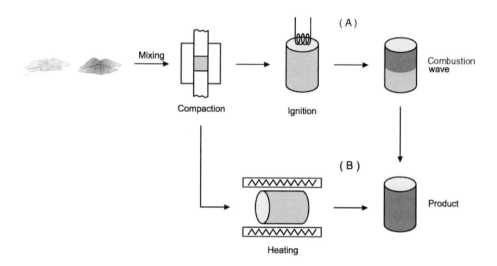

Figure 2-10. Schematic illustration of the self-propagating combustion mode (route A) and the thermal explosion mode (route B).

Figure 2-11. SHS before ignition (front), during reaction (middle) and after reaction (back) of the compacted sample.

- *Simultaneous combustion mode*, also referred to as thermal explosion: Once the entire sample has been heated to the ignition temperature (T_{ig}), reaction takes place simultaneously through the reactant mixture, rather than as a propagating combustion wave (Figure 2-10, route B).

An early application of combustion synthesis was the "thermite" reduction of metal oxide powders with aluminum powder yielding either metal or metal/alumina composites (▶ glossary). An example is given in Eq. 2-16. The heat generated by the exothermic reaction (849 kJ·mol^{-1}) is sufficient for welding railroad tracks, for example (Figure 2-12).

$$3\ Fe_3O_4 + 8\ Al \longrightarrow 9\ Fe + 4\ Al_2O_3 \tag{2-16}$$

In practice, the aligned ends of the rail are surrounded by a refractory mold (▶ glossary) and pre-heated to 600–1000 °C. A crucible containing the thermite mixture is placed over the butt joint, and the mixture is ignited. The liquid iron and alumina formed in the reaction separate due to their different density. The heavier iron flows in the butt joint through an opening in the bottom of the crucible.

Figure 2-12. A device for welding railroad tracks by the thermite process.

Combustion synthesis is a versatile method for the synthesis of a variety of technologically useful solid materials, such as binary and ternary metal borides, carbides, silicides, chalcogenides, nitrides or hydrides, as well as alloys (▶ glossary), composites (▶ glossary) or cemented carbides. Compared with conventional ceramic processing, the main advantages of the SHS process are:

- it requires less energy than conventional methods;
- the short reaction times result in low operating and processing costs and time savings;
- expensive processing equipment is not necessary;
- inorganic materials can be synthesized and consolidated into a final product in one step by utilizing the chemical energy of the reactants;
- the high reaction temperatures may expel volatile impurities and thus result in higher purity products; and
- the high thermal gradients and rapid cooling rates can give rise to non-equilibrium phases.

SHS reactions can be characterized by the adiabatic combustion temperature T_{ad}. This can be calculated by assuming that the enthalpy of the reaction heats up the products and that no energy is lost by heating the surrounding environment. Thus, T_{ad} is a measure of the exothermicity of the reaction and defines the upper limit for any combustion system. As a rule-of-thumb, if T_{ad} <1200 °C, combustion does not occur, and if T_{ad} >2200 °C, self-propagating combustion occurs. In the range 1200 °C < T_{ad} <2200 °C, a combustion wave cannot propagate but can made to do so by special techniques such as pre-heating of the reactants. For example, the reactions

$$Si + C \rightarrow SiC \ (T_{ad} = 1527\,°C, T_{ig} = 1300\,°C) \ or$$
$$Ti + Al \rightarrow TiAl \ (T_{ad} = 1245\,°C, T_{ig} = 640\,°C)$$

are not self-sustaining when $T_{start} = 25\,°C$, while the reaction

$$Ti + C \rightarrow TiC \ (T_{ad} = 2937\,°C, T_{ig} = 1027\,°C)$$

is self-sustaining. The reaction Ti + Al $\rightarrow$ TiAl can be made self-sustaining by preheating above 100 °C.

Combustion syntheses are difficult to control because of the high reaction rates. However, some control is possible by:

- addition of diluents, which do not take part in the reaction, but increase the thermal mass of the system and lower T_{ad}. The diluent may be the product (for example, the rate of the highly exothermic reaction of titanium and boron is reduced by addition of TiB_2), or an inert compound. The latter option results in composite materials (▶ glossary).

- an increase in particle size of the reactants. This leads to a decrease of the combustion wave velocity, and the reaction rate decreases, as expected for solid–solid reactions.

The reactions can take place both in solid–solid and solid–gas systems. Ignition can be achieved using an electrical arc, a laser pulse, a spark, a chemical reaction, etc.

Solid–solid systems. The reactant mixture is usually pelletized to increase an intimate contact between individual particles and placed in a refractory container (▶ glossary), degassed and ignited in a vacuum or in an inert atmosphere. Typical examples are the syntheses of refractory materials (▶ glossary), such as carbides, borides or silicides from the corresponding elements.

Solid–gas systems. Such reactions are often called "filtration combustion" processes, because the gas is supplied to the reaction front from the surrounding atmosphere by "filtration" through the porous mass. They will be discussed in Section 2.2.

The materials obtained by combustion synthesis vary from friable, slightly sintered powders to monolithic porous products which are formed when a fraction of the products is molten at the peak temperature. The inherently high porosity of the products originates both from the density difference between the reactants and the products, and from outgassing or evaporation due to the high reaction temperatures. Important parameters that control the final product composition and morphology are:

- reactant particle size distribution, morphology and purity;
- green density (▶ glossary), which influences the thermal conductivity of the reactant mixture;
- ignition techniques, initial temperature and pressure of the reacting mixture;
- the difference between the temperature of the initial reaction mixture and the completely reacted product; and
- size of the sample and reactor configuration.

The usual techniques can be applied to densify the materials after synthesis (see Section 2.1.4). However, an *in-situ* consolidation would provide an important economical method for manufacturing ceramic materials. In order to obtain a high degree of consolidation, it is advantageous to maintain one of the constituents in a partially molten condition. Simultaneous combustion synthesis and densification is currently used in the direct manufacture of hard-alloy (▶ glossary) components, such as rollers, drawing blocks, pressing equipment and cutting inserts. Densification can be achieved by applying a suitable consolidation pressure either during the reaction or immediately after the reaction is completed and while the products are still in a plastic state at the high combustion temperatures.

Classification of combustion synthesis reactions

Combustion synthesis reactions can be classified according to the kind of reactants.

The product is synthesized from its elements. Many carbides, silicides, borides, nitrides, oxides or hydrides were prepared by this method, including solid solutions and composite materials (▶ glossary). For example, the reaction between titanium and carbon (Eq. 2-17), one of the most widely studied combustion synthesis reactions, results in stoichiometric TiC, which is a valuable refractory (▶ glossary) and abrasive material. The reaction is highly exothermic, liberating 183 kJ·mol^{-1}. The synthesis of 20-kg batches was reported to last about 60–90 s, while cooling of the reactor afterwards took 1.5–2 h.

$$Ti + C \longrightarrow TiC \qquad\qquad (2\text{-}17)$$

Numerous intermetallic compounds were also produced using the SHS method, for example a variety of shape-memory alloys (▶ glossary; such as TiNi). Aluminides have also received some attention, particularly aluminides with nickel, zirconium and copper.

Thermite-type reactions. Thermite reactions are extensions of the Goldschmidt process (▶ glossary) for the reduction of an ore to its constituent metals. In these systems, the combustion synthesis reaction involves the reduction of a metallic compound. Most common is the reduction of an oxide by a strongly reducing metal, mainly Al and Mg. Magnesium is more desirable, because the resulting MgO by-product is easily leachable with hydrochloric acid.

There are two types of thermite reactions. The first involves the reduction of an oxide to the element, for example the well-known "thermite reaction" (Eq. 2-16) and the second the reduction of an oxide to the element followed by its reaction with another element to form a refractory compound (▶ glossary) such as borides, carbides, silicides and nitrides (for example, Eqs. 2-18 and 2-19).

$$SiO_2 + C + 2\,Mg \longrightarrow SiC + 2\,MgO \tag{2-18}$$

$$TiO_2 + B_2O_3 + 5\,Mg \longrightarrow TiB_2 + 5\,MgO \tag{2-19}$$

In each case, a metal oxide is formed as a by-product. The obtained ceramic–ceramic composites (▶ glossary) are often interesting by themselves. For the synthesis of the pure non-oxide powders, the oxide has to be chemically leached. This technique may be used when the cost of the elemental powder is too high (for example, Ta, Hf, B) or when the direct reaction of the element does not generate a sufficient amount of heat to be self-sustaining.

For example, the heat of formation of B_4C from boron and carbon is only $-39\ kJ\cdot mol^{-1}$. Therefore, the T_{ad} (about $730\,°C$) is not high enough for a self-propagating reaction, when the reactant mixture is ignited at room temperature. This problem can be solved by pre-heating or by the thermal explosion technique. However, since elemental boron is rather expensive, the use of the thermite reaction (Eq. 2-20) is more economic. The reaction enthalpy of this particular reaction is $-1135\ kJ\cdot mol^{-1}$, resulting in T_{ad} of $2470\,°C$. Ignition occurs when the reactant mixture is heated to $930\,°C$ in an inert gas (argon) atmosphere.

$$2\,B_2O_3 + C + 6\,Mg \longrightarrow B_4C + 6\,MgO \tag{2-20}$$

An interesting variation is the "centrifugal thermite process" which was developed to coat the inside of pipes. A mixture of iron oxide and aluminum is placed inside the pipes, rotated at high speeds, and ignited. A surface layer of alumina and an inner layer of Fe is formed in the pipe due to the difference in density between Fe and alumina. The composite pipe has the strength and toughness of a metal and corrosion and abrasion resistance of a ceramic.

Solid-state metathesis (SSM) reactions. Unlike the SHS reactions which employ metals, non-metals and oxides, this method involves rapid, low-temperature initiated solid-state exchange reactions. Examples are given in Eqs. 2-21 to 2-24. Particularly useful are reactions between (anhydrous!) metal halides and alkaline-metal (or alkaline-earth metal) main-group compounds (Eqs. 2-23 and

2-24, for example). The desired product is easily separated from the alkaline
metal halide by-product by washing with alcohol and/or water.

$$Ti_2O_3 + 2\ AlN \longrightarrow 2\ TiN + Al_2O_3 \tag{2-21}$$

$$NaBF_4 + 3\ NaN_3 \longrightarrow BN + 4\ NaF + 4\ N_2 \tag{2-22}$$

$$6\ ZrCl_4 + 8\ Li_3N \longrightarrow 6\ ZrN + 24\ LiCl + N_2 \tag{2-23}$$

$$2\ VCl_3 + 3\ MgB_2 \longrightarrow 2\ VB_2 + 3\ MgCl_2 + 2\ B \tag{2-24}$$

A number of technologically important materials have been prepared by
self-propagating SSM reactions, for example, superconductors ($\blacktriangleright$ glossary;
NbN, ZrN [Eq. 2-23]), semiconductors ($\blacktriangleright$ glossary; GaAs, InSb), insulators
(BN, ZrO_2), magnetic materials (GdP, SmAs), chalcogenides (MoS_2, NiS_2),
intermetallics ($MoSi_2$, WSi_2), pnictides (ZrP, NbAs), and oxides (Cr_2O_3).

2.1.4 Sintering

When a compacted powder is heated at an elevated temperature which is
below its melting point, powder particles fuse together, voids between the particles decrease, and eventually a dense solid body is obtained. This phenomenon is called sintering. Sintering processes have been used extensively for the
manufacture of ceramics and ironware for hundreds of years, and today sintering is still a very important process for the manufacture of a wide variety of
industrial materials.

In this section, we will only briefly touch technological issues of sintering,
which are more thoroughly treated in materials science textbooks. Rather, we
will point out some chemical issues related to sintering processes.

The driving force of sintering is the excess surface free energy of a powder
compact. When heated, the system tries to decrease its surface free energy by
decreasing its total surface area. This is reached by mass transport that joins the
powder particles together.

There is a chemical potential difference between surfaces of dissimilar curvature within the system. A concave surface has a negative free energy, and
convex surfaces a positive free energy. As a consequence, mass transport occurs
from the particle surface (convex) to the interparticle necks or pores (concave).
The greater the curvature, i.e., the finer the particle size, the greater the driving
force for sintering.

During sintering, mass transport can occur by solid-state, liquid-phase, and vapor-phase mechanisms individually, or in combination (Figure 2-13):

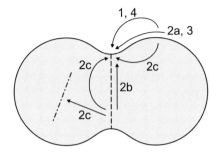

Figure 2-13. Diffusion paths during sintering. The numbers correspond to the numbers in the following paragraph.

- *Path 1: Evaporation–condensation.* Compared to a flat surface, the vapor pressure of a convex surface is higher and that of a concave surface is lower. As a consequence, a substance vaporizes at the particle surface and condenses at the necks and in the pores.

- *Path 2: Diffusion.* The driving force for diffusion are differences in vacancy concentration. The diffusion mechanism can be sub-classified into surface diffusion (2a), grain boundary diffusion (2b) and volume diffusion (2c). The driving force for surface diffusion is the difference in vacancy concentration between convex and concave areas. The vacancy concentration is smaller on a convex surface than that on a flat surface, and higher on a concave surface. As a consequence, vacancies flow from a neck (concave area) to a convex area and thus a mass flow occurs in the reverse direction. Mass flow through the volume of a grain (volume diffusion) may originate at convex surfaces, grain boundaries or dislocation in the grain matrix. Except in substances with high vapor pressure such as NaCl, the sintering of powders proceeds by the volume diffusion mechanism.

- *Path 3: Flow.* A perpendicular pressure pointing to the center of a powder particle acts on a convex surface, and a pressure pointing away from the center acts on a concave surface. When a substance is fluid, mass transport proceeds under the pressure difference.

- *Path 4: Dissolution–precipitation.* In the initial phase of sintering, the presence of a liquid phase which wets the solid phase allows the rearrangement of particle packing by gliding. Subsequently the substance dissolved from convex surfaces, where solubility is higher, is transported to concave surfaces, where solubility is lower, and precipitates out. Dissolution, precipitation or diffusion of a substance in the liquid phase can be rate-controlling.

Stages of sintering

The development of microstructures during sintering is rather complicated, but may be distinguished in three stages (Figure 2-14).

Initial stage. Initially, material is transported from higher-energy convex particle surfaces to the lower-energy, concave intersections between adjacent particles to form necks ("neck growth"). The powder particles fuse together and the area of contact increases gradually. Since mass is only transported from convex to concave areas, the total pore volume and the distance between the particle centers remain about constant, and shrinkage of the green body (▶ glossary) is only about 4–5 %. In this stage the relative density, which is the density of the powder compact divided by its theoretical density, is about 0.5–0.6.

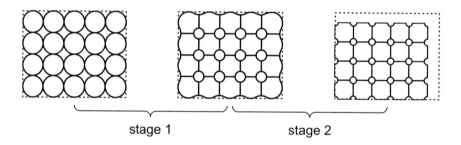

stage 1   stage 2

Figure 2-14. A two-dimensional sphere model illustrating the first two stages during sintering.

Intermediate stage. In this stage, interparticle necks grow, the area of grain boundaries (the interface plane shared by two grains) increases, interparticle contacts flatten, and the pore diameters decrease. The distance between the particle centers and the volume of the compact decreases (shrinkage of 5 % to 20 %), i.e. densification occurs. The relative density increases to about 0.95.

Final stage. When the relative density increases above 0.95, isolated spherical pores remain only at triple points (intersection lines where three grains (▶ glossary) meet) or inside the grain matrix. In the final stage, these pores are gradually eliminated and the relative density increases further.

Factors affecting sintering

The most important powder physical characteristics that can affect sintering are particle size, particle packing, and particle shape.

Particle size. Material transport occurs faster over shorter distances, and less material needs to be transported to fill small pores. Furthermore, very fine particles have high surface energies. Therefore, smaller powder particles speed up the sintering process and lower the sintering temperatures and pressure (see also Chapter 7). Due to the thermodynamic considerations discussed above, larger grains (▶ glossary) grow at the expense of smaller ones. Consequently, as sintering progresses, the average size of the grains increase ("coarsening"), and the size distribution becomes narrower. Since coarsening is much slower than sintering, grain growth can occur especially in the final sintering stage. Uncontrolled grain growth is usually detrimental to the ceramic's properties.

Particle packing. Improved particle packing increases the number of contact points between adjacent particles and the relative density of the compact. Consequently, densification occurs faster (better material transport) and with less volume shrinkage. One of the most important reasons for non-uniform particle packing is the formation of aggregates (see Section 3.3).

Particle shape. Irregular-shaped particles, which have a high surface area to volume ratio, have a higher driving force for densification and sinter faster than equiaxed particles. Particles that pack poorly sinter poorly.

Ceramics processing

Ceramic processing, i.e., the fabrication of ceramics from powders, involves a number of steps (Figure 2-15). Each of these steps must be carefully controlled because ceramics are basically flaw intolerant, and chemical and physical defects severely degrade properties. Mistakes in ceramic processing are cumulative, and generally cannot be corrected during sintering or by post-sintering processes.

We will only briefly describe the different steps of ceramic processing, not due to negligence of the importance of these topics, but in order to focus on the chemical issues. Readers interested in ceramic processing are referred to text-books on ceramic engineering.

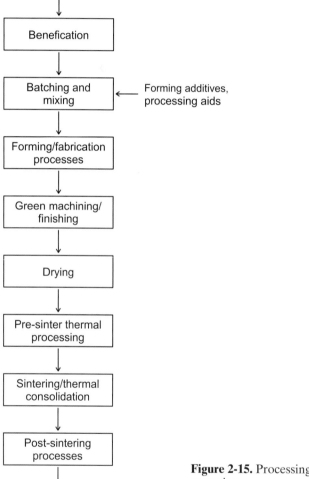

Figure 2-15. Processing steps in the fabrication of ceramics.

Benefication. These processes modify the chemical and/or physical properties of raw materials to render them more processable. They may include:

- particle size reduction by crushing, grinding or milling;

- purification by washing, extraction, chemical leaching (to extract contaminations), separation, etc.;
- sizing (sieving) and classification;
- calcination (▶ glossary) to effect decomposition of precursor compounds, to liberate gases (water) or to achieve structural transformations;
- preparation of a slurry by dispersion in liquids to wet the solid, break down agglomerates or stabilize the system to prevent agglomeration (flocculation); and
- granulation of powders to improve flow, handling, packaging, and compaction.

Batching and mixing. The constituents of a ceramic body are combined to produce a more chemically and physically homogeneous system for forming. Forming additives or processing aids are commonly used to render ceramic powders more processable.

- *Binders* are used to impart strength to a green (unfired) ceramic body (▶ glossary) for handling and machining. They are mostly organic polymers (poly(vinyl alcohol), poly(ethylene glycol), etc.) that adsorb to the particle surfaces and promote interparticle bridging or flocculation.
- *Plasticizers* are often added to binders to reduce crosslinking (▶ glossary) density and increase flexibility. Typical examples are water and ethylene glycol for vinyl binders, glycerin or ethylene glycol for clay bodies, or molten waxes or oils for thermoplastic polymers used in injection molding.
- *Lubricants* are added to lower friction between particles, and between particles and die surfaces. Stearic and oleic acid are good lubricants.
- *Deflocculants*, dispersants or anticoagulants are added to slurries to improve dispersion and dispersion stability. Deflocculants adsorb on particle surfaces and prevent the approach of particles either by electrostatic or steric stabilization (see Section 4.5).
- *Surfactants* or wetting agents can also be added to a slurry to improve dispersion. *Antifoams* (defoamers) are added to slurries to remove trapped gas bubbles from the liquid. Typical antifoams are fluorocarbons, silicones (see Section 5.2), stearates, and high molecular weight alcohols.

Forming/fabrication processes. A cohesive body of the desired shape and size is obtained by consolidation and molding of ceramic powders (Figure 2-16).

- *Dry forming*. Dry powders can be simultaneously compacted and shaped by pressing in a rigid die or flexible mold. Pressing is the most widely used forming process in ceramic manufacturing. Powders should be free-flowing, should have a high bulk density, should consist of deformable particles, and should not stick to the die. *Dry pressing* (also called die pressing) involves compacting a powder between two plungers in a die cavity (Figure 2-16a). *Isostatic pressing* allows the manufacture of complex shapes. Typically, a flexible (rubber) mold is filled with the powder, sealed, placed in a gas- or liquid-filled pressure chamber, and pressurized to compact the powder (Figure 2-16b). The gas or liquid serves to apply the pressure uniformly over the surface of the green body (▶ glossary).

- *Plastic forming*. Plastic ceramic bodies such as plastic clay bodies or ceramic bodies plasticized with organic binders deform inelastically without rupture under a compressive load. The cohesive strength required for plastic deformation is provided by capillary pore pressure and particle agglomeration. In the *extrusion* method (Figure 2-16c), the plastic body is ejected through an opening of the die to shape green bodies (▶ glossary) of simple geometries (for example, tubes, bricks, tiles, catalyst supports, electrodes). *Jiggering* involves shaping of a plastic clay body on a spinning porous plaster mold by a moistened shaping tool (Figure 2-16d). This is the most widely used forming technique used in manufacturing small, axial-symmetrical whiteware ceramics such as china or electrical porcelain. In *powder injection molding*, a hot ceramic/binder mixture is injected into a mold with a desired shape (Figure 2-16e). The binder can be removed easily from the green forms by thermal decomposition prior to sintering. It is possible to form fairly complicated shapes such as turbochargers by this method.

- *Paste forming*. Ceramic thick-film pastes, consisting of a ceramic powder, a sintering aid such as borosilicate glass, and organic additives, are used for decorating ceramic tableware or to form capacitors and insulation layers for microelectronic packaging.

- *Slurry forming*. Slurries with proper viscosity are prepared by dispersing powder in a solvent, usually water. A good slurry is well dispersed and free of air bubbles, and the particle setting rate should be low. A high solid concentration is desirable to minimize drying shrinkage and the time and energy required for forming.

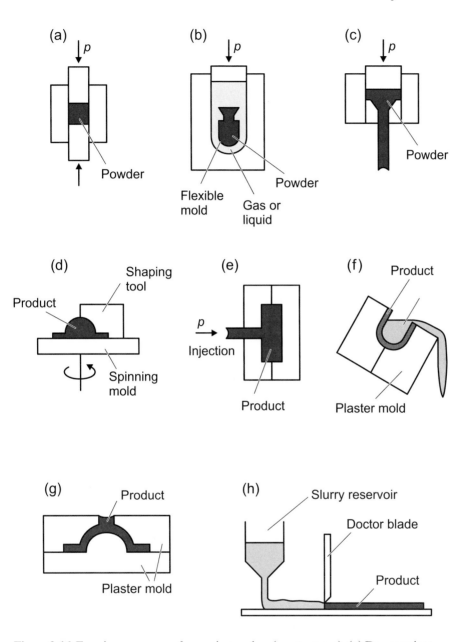

Figure 2-16. Forming processes of ceramic powders (p = pressure). (a) Dry pressing; (b) isostatic pressing; (c) extrusion; (d) jiggering; (e) injection molding; (f) drain casting; (g) solid casting; and (h) tape casting.

In the *slip-casting* method the slurries are poured in a porous mould having a cavity of the desired shape. The capillary suction of the porous mold draws the liquid from the slurry. A higher solids content cast is formed on the inner wall of the mold, its thickness increasing with the time. If the excess slurry is drained after a fixed time, hollow shapes are obtained ("*drain casting*", Figure 2-16f). Alternatively, the entire amount of slurry can be used to form a dense shape ("*solid casting*"; Figure 2-16g).

- *Gel casting* employs *in-situ* polymerization (▶ glossary) of organic monomers to produce a polymer that binds individual particles together to form a cohesive body. *Coagulation casting* is a related process in which the green body (▶ glossary) is formed by coagulation of particles. This is usually induced by a pH change. For example, a rise in pH can be achieved by adding urea to the slurry. Decomposition of urea by the enzyme urease liberates ammonia that increases the pH. *Tape casting* (doctor blading, knife coating) is used for producing ceramic sheets or tapes. A slurry is poured on a smooth supporting surface; its thickness is controlled by the doctor blade (Figure 2-16h).

Green finishing/machining. This is often required after forming in order to eliminate rough surfaces, smooth forming seams, and modify the size and shape of the green body (▶ glossary) in preparing for sintering.

Drying. Drying of the green body must be carefully controlled in order to avoid cracking, warping, and shape distortions. The different stages during drying are the same as discussed for the drying of gels later in this book (Section 4.5).

Pre-sinter thermal processing. In this step, organic or inorganic additives and impurities are burnt out or decomposed, and residual moisture is removed. This is achieved by heat treatment below the sintering temperature, or in a controlled series of ramps and isothermal holds.

Sintering. A ceramic body densifies during sintering. Thermally activated material transport (see above) transforms loosely bound particles into a dense, cohesive body. Porosity or void space between the particles is reduced, and interparticle contact or grain (▶ glossary) boundary area increases. The methods for sintering of powder compacts can be classified into two major categories. One is called *pressureless sintering*, which involves heating of the formed compact to about two-thirds of its melting temperature at ambient pressure and holding for a given time. *Pressure sintering* employs the simultaneous use of pressure and temperature, i.e., densification during sintering.

- *Pressureless sintering.* In *solid-state sintering*, densification occurs by solid-state, diffusion-controlled mass transport. Since mass transport occurs faster through a liquid than through a solid, densification occurs faster in the presence of a liquid phase. *Liquid-phase sintering* (LPS) requires a liquid phase (melt) that is chemically compatible, that wets the particles, and that dissolves some of the solid. In a system containing multiple phases, such as cermets or porcelain, the sintering temperature must be between the melting point of the phase that forms the melt and that of the other phase(s). Many non-oxide powders with covalent bonds between the constituting elements are difficult to sinter by solid-state sintering because mass transport by diffusion is low, even at high temperatures. In such cases, sintering aids are added that form a liquid phase. For example, the addition of a few percent of certain oxides to Si_3N_4 leads to liquid silicate phases during sintering.

- *Pressure sintering.* Compared to pressureless sintering, pressure sintering is more expensive, does not allow to produce complicated shapes, and is not suited for mass production. On the other hand, pressure sintering lowers the sintering temperature or allows faster sintering at the same temperature than pressureless sintering. Pressure sintering is usually used in manufacturing ceramics that are difficult to sinter without pressure such as non-oxide ceramics. The most common technique is called *hot pressing*. A powder is compacted in a die and pressed uniaxially between two forming rams while heated to high temperatures. The commonly used graphite dies allow temperatures up to 2500 °C, but limit the maximum pressure to about 100 MPa. *Hot isostatic pressing* (HIP) is similar to cold isostatic pressing used in forming of the green body (▶ glossary), except that the pressing is done at the sintering temperature. The "flexible" mold is then made of metals or glasses that soften at high temperatures and transmit the applied pressure uniformly to the ceramic body. The pressuring medium is typically a gas (argon). Complex shapes can be formed by HIP. Si_3N_4 can be sintered by HIP under a nitrogen gas pressure of 1–5 MPa, because high-pressure nitrogen can suppress the decomposition reaction in elemental silicon and nitrogen at temperatures above about 1800 °C.

Postsintering processes. These may involve machining to meet dimensional tolerances, surface coating or modification, etc.

2.2 Solid–Gas Reactions

The use of a combination of solid and gaseous reactants allows the problem of poor contact between two solids to be overcome. Furthermore, the unfavorable equilibria involving gaseous species (for example, the thermal decomposition of nitrides into the elements) will be shifted towards the products because there is a large excess of the gaseous reactant.

We will discuss some important process options for an industrially important solid–gas reaction, the "direct nitridation" of elemental silicon to give silicon nitride (Si_3N_4). Silicon nitride ceramics have high reliability, thermal shock and wear resistance, strength and hardness (▶ glossary) as well as good corrosion and oxidation resistance even up to 1200–1300 °C. The excellent thermomechanical properties combined with the low density make Si_3N_4 suitable for applications such as moving automotive parts (valves, turbocharger rotors, piston heads, etc.) or gas turbine rotors. The excellent wear resistance is utilized in metal machining inserts or ultra-high speed bearings.

There are two hexagonal modifications of crystalline Si_3N_4, the α- and the β-phase (Figure 2-17). In both modifications, three-dimensional networks are formed from corner-sharing SiN_4 tetrahedra. The covalent Si–N bonds are the reason for the high hardness (▶ glossary) and strength (the same is true for SiO_2 or SiC, because of the covalent Si–O and Si–C bonds).

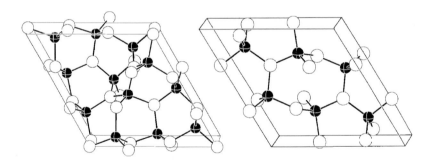

Figure 2-17. The crystal structures of α- (left) and β-Si_3N_4 (right).

The α-phase is the low-temperature modification that transforms to the β-phase at temperatures >1650 °C. The α/β ratio is an important parameter for the characterization of Si_3N_4 powders. The α-phase has a better sintering behavior.

There are three technically important methods for the preparation of Si_3N_4 powders:

1. Carbothermal nitridation of SiO_2 (see Section 2.1.2).
2. Ammonolysis of $SiCl_4$ either in the liquid state (see Section 5.6) or in the gas phase (see Section 3.3).
3. Nitridation of silicon powder.

The latter method, although nearly a century old, is still the dominating technical process. The overall process is represented by Eq. 2-25.

$$3\ Si + 2\ N_2 \longrightarrow Si_3N_4 \tag{2-25}$$

The process can be run continuously or discontinuously. The purity of the silicon employed determines the purity of the obtained Si_3N_4 powder. Therefore, semiconductor-grade (▶ glossary) silicon is required to obtain highly pure Si_3N_4 powders. Reaction with elemental nitrogen proceeds with sufficient reaction rates only above 1100 °C. The rate not only depends on the particle size of the silicon powder employed, but also on the presence of impurities. For example, traces of iron catalyze the nitridation reaction. Owing to the exothermic nature of the reaction ($\Delta H = -746$ kJ $\cdot$ mol^{-1}), the process must be carefully controlled. The temperature must be high enough to obtain complete conversion, but too high temperatures favor formation of the undesired β-phase and of aggregates. Optimum results are obtained at about 1250–1350 °C with the nitridation gas diluted by 20–40 % H_2 and the silicon powder diluted by α-Si_3N_4. Both measures help to moderate the exothermic nature of the direct nitridation reaction. However, the tight temperature control requires extended periods of reaction times (several hours). Intermediate grinding is needed to expose unreacted silicon to the nitridation atmosphere.

When Si_3N_4 is prepared in a combustion mode, high gas pressures are necessary to reach self-sustaining conditions. Otherwise gas is depleted at the reaction front. Ignition does not occur if the nitrogen pressure is below 50 MPa. At higher pressures, however, the reaction becomes self-propagating. The high pressure is also useful to suppress the thermal decomposition of Si_3N_4. In contrast to the direct nitridation, the reaction is extremely rapid, and since temperatures >1700 °C are reached, a product with a high portion (>95 %) of the β-phase is obtained. As was the case with the controlled nitridation, the addition of Si_3N_4 as a diluent is an important process variable.

Many oxides, hydrides, nitrides, and oxynitrides are formed by heating a solid reagent in an oxygen (air), hydrogen, or nitrogen atmosphere, respectively. Ammonia gas is also used as a nitrogen source for the preparation of nitrides and oxynitrides (see also Chapter 3).

As discussed above for Si_3N_4, conventional methods of solid–gas syntheses involve heating a solid in a gas atmosphere below its ignition point. The reaction rates are low at these conditions because metal ions must diffuse through the product layer to the product/gas interface, or gas molecules through the product layer to the metal/product interface.

Only when the ignition conditions (temperature and gas pressure) are met, the process can proceed in a self-sustaining regime (see Section 2.1.3). High gas pressures must often be employed in order to avoid depletion of the gas at the reaction front. The reactant gas flow can be either co-current (Figure 2-18, left) or counter-current (Figure 2-18, right) to the direction of the propagating combustion front.

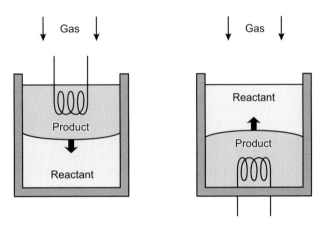

Figure 2-18. Gas flow in self-sustaining gas/solid reactions.

Another option to avoid gas depletion is the addition of solid compounds that cleave the reactant gas upon heating. For example, complete conversion of the metal to the corresponding nitride was achieved in some cases by compacting the metal powder with a stoichiometric portion of NaN_3. An oxidizer (nitrates, for example) may be added when oxides are prepared. A fuel (e.g., hydrazine, glycine, urea) can also be helpful to obtain spontaneous combustion, especially in combination with oxidizers.

When a pure metal burns in a gas atmosphere, the heat generated during the reaction is so high that melting of the reacting metal or the product and/or decomposition of the product may occur (for example, transition metal hydrides are unstable at temperatures between 500–1000 °C). Excessive melting of the solid reactant can cause a significant decrease in the gas permeability (▶ glossary), and hence incomplete reaction. Dilution of the metal reactant with an inert compound (usually the product of the reaction) can reduce the reaction temperature and therefore the extent of melting.

2.3 Decomposition and Dehydration Reactions

Another case involving both a solid and a gas phase are thermal decomposition reactions. While in the previously discussed solid–gas reactions the gaseous compound is a reactant and has to be carried *to* the reaction front, it is a product in thermal decomposition reactions and has to be carried *away*. Typical salts that decompose by generation of a gaseous by-product when heated to high temperatures are carbonates, oxalates, nitrates, and hydroxides. The use of these compounds to generate metal oxides *in situ* in solid–solid reactions was mentioned earlier in Section 2.1.1. Thermal decomposition can, of course, also be used to prepare bulk quantities of the corresponding binary metal oxides. A technically applied example is the preparation of γ-Al_2O_3 by dehydration of boehmite (AlOOH) (see also Section 4.4).

In general, thermal decomposition of solids initiates at some structural defects such as surfaces, grain boundaries (▶ glossary), or dislocations. Small nuclei are formed that subsequently grow with time. The most common type of thermal decomposition follows a sigmoid rate law, and is represented by the Avrami–Erofeyev equation (Eq. 2-2).

A solid-state reaction is said to be topochemically controlled when the reactivity is controlled by the crystal structure rather than by the chemical nature of the reactants. The products obtained in many solid-state decomposition reactions are determined by topochemical factors, particularly when the reaction occurs within the solid without the separation of a new phase. In topochemical solid-state reactions, only little structural rearrangement of the atoms leads to new phases (see also Section 2.1.1). They allow the preparation of metastable solids (▶ glossary) not accessible by other approaches. The conditions for such reactions are usually much milder than conventional solid-state syntheses. Therefore, they are considered as "soft chemistry".

One of the best investigated examples is the formation of metastable MoO_3 and WO_3 phases by dehydration of $MoO_3 \cdot {}^1/_3\, H_2O$ and $WO_3 \cdot {}^1/_3\, H_2O$ at 300 °C. The thermodynamic stable structures of WO_3 (ReO_3-type monoclinic structure with [WO_6] octahedra sharing all edges with neighboring octahedra) and MoO_3 (layered orthorhombic structure) are shown schematically at the right-hand side of Figure 2-19.

The basic structural element of the hydrates is a plane of corner-sharing octahedra forming six-membered rings in the crystallographic (001) plane (left-hand side of Figure 2-19; only one ring per plane is shown). The planes are stacked parallel to each other, with every other layer shifted by half the ring diameter (vertically in Figure 2-19; i.e., the central octahedron in the structure plots of $MO_3 \cdot {}^1/_3$ H_2O belongs to the layer below).

In the metastable hexagonal WO_3 phase formed upon heating $WO_3 \cdot {}^1/_3$ H_2O, the same layers of six-membered rings of $[WO_6]$ octahedra are found as in the precursor hydrate. However, they are stacked above each other, leading to tunnels along perpendicular to the (001) planes. Dehydration of $MoO_3 \cdot {}^1/_3$ H_2O at 300 °C results in a (metastable) monoclinic MoO_3 modification with a ReO_3-type structure (the same as the thermodynamically stable modification of WO_3).

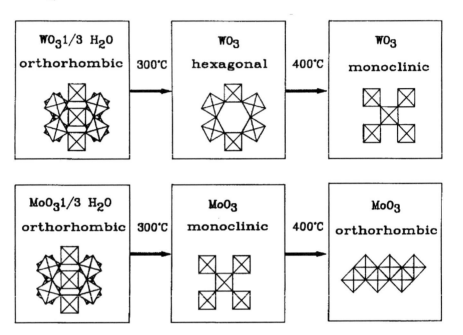

Figure 2-19. Structural relations in the dehydration of $MoO_3 \cdot {}^1/_3 H_2O$ and $WO_3 \cdot {}^1/_3$ H_2O and in the phase transformations of MoO_3 and WO_3. Each crossed square represents an octahedron viewed from above.

In both cases, dehydration proceeds by oriented nucleation (▶ glossary) of the metastable (▶ glossary) phase on a crystallographic plane of the hydrate, and of oriented growth with displacement of the hydrate/oxide interface boundary. The difference arises from nucleation (▶ glossary) at different crystallographic planes. WO_3 nucleates at the (001) plane of $WO_3 \cdot {}^1/_3 H_2O$ (the plane of the six-membered rings), while the nucleation energy for MoO_3 is lower at the (100) plane of $MoO_3 \cdot {}^1/_3 H_2O$ (the trace of the plane being horizontal in Figure 2-19). Inspection of the crystal structures of $MoO_3 \cdot {}^1/_3 H_2O$

and MoO_3 shows that the oxygen arrangement in the (100) plane is indeed similar.

Other examples of topochemical reactions mentioned in this book are the annealing of $YBa_2Cu_3O_{7-y}$ in an oxygen atmosphere to give superconducting (▶ glossary) $YBa_2Cu_3O_{7-x}$ (y > x) (see Section 2.1.1), or the solid–state polymerization (▶ glossary) of cyclic S_2N_2 to $[SN]_n$ mentioned in Section 5.7 (see Figure 5-17).

2.4 Intercalation Reactions

The most important group of topochemical reactions in inorganic chemistry are intercalation and deintercalation reactions.

The term *intercalation* describes the insertion of guest species (atoms, molecules or ions), called intercalates, into a crystalline host lattice that retains its general structural features. However, during intercalation reactions the guest and host may undergo perturbations in their geometrical, chemical, and electronic environment. Intercalation results in the modification of the properties of the host material, giving rise to materials with new properties. Intercalation reactions are usually reversible, and typically occur near room temperature.

After some general remarks on layered solid compounds and a technically important example for the use of intercalaction compounds – the lithium ion battery – we will briefly touch on the mechanistic aspects of intercalation reactions, and then turn to methods to prepare intercalation compounds (Section 2.4.2). At the end of this section, two new promising topics of intercalation chemistry will be introduced: the preparation of inorganic–organic hybrid materials and of porous solids by "pillaring" (Section 2.4.3).

2.4.1 General Aspects

Two classes of intercalation systems can be distinguished with respect to the dimensionality of the host lattice:

1. Three-dimensional lattices with parallel channels or interconnected channel systems, as in zeolites (see Section 6.4). The uptake of guests is restricted to those having smaller diameters than the pores ("molecular sieves").

2. Low-dimensional host lattices (layered lattices, or chain structures consisting of one-dimensional stacks). The spacings between the layers or chains can adapt to the dimensions of the guest species.

The remainder of this section is restricted to layered compounds, the most important group in terms of chemical reactivity and applications. The ability to adjust the interlayer separation is responsible for the wider occurrence of intercalation compounds for layered structures.

A large variety of guest species can be intercalated, ranging from protons and metal atoms or ions up to large molecular species and polymers. The separation of two host layers can vary over wide ranges, depending on the size and orientation of the host. This is illustrated by some graphite intercalation compounds. Graphite is probably the best investigated host material, and forms intercalation compounds with several hundred chemical species. The graphite layers are only one atom thick, and therefore relatively flexible materials are obtained after intercalation. The interlayer distance (335 pm) is only slightly enlarged (to 371 pm) when lithium is intercalated. Intermediate cases are, for example, potassium (535 pm; Figure 2-20) or AsF_5 (815 pm) as intercalates. Intercalation of KHg leads to an interlayer separation of 1022 pm with potassium and mercury arranged in three layers in between the graphite layers.

In general, layered compounds are characterized by strong *intra*layer bonds and weak *inter*layer interactions. The layers may be electrically neutral or charged:

- *Neutral layers*, held together by van der Waals interactions: for example, graphite, oxychlorides (FeOCl), transition metal disulfides.
- *Negatively charged layers*, separated by mobile (exchangeable) cations: for example, clays, transition metal phosphates.
- *Positively charged layers*, separated by anions: layered double hydroxides, $[M_{1-x}{}^{II}M_x{}^{III}(OH)_2]^{x+}$.

Examples for the different types of layered compounds will be given below. With respect to the *electronic properties* of the host lattice two qualitatively different types are found:

1. For *insulator host lattices* (zeolites, layered aluminosilicates, metal phosphates, etc.) the basic physical properties of the host lattice are not affected by intercalation. They are interesting as catalysts or catalyst supports, adsorbents, ion exchangers, etc.
2. Host *lattices which can reversibly uptake electrons from or* (less often) *donate electrons to the host* (graphite, metal dichalcogenides, metal oxy-

halides, etc.) may be reduced or oxidized during intercalation. This results in strong changes of the physical properties of the host matrix, i.e., the electric conductivity. Such intercalation compounds are of interest as electrode materials for batteries, as electrochromic systems, sensor materials, etc.

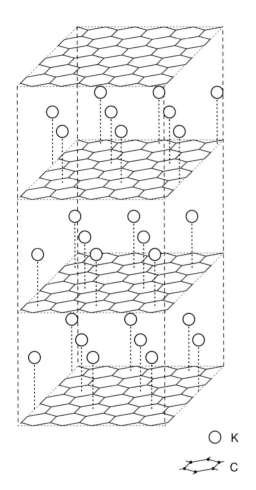

○ K

<svg>⬡</svg> C

Figure 2-20. Model of the C_8K structure.

A prototypical example for the latter case is the reversible reaction of lithium with TiS_2 (Eq. 2-26).

$$Li + TiS_2 \longrightarrow Li^+[TiS_2]^- \qquad (2\text{-}26)$$

Many of the metal dichalcogenides MX_2 (M = Ti, Zr, Hf, V, Nb, Ta, Mo, W, Sn; X = S, Se) possess layered structures. The layers consist of two layers of close-packed chalcogenide ions between which cations reside in octahedral or trigonal prismatic sites. Thus, the sequence of atoms perpendicular to the layers is X-M-X···X-M-X··· (Figure 2-21).

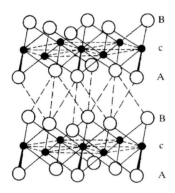

Figure 2-21. The structure of TiS_2. The open circles (A and B layers) represent the sulfur atoms, the full circles (c layers) the metal atoms.

During the intercalation of lithium into TiS_2, Ti^{4+} is reduced to Ti^{3+}, and lithium cations are inserted between the sulfide layers. The van der Waals forces between the layers of TiS_2 are replaced by Coulomb interactions between the negatively charged layers and the Li^+ cations. The interlayer spacing of the final product, $LiTiS_2$, is expanded relative to the parent TiS_2.

Several other transition metal oxides or chalcogenides are capable of reversible lithium insertion. Apart from protons, no other cations than Li^+ can penetrate so easily into solids. This is the chemical basis for ambient-temperature rechargeable batteries ("lithium batteries") used in cellular phones, portable computers, etc. Lithium batteries offer several advantages, such as higher voltages, higher energy density, and longer shelf-life compared to other systems. The operating principle of a typical battery is shown in Figure 2-22.

In lithium ion transfer cells both electrodes are capable of reversible lithium insertion. Because of the difference in chemical potentials of Li in the two electrodes, the transfer of Li^+ from the cathode through the electrolyte to the anode (discharge) delivers energy, while the reverse process (charge) consumes energy (Figure 2-22). Such cells are also called "rocking-chair" cells, because the Li^+ ions shuttle between the cathode and anode host during the discharge-charge cycles.

The majority of commercially available cells use layered $LiCoO_2$ as the cathode, a graphitic carbon as the anode, and liquid organic (lithium salt dissolved in an aprotic organic solvent) or polymer electrolytes (Eq. 2-27). Other cathode materials are $LiNiO_2$, the spinel (▶ glossary) $LiMn_2O_4$, and V_2O_5. Use of the fully lithiated oxides has the advantage that the cells can be manufactured in the discharged state (i.e., from $LiCoO_2$ and graphite).

cathode: $Li_{0.35}CoO_2 + 0.65\ Li^+ + 0.65\ e^-$ $\underset{charge}{\overset{discharge}{\rightleftharpoons}}$ $LiCoO_2$

anode: C_6Li_x $\underset{charge}{\overset{discharge}{\rightleftharpoons}}$ $C_{graphite} + x\ Li^+ + x\ e^-$

overall: $Li_{0.35}CoO_2 + C_6Li$ $\rightleftharpoons$ $LiCoO_2 + graphite$
        charged   *discharged*

$$(2\text{-}27)$$

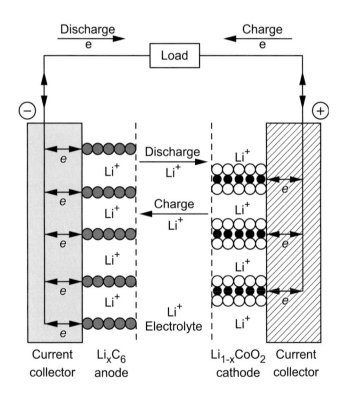

Figure 2-22. Schematic illustration of the charge/discharge process in a rechargeable lithium ion battery.

The carbon material used for the anode is polycrystalline and consists of aggregates of graphite crystals. At ambient pressure, a maximum loading of C_6Li can be reached for highly crystalline graphitic carbon.

The structure of $LiCoO_2$ is derived from the layered metal dichalcogenide structures MX_2 discussed above (Figure 2-21). The MO_2 equivalents are not stable, due to the higher electronegativity and lower polarizability of oxygen compared to sulfur or selenium. This leads to more ionic M–O bonds, and a stronger repulsion between adjacent oxygen layers. However, alkali metal ions (M′) between the layers stabilize the layered structures $(M′_xMO_2)$, because they screen the anionic repulsion. The Li^+ ions have a high two-dimensional mobility between the CoO_2 layers. Because of the instability of a layered CoO_2 structure, not all Li^+ is electrochemically removed from $LiCoO_2$.

Mechanistic aspects of intercalation reactions

Intercalation reactions in general involve breaking the interlayer interactions in the host lattice and the formation of new interactions between the guest and host. Therefore, layered host lattices with stronger interlayer bonding are more difficult to intercalate. This effect can be observed qualitatively in the intercalation reactions of the layered transition metal dichalcogenides (MX_2). The metal disulfides intercalate guest molecules under milder conditions than the equivalent isomorphous diselenides (MSe_2), and there is no report of intercalation of a layered metal ditelluride (MTe_2). This trend is attributed to increased covalent bonding between the dichalcogenide layers in the series S < Se < Te.

The difficulty of breaking the interlayer interactions can be minimized by the phenomenon of staging. Staging refers to situations in which certain interlayer regions are *totally vacant* while others are *partially or totally occupied*. The order of the staging is given by the number of layers between successive filled or partially filled layers (Figure 2-23).

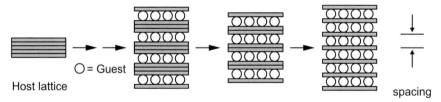

Figure 2-23. Representation of nth stage intercalates (n = 1–3).

Two extreme staging behaviors are observed in the reaction between guest and host species. A successive sequence of stages, from higher stages down to the 1st stage is observed, when potassium is reacted with graphite, for example. Each stage corresponds to a specific stoichiometry (for example, 1st stage: C_8K, Figure 2-20) and can be converted to the (n–1)th stage by reaction with additional potassium. The staging behavior is a general phenomenon in graphite intercalation compounds.

An example for the other staging behavior is Li_xTiS_2 ($0 \leq x \leq 1$), where a single first stage phase (i.e., a certain amount of Li ions between each pair of successive layers) is formed over the entire composition range. In reality, most intercalation reactions lie somewhere between.

One model explains staging in terms of a domain structure with all interlayers being involved, but with local concentrations of the guest species in certain regions (Figure 2-24).

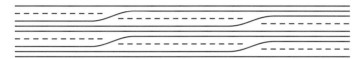

Figure 2-24. Model of a stage 3 compound with the intercalated compound $(-)$ between all layers.

The kinetics of incorporation of guest species into host lattices is complex. The reaction seems to be initiated at defects on the host surface. Therefore, variations in reactivity often arise from one batch to another of the same host due to microcrystalline or particle size differences.

2.4.2 Preparative Methods

The products of intercalation reactions are insoluble and therefore cannot be purified or separated.

Types of intercalation reactions

Direct reaction. The simplest and most widely used method is the direct reaction of the guest with a potential host. For example, the first alkali metal intercalates of the metal dichalcogenides were prepared by the reaction of the solid host with the metal vapor at temperatures of 600–800 °C (as in Eq. 2-26). An alternative is the use of alkaline metals dissolved in liquid ammonia. This offers many advantages, but ammonia is often cointercalated. Heating *in vacuo* to remove the ammonia may result in undesired side reactions.

Butyllithium (LiBu) is a simple and generally applicable reagent for lithium intercalation, giving products of high purity (Eq. 2-28).

$$x \, LiBu + MX_n \longrightarrow Li_xMX_n + x/2 \, C_8H_{18} \tag{2-28}$$

Liquid compounds or low-melting solids can be used neat, whereas solid organic and organometallic guests are usually dissolved in polar organic solvents. Although highly polar solvents often accelerate the intercalation process, they can complicate the product distribution by cointercalating with the guest molecules, as mentioned above for ammonia.

Examples for the direct reaction are given in Table 2-3. Note that there are examples for reactions proceeding with and without electron transfer.

Table 2-3. Examples for the direct reaction.

Reactants (host/guest)	Products
TaS_2/pyridine	$TaS_2(C_5H_5N)_{0.5}$
VSe_2/Cp_2Co	$VSe_2(Cp_2Co)_{0.25}$
TiS_2/Na(naphthalide)	$NaTiS_2$
α-$Zr(HPO_4)_2 \cdot H_2O$/RNH_2	$Zr(HPO_4)_2(RNH_2)_2$
$V(O)PO_4$/EtOH	$V(O)PO_4(EtOH)_2$
graphite/K	C_8K
graphite/H_2SO_4	$C_{24}{}^+HSO_4{}^- \cdot 2 \, H_2SO_4$
graphite/OsF_6	$C_8(OsF_6)$
graphite/$AlBr_3$/Br_2	$C_9(AlBr_3)(Br_2)$
FeOCl/pyrrole	$FeOCl(polypyrrole)_{0.34}$
FeOCl/Cp_2Fe	$FeOCl(Cp_2Fe)_{0.16}$
$NiPS_3$/LiBu	$Li_{4.4}NiPS_3$
MoO_3/H^+/reductant	$H_{1.7}MoO_3$
V_2O_5/LiI	$Li_xV_2O_5$

Almost all organometallic compounds known to form intercalates are good reducing agents, i.e. they form stable cations. Metallocenes, Cp_2M, are the most prominent examples. There must be a match between the reducing power of the organometallic guest and its ability to intercalate specific host lattices. For example, cobaltocene, Cp_2Co, with a first ionization potential of 5.5 eV, will intercalate into both TaS_2 and $FeOCl$, but ferrocene, Cp_2Fe, which is much less reducing (first ionization potential 6.88 eV) will only intercalate into the more oxidizing host lattices of $FeOCl$ or α-$RuCl_3$. The $FeOCl$ structure type (e.g., $FeOCl$, $TiOCl$, $VOCl$) consists of neutral layers of edge-sharing distorted *cis*-$[FeCl_2O_4]$ octahedra, with the chlorine atoms directed towards the interlayer space (Figure 2-25).

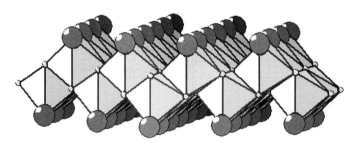

Figure 2-25. One layer of the FeOCl structure. The large spheres represent the chlorine atoms directed towards the interlayer space.

Graphite is the only host lattice known to be able to undergo intercalation reactions by either reduction (e.g., formation of C_8K by reaction with metallic potassium) or oxidation of the layers (e.g., formation of C_8Br by reaction with bromine or of $C_{24}{}^+NO_3{}^- \cdot 3\ HNO_3$ by reaction with nitric acid).

Metal phosphorus trisulfides, MPS_3, or more precisely $M_2(P_2S_6)$ (M = Mn^{2+}, Cd^{2+}, and many other divalent cations), are structural analogues of the metal dichalcogenides (see Figure 2-21), if one considers half of the metal atoms replaced by P_2 pairs. They exhibit two types of intercalation reactions. The first are redox intercalations with reducing guests (see the reaction of $NiPS_3$ with LiBu in Table 2-3 as an example). The second type is unique for the MPS_3 compounds and involves intercalation of cations with concomitant loss of the M^{2+} cations from the layers into solution to maintain charge balance (see the reaction of $MnPS_3$ with KCl in Table 2-4 as an example).

Intercalation reactions sometimes take weeks or even months for completion, depending on the host, and sometimes require elevated temperatures. Sonification of reaction mixtures increases the rates of intercalation in many cases. This is not due to improvement of mass transport, but in a significant decrease in the particle size and surface damage of the host. As a matter of fact, the crystallinity of the final product is much inferior compared to the material obtained using conventional methods.

Ion exchange. Intercalated guest ions can often be replaced by immersing the material in a concentrated solution containing another potential guest ion. Some illustrative examples are given in Table 2-4. In some cases, pre-intercalating with an alkali metal ion and then ion exchanging the alkali metal cation with the final guest cation is a useful strategy for intercalating large cations that do not intercalate directly (see intercalation of the tris(2,2'-bipyridyl)ruthenium dication into $MnPS_3$ in Table 2-4).

Intercalation reactions involving silicate hosts (zeolites, clays, pyrochlores, etc.) generally proceed by ion-exchange processes.

Table 2-4. Examples for ion-exchange reactions.

Reactants	Products	Solvent
$Na_xTiS_2/LiPF_6$	Li_xTiS_2	dioxolane
$K[Al_2(OH)_2(Si_3AlO_{10})]/Na^+$	$Na[Al_2(OH)_2(Si_3AlO_{10})]$	H_2O
$Na_{0.33}(H_2O)_{0.66}TaS_2/[Cp_2Co]I$	$TaS_2(Cp_2Co)_{0.2}$	MeOH
$MnPS_3/KCl$	$Mn_{0.8}PS_3(K)_{0.4}(H_2O)_{0.9}$	H_2O
$Mn_{0.8}PS_3(K)_{0.4}(H_2O)_y/[Ru(bipy)_3]^{2+}$	$Mn_{0.8}PS_3[Ru(bipy)_3]_{0.2}$ *	H_2O

* bipy = 2,2'-bipyridyl

Long before there was a scientific understanding of their properties, clays enjoyed wide practical us as clarifiers for alcoholic beverages and as decolorizers for edible oils. The practice of mixing raw wool with an aqueous slurry of clays to remove grease dates back to antiquity. *Smectic clays* consist of negatively charged layers that are compensated and bound together by readily exchangeable interlayer cations (normally hydrated Na^+ or Ca^{2+}). A charged layer contains a central sublayer of edge-sharing octahedra sandwiched by two sublayers of linked tetrahedra ("2:1 layer"). The cations in the tetrahedral layer are normally Si^{4+}, and those in the octahedral layer divalent or trivalent cations (for example, Al^{3+}, Fe^{3+}, Mg^{2+}, or Fe^{2+}). Substitution of metal cations with similar size but lower valency (i.e., substitution of Si^{4+} by Al^{3+} in the tetrahedral sublayers, or Al^{3+} by Mg^{2+} or Mg^{2+} by Li^+ in the octahedral sublayer) creates a net negative charge for the layers. This is compensated by the interlayer cations. An example is the mineral muscovite $K[Al_2(OH)_2(AlSi_3O_{10})]$ (see left side of Figure 2-27). One quarter of the silicon atoms in the tetrahedral layer are replaced by Al^{3+} ions. The thus created negative layer charge is balanced by interlayer K^+ cations. The amount and the site of the substitution influences the properties of the clays, since they determine the charge density and the strength of the interaction between the layers and the interlayer cations. Their swelling property makes smectic clays particularly attractive. By dispersion in water, such clays form a stable thixotropic gel (▶ glossary).

Another class of clay minerals is represented by kaolinite, $Al_2(OH)_4(Si_2O_5)$ consisting of one tetrahedral silicate layer and one layer of edge-sharing octahedra with central Al^{3+} cations. The 1:1 layers are electrically neutral and are linked head-to-tail via hydrogen bridges. Kaolinite is one of the few clay minerals capable of intercalating neutral guest species. However, the guest species must be able to form strong hydrogen bonds with the host's layers (e.g., urea, acetamide).

Exfoliation and reflocculation. Layered lattices are able to disintegrate under appropriate reaction conditions to give colloidal solutions (▶ glossary) of the layers. Among the oldest examples for this behavior are smectic clays. Complete delamination (exfoliation) to colloid suspensions is achieved under appropriate conditions. Another example is the hydrated alkali metal dichalcogenides ($Na_x(H_2O)_yTaS_2$, for example) which can be completely dispersed in solvent with high dielectric constants.

An interesting aspect of colloid (▶ glossary) formation is the possibility of reconstructing the layered structure in the presence of a guest. Large molecules, such as cyanines, substituted ferrocenes, hydrocarbons, or polymers can be inserted in the host lattices by reacting them with the colloid solution. Subsequent precipitation of the intercalated systems (reflocculation) is achieved by removal of the solvent or increasing the electrolyte concentration. The structure of thus obtained polyaniline/MoS_2 is shown in Figure 2-26, as an example.

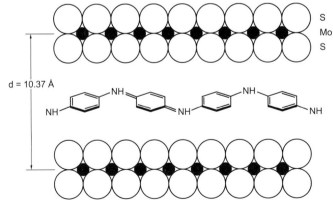

$d = 10.37$ Å

S
Mo
S

Figure 2-26. Schematic structure of polyaniline/MoS_2.

Electrointercalation. Electrointercalation has been discussed earlier in the case of lithium batteries. The host lattice serves as the cathode of an electrochemical cell. Some of the advantages of the electrolysis method over conventional techniques are its simplicity, ease of control of stoichiometry, and fast reaction rate at room temperature.

Important examples of electrointercalation is the oxidation and reduction of various compounds $A_xB_yO_z$ whose structures are derived from perovskite (▶ glossary) ABO_3 (see Section 2.1.1). Oxidation and reduction can be performed either electrochemically, or by reaction with oxygen or hydrogen. In the reduced state, the compounds belong essentially to the $A_nB_nO_{3n-1}$ and $A_{n+1}B_nO_{3n+1}$ series, with A = La, Nd, Sr or Ba, and B = Fe, Co, Ni or Cu. The redox reaction can be summarized as in Eq. 2-29, δ representing the amount of inserted oxygen. Typical examples in the $A_nB_nO_{3n-1}$ series are $Sr_2M_2O_5$ (M = Co and/or Ni), resulting in metallic $SrMO_3$ after oxidation. Examples of the $A_{n+1}B_nO_{3n+1}$ series include La_2CuO_4, M_2NiO_4 (M = Nd or La) and $La_3Ni_2O_7$. Their oxidation gives superconducting (▶ glossary), semiconducting (▶ glossary) or metallic compounds, respectively. In the $A_{n+1}B_nO_{3n+1}$ series, the compounds are only partially oxidized (δ typically 0.1 to 0.25).

$$A_xB_yO_z + 2\delta\ OH^- \longrightarrow A_xB_yO_{z+\delta} + \delta\ H_2O + 2\delta\ e^- \qquad (2\text{-}29)$$

The structures of the compounds in the reduced state can be considered as oxygen-deficient perovskites (▶ glossary) with channels perpendicular to the z axis. Upon oxidation, oxygen diffuses in and either partially or completely fills the vacant sites in these channels. Note that annealing of $YBa_2Cu_3O_6$ (Section 2.1.1) is a related topochemical process.

Intercalation of polymers in layered systems

Considerable interest has focused upon the synthesis of inorganic-organic multilayer composites (▶ glossary) by insertion of macromolecules into layered host lattices. Extended studies have been performed with iron oxychloride (FeOCl), metal phosphates, layered silicates, and transition-metal dichalcogenides. Examples of polymers that have been inserted are polyaniline, polypyrrole, polyfurane, polycaprolactame, and poly(ethylene oxide). Due to the size of the polymer molecules there is a smooth transition between truly intercalated polymers and exfoliated sheets dispersed in the polymer matrix.

There are several preparative strategies:

- Direct intercalation of polymer chains into the host lattice. A disadvantage is the slow kinetics for the transport of macromolecules in the interlayer space. An example is the formation of poly(ethylene oxide) in mica type sheet silicates by reaction of the polymer melt with Na^+ or alkylammonium exchanged host lattices. Another approach is first to make the silicates organophilic by exchanging the interlayer cations with organic cations and also swelling the lattice in the presence of a polar solvent. The enlarged interlayer spacing then allows polymer molecules to intercalate.
- Delamination of the host lattice into a colloid system (▶ glossary) in appropriate solvents, addition of the polymer and subsequent re-precipitation of one-dimensionally disordered intercalates (see above).
- Intercalation of monomer molecules with subsequent *in-situ* polymerization (▶ glossary).

2.4.3 Pillaring of Layered Compounds

The development of systems with nanometer-sized rigid pores is of considerable interest in shape-selective catalysis, adsorption, ion exchange, and the design of new structural and functional materials (see Chapter 6). One approach is based on the transformation of *layered* systems into porous framework materials by introducing "pillars" into the interlayer space (Figure 2-27).

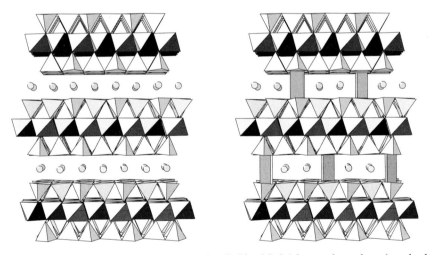

Figure 2-27. The layer structure of muscovite (left) with 2:1 layers (see above) and schematic representation of the clay with pillars (right).

The most extensively studied and best understood materials are alumina pillared clays, which are stable in both oxidizing and reducing atmospheres and offer high thermal stability and high surface areas. They may serve as an example of how the pillared materials are prepared.

The cation $[Al_{13}O_4(OH)_{24}(H_2O)_{12}]^{7+}$ (abbreviated Al_{13}) consists of 12 aluminum octahedra surrounding a central aluminum tetrahedron (Figure 2-28) and is approximately 860 pm in diameter. Al_2O_3-pillared clays have been obtained by exchanging the ions in the interlayer space of smectic clays with solutions containing the Al_{13} ion. Upon calcining the intercalates at about 500 °C, the Al_{13} ions dehydrate and convert to aluminum oxide particles, propping up the layers as pillars and generating interlayer space of molecular dimensions, i.e., a two-dimensional porous network. The alumina pillars are anchored to the tetrahedral aluminosilicate layers of smectic clays by chemical reactions. These pillared clays have interlayer free spacings of 700–800 pm, corresponding to the size of the Al_{13} unit, and surface areas of ~300 $m^2 \cdot g^{-1}$.

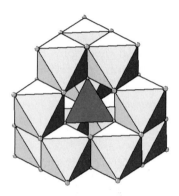

Figure 2-28. The structure of $[Al_{13}O_4(OH)_{24}(H_2O)_{12}]^{7+}$ ("Al_{13}")

In general, the micropore structure is tailored by the nature of the host material and of the pillaring species, and may have pore sizes larger than those of zeolites and related materials.

The ion-exchange technique described for $[Al_{13}O_4(OH)_{24}(H_2O)_{12}]^{7+}$ can similarly be employed for other cationic metal compounds, such as other hydroxo or oxo cluster ions (for example, $[Fe_{13}O_4(OH)_{24}(H_2O)_{12}]^{7+}$ or $[Zr_4(OH)_8(H_2O)_{16}Cl_z]^{(8-z)+}$), cationic metal–organic complexes, cationic polynuclear halide compounds or positively charged colloidal particles (▶ glossary).

Other layered host materials than smectic clays show limited swellability. Hence, they usually cannot exchange directly with the pillaring species, and pre-swelling of the layered substance by amine is performed prior to the insertion of pillaring species. Amines intercalate readily into these hosts to form alkylammonium intercalates. The expanded layers permit the subsequent ion-exchange reaction of the interlayer alkylammonium ions with large cationic pillar precursors. This method was applied successfully to the synthesis of Al_2O_3- or Cr_2O_3-pillared phosphates, for example.

The majority of layered host materials consists of neutral or negatively charged layers. The few compounds with positively charged layers, i.e. anion-exchange properties, are the so-called layered double hydroxides. Their structures consist of *positively* charged layers $[M_{1-x}^{II}M_x^{III}(OH)_2]^{x+}$ ($M^{II} = Mg^{2+}$, Ni^{2+}, etc.; $M^{III} = Al^{3+}$, Fe^{3+}, etc.). The structure of the layers is similar to that of the metal dichalcogenides, with the OH groups instead of the chalcogenide ions. They contain exchangeable anions and water in the interlayer spaces and can be pillared by using *negatively* charged polyoxometalate anions such as $V_{10}O_{28}^{6-}$, $[SiV_3W_9O_{40}]^{7-}$ or $PW_{12}O_{40}^{3-}$. For layered double hydroxides that do not swell in water, the pre-swelling approach can also be used: the hydroxides are first intercalated with large organic anions prior to polyoxometalate exchange.

Microporous zirconium phosphonate materials with organic pillars were derived from layered metal phosphates or phosphonates. The best known members of this group are zirconium phosphate, $Zr(HPO_4)_2 \cdot H_2O$ (Figure 2-29) and other M^{IV} phosphates and arsenates (M^{IV} = Zr, Ti, Ge, etc.).

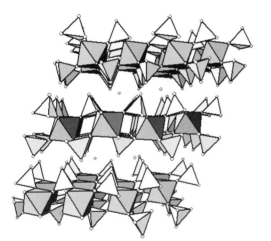

Figure 2-29. The crystal structure of α-$Zr(HPO_4)_2 \cdot H_2O$. The tetrahedra represent the phosphate groups, and the octahedra the coordination polyhedra of the zirconium atoms. The small circles in the interlayer space represent water molecules.

The Zr atoms of α-$Zr(HPO_4)_2 \cdot H_2O$ atoms are octahedrally coordinated by the oxygen atoms of six hydrogen phosphate anions. Each HPO_4^{2-} anion bridges three Zr^{4+} cations. The fourth corner of each phosphate tetrahedron is occupied by an OH group which points towards the interlayer space. The proton can be exchanged by other cations.

The corresponding metal phosphonates have the same general structure, the OH group being replaced by an organic group R (i.e., the composition is $Zr(O_3PR)_2$). The organic groups thus also point towards the interlayer space. Due to their size, the interlayer spacing is expanded, e.g. from 756 pm in α-$Zr(HPO_4)_2 \cdot H_2O$ to 1470 pm in $Zr(O_3PPh)_2$. Compounds with organic pillars were obtained by using bisphosphonic acids of the type $HO_3P-X-PO_3H$. While the two PO_3 groups are incorporated into adjacent layers, the interlayer distance can be adjusted by the size if the organic group X that bridges the layers.

2.5 Further Reading

S. B. Bhaduri, S. Bhaduri, "Combustion synthesis", in *Non-Equilibrium Processing of Materials*, Pergamon, Amsterdam, 1999, Ed. C. Suryanarayana, pp. 289–309.

C. R. Bowen, B. Derby, "Selfpropagating high temperature synthesis of ceramic materials", *Br. Ceramic Trans.* **96** (1997) 25–31.

M. S. Dresselhaus (Ed.), *Intercalation in Layered Materials*, Plenum, New York, 1987.

M. S. Dresselhaus, "New trends in intercalation compounds", *Mater. Sci. Engin.* **B1** (1988) 259–277.

M. S. Dresselhaus, G. Dresselhaus, "Intercalation compounds of graphite", *Adv. Phys.* **30** (1981) 139–326.

E. G. Gillan, R. B. Kaner, "Synthesis of refractory ceramics via rapid metathesis reactions between solid-state precursors", *Chem. Mater.* **8** (1996) 333–343.

J. Gopalakrishnan, "Chimie douce approaches to the synthesis of metastable oxide materials", *Chem. Mater.* **7** (1995) 1265–1275.

F. F. Lange, "Powder processing science and technology for increased reliability", *J. Am. Ceram. Soc.* **72**(1) (1989) 3–15.

H. Lange, G. Wötting, G. Winter, "Silicon nitride – from powder to ceramic materials", *Angew. Chem. Int. Ed. Engl.* **30** (1991) 1579–1597.

Z. A. Munir, J. B. Holt (Eds.), *Combustion and Plasma Synthesis of High Temperature Materials*, VCH, Weinheim, 1990.

A. Manthiram, J. Kim, "Low temperature synthesis of insertion oxides for lithium batteries", *Chem. Mater.* **10** (1998) 2895–2909.

J. J. Moore, H. J. Feng, "Combustion synthesis of advanced materials", *Progr. Mater. Sci.* **39** (1995) 243–273 (part I), 275–316 (part II).

D. O'Hare, "Inorganic intercalation compounds" in *Inorganic Materials*, Wiley, 1992, Eds. D. W. Bruce, D. O'Hare, pp. 165–235.

K. Ohtsuka, "Preparation and properties of two-dimensional microporous pillared interlayered solids", *Chem. Mater.* **9** (1997) 2039–2050.

G. Ouvrard, D. Guyomard, "Intercalation chemistry", *Curr. Opin. Solid State Mater. Sci.* **1** (1996) 260–267.

C. N. R. Rao, "Chemical synthesis of solid inorganic materials", *Mater. Sci. Eng.* **B18** (1993) 1–21.

J. Rouxel, "Design and chemical reactivity of low-dimensional solids – some soft chemistry routes to new solids", *Am. Chem. Soc. Symp. Ser.* **499** (1992) 88–113.

P. A. Salvador, T. O. Mason, M. E. Hagerman, K. Poeppelmeier, "Layered transition-metal oxides and chalcogenides", in *Chemistry of Advanced Materials – an Overview*, Wiley-VCH, New York, 1998, Eds. L. V. Interrante, M. Hampden-Smith, pp. 449–498.

R. Schöllhorn, "Intercalation systems as nanostructured functional materials", *Chem. Mater.* **8** (1996) 1747–1757.

D. Segal, "Chemical synthesis of ceramic materials", *J. Mater. Chem.* **7** (1997) 1297–1305.

J. Subrahmanyam, M. Vijayakumar, "Self-propagating high-temperature synthesis", *J. Mater. Sci.* **27** (1992) 6249–6273.

A. W. Weimer (Ed.), *Carbide, Nitride and Boride Materials – Synthesis and Processing*, Chapman & Hall, London, 1997.

M. S. Whittingham, A. J. Jacobson (Eds.), *Intercalation Chemistry*, Academic Press, New York, 1982.

M. Winter, J. O. Besenhard, M. E. Spahr, P. Novak, "Insertion electrode materials for rechargeable lithium batteries", *Adv. Mater.* **10** (1998) 725–763.

H. Yanagida, K. Koumoto, M. Miyayama, *The Chemistry of Ceramics*, Wiley & Sons, Chichester, 1996.

H. C. Yi, J. J. Moore, "Self-propagating high-temperature (combustion) synthesis (SHS) of powder-compacted materials", *J. Mater. Sci.* **25** (1990) 1159–1168.

3 Formation of Solids from the Gas Phase

This chapter deals with reactions, in which solid products are obtained from gaseous precursors or intermediates. The product phases may be formed by gas phase reactions (Section 3.3) or by thermal reaction of the gaseous precursors after adsorption to surfaces in chemical vapor deposition (CVD) processes (Section 3.2). In chemical transport reactions (Section 3.1), gaseous intermediates are in thermodynamic equilibrium with the solid phase. Their diffusion results in an overall mass transport of the solid.

3.1 Chemical Vapor Transport

In this section we will first deal with the general principle of chemical vapor transport phenomena. We will then demonstrate their important role in halogen lamps, before discussing some of the chemical reactions involved.

The general principle behind transport reactions is that a non-volatile solid (A) *reversibly* reacts with a gaseous agent (B) to form a gaseous product (AB). The equilibrium constant associated with this reversible reaction is temperature-dependent, and therefore the equilibrium concentration of AB changes when the temperature is changed. Thus, a concentration gradient of AB develops when the system is exposed to a temperature gradient. This provides the driving force for an overall mass transport by diffusion from one site at the temperature T_1 (the source), where AB is formed, to another site at the temperature T_2 (the sink), where the chemical equilibrium position is shifted towards the side of A and B (Figure 3-1).

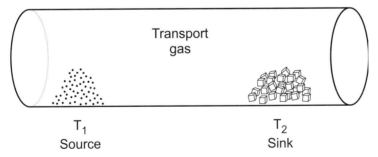

Figure 3-1. The principle of chemical transport reactions. The reaction vessel can be open or closed (see text).

The equilibrium positions at the two temperatures should not be too far on either side for an efficient transport reaction. Exothermic reactions cause migration into the region of higher temperature ($T_1 < T_2$), and endothermic reactions in the opposite direction ($T_1 > T_2$). The intermediate formation of gaseous compounds distinguishes chemical transport from sublimation.

Chemical transport is most often used for the purification of solids and particularly for the growth of single crystals. This is possible because only one chemical species is transported and thus the very pure compound is deposited at the sink. However, chemical transport can also be used for the preparation of new compounds. In this case the transport reaction is coupled with a subsequent reaction at the sink (see below).

Transport reactions can be performed in open systems with a continuous stream of the transport agent, or in closed systems. An example for the first is found in nature. In the presence of volcanic gases, Fe_2O_3 is volatilized and deposited at another site. At hot sites in a volcano, Fe_2O_3 reacts with HCl (the transport agent), and gaseous $FeCl_3$ and water vapor is formed. $FeCl_3$ is transported with the gases and deposited at cooler sites by back-reaction (Eq. 3-1).

$$3\ Fe_2O_3(s) + 6\ HCl(g) \rightleftharpoons 2\ FeCl_3(g) + 3\ H_2O(g) \tag{3-1}$$

A technical example for the chemical transport in a continuous stream of the transport agent is the Mond process in which elemental nickel in high purity (99.9–99.99 %) is produced (Eq. 3-2). Nickel is purified by reversible reaction with CO. $Ni(CO)_4$ is formed in an exothermic reaction at 50–80 °C and atmospheric pressure, transported out of the solid mixture formed in the metallurgical process and then decomposed at 230 °C.

$$Ni + 4\ CO \underset{230\ °C}{\overset{50\ °C}{\rightleftharpoons}} Ni(CO)_4 \tag{3-2}$$

Halogen lamps

The most important technical application of chemical transport reactions in closed systems is in halogen lamps. Since Edison made the first incandescent lamp in 1879, with a carbon filament, the principle has not changed very much. When an electric current is passed through a solid body, it is heated to a high temperature and electromagnetic radiation is emitted, mostly as infrared radiation (heat). The wavelength of the radiation decreases with increasing temperature. In incandescent lamps, only a few percent of the electrical energy are converted in visible light, and most of the energy is lost as heat (Figure 3-2). For example, the yield of light of a heated tungsten filament is only 1.4 % at 2400 °C.

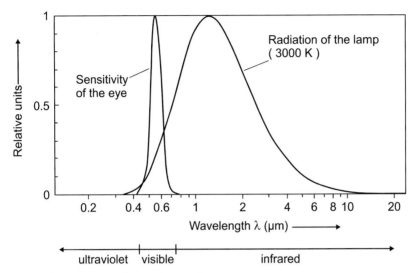

Figure 3-2. The spectrum of incandescent lamps.

The major improvements since Edison's invention have been the replacement of carbon by tungsten, the use of an inert-gas filling instead of evacuating the bulb, the coiling of the filament, and finally the addition of halogens.

Tungsten filaments. The great advantage of tungsten compared with other filament materials is that its vapor pressure is low – so that loss of the filament material by evaporation is slow – and the melting point is high (3400 °C). Other advantages of tungsten are its selective emissivity and its great strength at elevated temperatures. Improving the yield of visible light by a higher temperature of the tungsten filament is limited, because the higher the temperature, the greater the rate at which the filament evaporates, resulting in a shorter life time. For example, in a lamp whose life is completely determined by vapor transport of tungsten, a 70 % increase in efficacy, achieved by raising the filament temperature from 2527 to 2927 °C, will shorten the life time by a factor of about 100.

By increasing temperature of the tungsten, the vapor pressure of the filament material is increased. The vaporized metal condenses at the colder regions of the bulb. This has two consequences: the tungsten wire becomes thinner and eventually "blows". Furthermore, the bulb blackens and thus the yield of light is reduced. In ordinary incandescent lamps, the size of the bulb therefore is large to keep blackening of the bulb, and hence light losses, within reasonable bounds.

Inert gas filling. The inert gas filling reduces the rate of tungsten evaporation by inhibiting the transport of tungsten. The rate of diffusion is inverse proportional to the total pressure. The gas should preferentially have a high molar mass because thermal conductivity is inversely proportional to the molar mass of the gas. An economic compromise is the use of argon with a low portion (7 %) of nitrogen in conventional lightbulbs.

Halogens. With the addition of a small amount of iodine (about 0.1 mg per ml bulb volume) as transporting agent (halogen lamps) the vaporized tungsten is transported back to the hot filament (Figure 3-3).

Figure 3-3. Deposition of tungsten crystals on the tungsten filament of a halogen lamp (130-fold magnification).

A simplified picture is that the evaporated tungsten from the hot filament diffuses through the inert gas towards the bulb wall, reacting in the cooler zones with the transport medium to form a volatile compound. The reaction products diffuse backwards, i.e., in the direction of the filament. The volatile compounds dissociate at the filament, and the tungsten is released. Iodine as the transport medium is then available again for renewed reaction in cooler zones.

The "back reaction" of the deposited tungsten with iodine requires a certain minimal temperature of the glass wall. For this reason, the size of halogen bulbs is much smaller than that of conventional lightbulbs. The filament heats the glass wall to the required temperature (about 600 °C). The lower volume of the halogen bulbs allows the use of the more expensive xenon or krypton as gas filling instead of argon. Furthermore, because of their greater mechanical strength, these bulbs can be filled with inert gas at a higher pressure. Both measures reduce the transport of tungsten and thus permit a higher filament temperature.

The chemical processes in halogen lamps are less straightforward than it may appear. For example, it was found that low concentrations of oxygen are necessary for a good performance of the lamps. The compound responsible for the transport of tungsten is WO_2I_2 (instead of WI_2, as originally thought). This is formed at about 600 °C at the glass wall by reaction of the deposited tungsten with iodine and oxygen (or water) in an exothermic reaction (Eq. 3-3).

$$W + O_2 + I_2 \xrightleftharpoons{600\ ^\circ C} WO_2I_2 \qquad (3\text{-}3)$$

When WO_2I_2 migrates towards the hot filament (about 3000 °C), it dissociates sequentially into WO_2, WO, and eventually tungsten atoms (close to the filament) (Eq. 3-4). The tungsten atoms are then deposited at the hot filament. Thus, the process at the filament surface is not a chemical reaction, but a condensation process.

$$WO_2I_2 \underset{}{\overset{-I_2}{\rightleftharpoons}} WO_2 \underset{}{\overset{-1/2O_2}{\rightleftharpoons}} WO \rightleftharpoons W + 1/2\,O_2$$

increasing temperature →

$$(3\text{-}4)$$

As a consequence, the tungsten atoms are deposited at the coldest (i.e., thickest) parts of the filament. The transport reaction therefore is not a self-healing process, since tungsten preferentially evaporates from the hottest parts of the filament but is redeposited at the coldest parts. Tungsten evaporates from so-called "hot spots" at a faster rate until the lamp fails through localized melting or fracture of the filament. The "hot spots" are small regions in the filament where the temperature is higher because of local inhomogeneities. The development of a truly regenerative cyclic process, which would increase the life of bulbs considerably, is therefore still a challenging task.

Another type of lamp where transport reactions play an important role is the metal halide lamp. A typical construction of such a lamp is shown in Figure 3-4. The arc tube of a metal halide lamp is made from quartz glass or poly-crystalline alumina with two tungsten electrodes. The arc tube contains mixtures of metal iodides and/or bromides which vaporize during operation of the lamp.

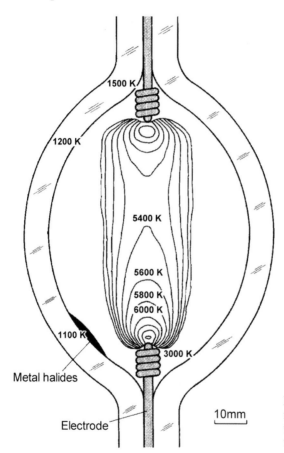

Figure 3-4. Arc tube of a metal halide lamp with temperature distribution under operating conditions.

Of particular interest are alkali halide/lanthanide halide systems. DyI_3, TmI_3 and HoI_3 are important for light generation, because these metals display many lines in the visible spectrum when excited in the high-temperature arc column of the lamp. The intensity of the light emission increases if the concentration of the rare earth metal in the arc column is raised. This is possible by the formation of gaseous heterometal complexes by chemical vapor transport reactions. For example, $NaDyI_4$ and Na_2DyI_5 have been identified in the NaI/DyI_3 system. In the high-temperature arc core, the metal complexes dissociate into the elements or ionized species, which are then transported outward by convection and diffusion to the lower temperature mantle where they recombine again.

Transport reactions

Under laboratory conditions, transport reactions are mostly carried out in closed tubes typically having a length of 10–20 cm and a diameter of 1–2 cm. The sample is placed into the ampoule, and the transport agent is added. The tube is then evacuated at different pressures, depending on the kind of transport agent, sealed and placed in a tubular furnace to which a temperature gradient (Figure 3-1) is applied. The transport agent reacts at the source, but is liberated during the back reaction and diffuses back to the source. It therefore remains in the system and is available for further reactions. Thus, small amounts of the agent can transport large amounts of the solid.

In general, transport reactions in closed systems can be described in four steps:

1. The chemical reaction at the source, leading to an equilibrium between the condensed phase and the gas phase.
2. Mass transport by diffusion of the gaseous compounds from the source to the sink.
3. Deposition of the condensed phase at the sink.
4. Diffusion of the gaseous transport agent back to the source.

The overall reaction is in most cases diffusion-limited. When a gaseous compound is produced at one site of the reaction vessel and consumed at another, the different partial pressures of the gas give rise to a mass flux between the two temperature regions. It is obvious that there is faster transport if the partial pressure gradient is large. On the other hand, a high total pressure slows down the balancing of the partial pressures. Therefore, the total pressure in the ampoules is usually kept at around 1 bar at the reaction temperature by addition of the appropriate amount of the transport agent.

A large number of chemical transport reactions are known and some typical examples are given for chromium compounds in Table 3-1. In most cases, the reactions are not as simple as the chemical equations may indicate, because very often different gaseous species are involved. In principle, it is possible to calculate the best choice of the transport agent from the thermodynamic data of all species involved in the reaction.

Table 3-1. Transport reactions of chromium compounds

Transported compound	Transport agent	Gas-phase species
Cr	I_2	CrI_2
CrX_3 (X = Cl, Br, I)	X_2	CrX_4
$CrCl_3$	$AlCl_3$	$CrAl_3Cl_{12}$
Cr_2O_3	O_2/Cl_2	CrO_2Cl_2
Cr_2O_3	H_2O/Cl_2	$CrO_2Cl_2 + HCl$
Cr_2O_3	$H_2O/HgCl_2$	$CrO_2Cl_2 + Hg + HCl$
$CoCr_2O_4$ (spinel)	Cl_2	$CrO_2Cl_2 + CoCl_2$
$Cr_2(SO_4)_3$	HCl	$CrO_2Cl_2 + SO_2 + H_2O$
$Cr_2P_2O_7$	I_2/P_4	$CrI_2 + P_4O_6$
$CrSi_2$	I_2	$CrI_2 + SiI_4$

Halogens are very often used as the transport agent for elemental metals, as has been discussed earlier for halogen lamps. Another example is the van Arkel–de Boer process, the first deliberately applied transport reaction. This process is used for the purification of metals (such as Ti, Cr, or V), and makes use of the exothermic reaction between metal and iodine to form a volatile iodide (for example, CrI_2 from Cr and I_2). The metal is redeposited at a hot filament, thus allowing the metals to be extracted from their oxides, nitrides, carbides, etc. The volatile metal halide can also be formed by reaction with other halogen sources, such as hydrogen halides or metal halides.

Transport of metal oxides is often achieved with hydrogen halides as transport agents. The formation of water as the by-product shifts the equilibrium for the formation of the metal halide to the right (for example, Eq. 3-1). Elemental halogens are less often used for the transport of oxides, because the equilibrium concentration of binary metal halides is usually very low. However, more volatile oxo halides may be formed, as shown in Table 3-1 for chromium oxides.

Many metal halides can be transported with $AlCl_3$, which forms volatile complexes (for example, Eq. 3-5). Some metal halides migrate without an additional transport agent ("auto transport"). Volatile metal halides acting as transporting agent are formed by partial disproportionation.

Two substances can be transported in different directions along the temperature gradient if one reaction is exothermic and the other endothermic.

An example is given in Eq. 3-6. CuCl forms exothermically from Cu_2O with HCl and endothermically from Cu. This allows the separation of Cu and Cu_2O, as Cu_2O is redeposited at higher temperature and Cu at a lower temperature.

$$CoCl_2(s) + Al_2Cl_6 \rightleftharpoons \underset{Cl}{\overset{Cl}{Al}} \underset{Cl}{\overset{Cl}{Co}} \underset{Cl}{\overset{Cl}{Al}} \overset{Cl}{\underset{Cl}{}} \quad (g)$$

$$(3\text{-}5)$$

$$Cu_2O(s) + 2\,HCl(g) \underset{900\,°C}{\overset{500\,°C}{\rightleftharpoons}} 2\,CuCl(g) + H_2O(g)$$

$$(3\text{-}6)$$

$$Cu(s) + HCl(g) \underset{500\,°C}{\overset{600\,°C}{\rightleftharpoons}} CuCl(g) + 1/2\,H_2(g)$$

For use as a preparative method, the transport reaction is coupled with a subsequent reaction at the sink. For example, metallic niobium and silica do not react with each other if heated together at 1100 °C in vacuum. However, in the presence of small amounts of H_2, gaseous silicon monoxide, SiO, is formed which migrates to the niobium and gives the silicide (Eq. 3-7). The role of SiO for the transport of silicon has already been discussed in Section 2.1.2.

$$SiO_2(s) + H_2 \rightleftharpoons SiO(g) + H_2O \quad \text{(transport reaction)}$$

$$(3\text{-}7)$$

$$3\,SiO(g) + 8\,Nb \longrightarrow Nb_5Si_3 + 3\,NbO$$

3.2 Chemical Vapor Deposition

Chemical vapor deposition (CVD) is one of the most important methods to prepare thin films and coatings of inorganic materials of various composition. Thin films are of increasing technological importance because they can add new chemical and physical properties to a substrate material. Unique combinations of properties can thus be obtained through particular combinations of substrate and coating.

After an introduction into general aspects of CVD, such as precursor chemistry, the different steps occurring during CVD, equipment and techniques, we will discuss the deposition of metals (Section 3.2.2), diamond (Section 3.2.3), metal oxides (Section 3.2.4), metal nitrides (Section 3.2.5) and compound semiconductors (▶ glossary; Section 3.2.6). These classes of compounds were selected because they allow an elaboration to be made on the different chemical processes that occur during the thermal decomposition of different precursor compounds on the substrate surface.

3.2.1 General Aspects

Chemical vapor deposition (CVD) is a process where one or more volatile precursors are transported via the vapor phase to the reaction chamber, where they decompose on a heated substrate. This results in the deposition of a solid thin film Materials deposited by CVD include metals and different multi-element materials, such as oxides, sulfides, nitrides, phosphides, arsenides, carbides, borides, silicides, etc.

A first example is given in Eq. 3-8. TiB_2, which has outstanding properties as an armor material, melts at 3325 °C. Therefore, the preparation of thin films by melting TiB_2 is practically impossible. However, CVD of TiB_2 films according to Eq. 3-8 already occurs at about 1000 °C, i.e., substantially below the material's melting point.

$$TiCl_4 + 2\ BCl_3 + 5\ H_2 \longrightarrow TiB_2 + 10\ HCl \qquad (3\text{-}8)$$

Initial applications of CVD were largely hard coatings to improve the life and performance of cutting tools. This is still a large market, together with coatings for aircraft engine components. More recently, CVD is used heavily in the microelectronics industry, as intricate three-dimensional structures can be built on substrates to perform complex tasks in electronic components (see Section 3.2.2). In the glass industry, CVD is used to coat large panels of glass with SnO_2, TiN or SiO_2. Typically, the glass panels are formed on a float line in which the hot glass is transported on top of molten tin. The panels pass through several CVD reactors in which reactants are directed toward the hot substrate, where they form coatings. Other applications are layers in solar cells, coatings for catalysis, membranes, or optical layers in waveguides. "Synthetic gold" coatings (non-stoichiometric TiN) are deposited on large volumes of personal jewelry.

CVD processes involve a series of steps (Figure 3-5):

- Transport of the reagents ($TiCl_4$, BCl_3 and H_2 in the above example) in the gas phase, often in a carrier gas, to the deposition zone.

- Diffusion or convection of gaseous precursors through the boundary layer (hot layer of gas immediately adjacent to the substrate).
- Adsorption of film precursors onto the growth surface.
- Surface diffusion of precursors to growth sites. The probability that a precursor molecule reacts directly at the first point of contact with the surface should not approach unity, because this may result in rough surface topologies, i.e., some surface mobility of the precursors to growth sites is desirable.
- Surface chemical reactions leading to deposition of a solid film. (TiB_2 in the above example) and the formation of by-products (HCl in the above example).
- Desorption of by-products.
- Transport of gaseous by-products out of the reactor.

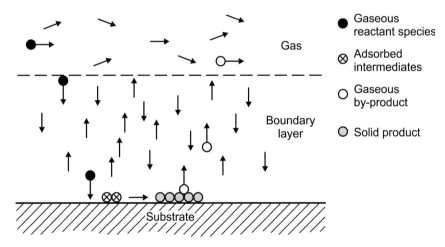

Figure 3-5. Schematic representation of the steps in CVD processes.

In the majority of CVD processes, gas-phase reactions (mostly in the hot boundary layer) prior to adsorption of the precursor on the surface are undesirable, because they can result in particle formation, incomplete reaction of the precursor, and depletion of the precursor concentration at the surface. However, there are examples (see Section 3.2.6 for CVD of GaAs) where gas-phase reactions are beneficial for the film quality. The total pressure in the reactor controls the degree of gas-phase reactions, as when it is lowered the probability of gas–gas collisions becomes smaller.

Once the precursor has been adsorbed onto the substrate and has diffused to growth sites it should undergo surface-initiated chemical reactions to deposit the desired material. The nature of the ligand and the type of metal–ligand bonds determine and control the decomposition pathway.

An ideal precursor should:

- be liquid rather than solid or gaseous;
- have a good volatility;
- have good thermal stability in the delivery system, and during its evaporation and transport in the gas phase;
- decompose cleanly and in a controlled manner on the substrate, without incorporation of contaminations;
- give by-products which are stable and readily removed from the reaction zone;
- have a high purity or should be readily purified;
- be readily available in consistent quality and quantity at low cost; and
- be non-toxic and non-pyrophoric.

In reality, precursors rarely meet all criteria, and compromises have to be made. In particular, the requirement of long-term thermal stability at ordinary temperatures (shelf-life) on one hand, and a high reactivity at higher temperatures (to achieve high deposition rates) on the other hand, often results in a narrow thermal stability window.

Most industrially important CVD processes utilize simple precursors such as metal hydrides (SiH_4, AsH_3), metal alkyls (Al^iBu_3, $GaEt_3$), or volatile metal halides (WF_6, $TiCl_4$). Metal halides generally require the presence of a reducing agent, and are often corrosive or liberate corrosive by-products (HX, X_2). Metal–organic and organometallic compounds are usually more volatile than most inorganic compounds. However, they often thermally decompose at relatively low temperatures, which can lead to carbon or other contaminations of the substrate (see below).

Volatility is enhanced by minimizing all types of interactions between the precursor molecules in the condensed state (e.g., hydrogen bonds, dipole–dipole interactions, van der Waals interactions). Fluorination of ligands or substituents helps to reduce van der Waals interactions between the molecules. Since smaller non-polar compounds have a higher vapor pressure, oligomerization or aggregation of the precursors should be prevented. For this reason, β-diketonate ligands are often used in metal-containing precursors. They are generally chelating (▶ glossary) and thus occupy two coordination sites at the metal center. Therefore, β-diketonate complexes of metals with low coordination numbers are often monomeric and volatile.

Volatility problems arise with metals that prefer higher coordination numbers. This results in oligomerization to satisfy the metals coordination number, mostly by the formation of ligand bridges. There are two strategies to prevent oligomerization:

1. Employment of sterically demanding ligands. They limit the steric accessibility of ligands of an adjacent molecule to the metal center. However, this approach may result in two problems:
   (a) The metal center is generally very reactive, due to its low coordination number. Therefore, small molecules (such as O_2 or H_2O) can penetrate the ligand shell and thus can make the compounds difficult to handle from a practical viewpoint.
   (b) The large ligands result in a relatively high molecular weight of the compound, which tends to decrease the volatility.
2. Introduction of chelating (▶ glossary) multidentate ligands into the coordination sphere of the metal, which can satisfy the preferred higher coordination number of the metal while preventing oligomerization. The problems associated with this approach are:
   (a) the multidentate ligands may prefer to bridge two or more metal centers rather than chelate; or
   (b) they may dissociate prior to or during transport, which can lead to oligomerization.

Chemical solutions for this problem will be discussed in Section 3.2.4 for alkaline earth precursors for CVD of high-temperature superconductors (▶ glossary).

A general problem associated with the use of organometallic compounds is the carbon incorporation into the deposited film. To avoid this, precursors must be designed that completely eliminate the supporting organic ligands during the surface reactions. The volatile products must additionally be removed rapidly from the deposition zone in order to prevent contamination of the film by unwanted decomposition reactions. This is possible when the ligand is itself a stable molecule, e.g., CO, alkenes, or when a thermal reaction results in a stable by-product by a facile reaction, such as β-hydrogen elimination. In many cases, additional reagents are used to oxidize the ligands or the metal, to reduce the metal, or to protonate the ligands. Examples will be given in the following sections.

There are two approaches for the CVD of multi-element materials, exemplarily shown in Eqs. 3-9 and 3-10 for the formation of GaAs (see also Section 3.2.6):

- Two (or more) individual precursors (*multiple-source precursors*) may be used that decompose individually on the substrate and eventually give a film of the desired multi-element composition. In the case of GaAs, the most commonly employed precursors are $GaMe_3$ and AsH_3 (Eq. 3-9).

$$GaMe_3 + AsH_3 \longrightarrow GaAs + 3\ CH_4 \tag{3-9}$$

- *Single-source precursors* contain the elements desired in the final film in one component, preferably in the ratio also required in the film. An example for GaAs is the cyclic compound $[Et_2Ga\text{-}As^tBu_2]_2$ (Eq. 3-10). In single-source precursors, the bonds between the film-forming elements must be stronger than those to the supporting ligands or substituents. The latter can be then removed cleanly by selective bond breaking. Otherwise, the advantages of a single source are lost.

$$
\begin{array}{c}
Bu^t \diagdown \qquad \diagup {}^tBu \\
Et \diagdown \quad As \quad \diagup Et \\
\quad Ga \qquad Ga \\
Et \diagup \quad As \quad \diagdown Et \\
Bu^t \diagup \qquad \diagdown {}^tBu
\end{array}
\longrightarrow
GaAs + 4\ C_2H_4 + 4\ CH_2{=}CMe_2 + 4\ H_2
$$

$$(3\text{-}10)$$

One problem associated with the multiple-source precursor approach is the reproducible control of film stoichiometry. This problem arises because the precursors or the intermediates formed during film deposition have different volatilities and reactivities. A typical example is the deposition of lead-containing films such as $PbTiO_3$, which can result in formation of lead-deficient materials due to desorption of volatile PbO.

Some of the problems associated with multiple-source precursors are circumvented by single-source precursors. Their use additionally allows a simplification of the precursor delivery system. However, single-source precursors have higher molecular weights than multiple-source precursors. They therefore are generally less volatile. It is also difficult to deposit films with non-integral stoichiometry, doped films, or materials with two or more metals. Furthermore, the stoichiometry of the precursor is not always retained in the film.

Equipment

A schematic of a typical CVD reactor is shown in Figure 3-6. Features common to all CVD reactors include a precursor delivery system, the reactor, and an exhaust system to remove by-products. In conventional CVD processes, the substrate temperature typically varies between 200 and 800 °C, depending on the nature of the layer, and the pressure in the reactor cell between 0.1 mbar and 1 bar.

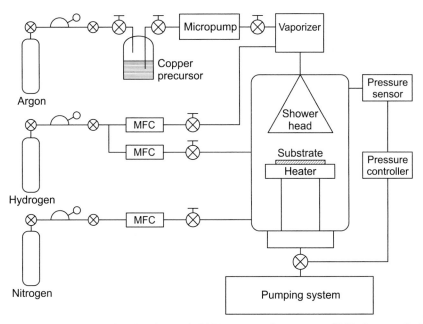

Figure 3-6. Schematic of a thermal CVD reactor for copper CVD from a Cu(II) precursor (see Eq. 3-19) (MFC = mass flow controller).

The *precursor delivery method* plays a critical role because the overall deposition rate can be limited by the feed rate of precursor delivery into the reactor (see below). The transport of liquid or solid precursors with relatively low vapor pressures is controlled by bubbling a carrier gas through the precursor. The vapor pressure of liquids should be >0.01 bar at 25 °C for efficient industrial utilization. If the volatility of the precursors at room temperature is not sufficient, the storage vessel must be heated. Conventional delivery methods are unsuitable if the precursors are thermally unstable. Alternative methods of delivery must then be used such as liquid delivery, aerosol (▶ glossary) delivery, spray pyrolysis, and supercritical fluid delivery.

Two main *reactor types* can be distinguished in conventional CVD systems: hot- and cold-wall reactors (Figures 3-7 and 3-8).

In *hot-wall reactors* (Figure 3-7), the substrate and the chamber walls are maintained at the same temperature. The advantages are that:

- they are simple to operate;
- they can accommodate several substrates;
- obtaining a uniform substrate temperature is easy;
- they can be operated under a range of pressures and temperatures; and
- different orientations of the substrate relative to the gas flow are possible.

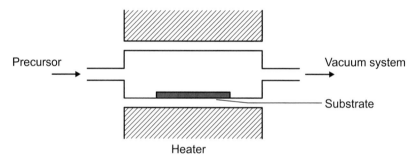

Figure 3-7. Schematic drawing of a hot-wall reactor.

The major problems are that:

- deposition occurs not only on the substrate but also on the reactor walls (these deposits can eventually fall off the reactor walls and contaminate the substrate);
- the large consumption of the precursor makes control of the gas composition more difficult and can result in feed-rate-limited deposition (see below); and
- gas-phase reactions in the heated gas can occur.

For these reasons, hot-wall reactors are used primarily at the laboratory scale, and in industry for CVD of semiconductors (▶ glossary) and oxides, with precursors having high vapor pressures. They are also often used to determine product distributions because the large heated surface area can consume the precursor completely and provide high yields of the reaction products.

Most metal CVD in industry is carried out in *cold-wall reactors* (Figure 3-8). The substrate is maintained at higher temperature then the reactor walls. The advantages are that:

- pressure and temperature can be controlled;
- plasmas can be used;
- deposition does not occur on the reactor walls;
- gas-phase reactions are suppressed; and
- higher deposition rates can be obtained because deposition occurs only on the heated substrate (higher precursor efficiency).

A disadvantage is that the steep temperature gradients near the substrate surface may lead to severe convection. This can result in non-uniform coatings.

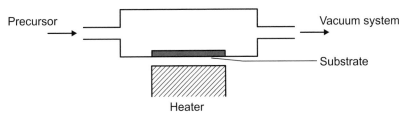

Figure 3-8. Schematic drawing of a cold-wall reactor.

CVD is a complex process with a large number of variables that influence the film properties: reactor geometry, reactant delivery, total pressure, gas and substrate temperature, substrate environment, gas composition and homogeneity, flow rate, substrate surface, time of deposition, gas flow behavior, deposition rate, nucleation (▶ glossary) density, thermodynamic and kinetic properties of all involved species, particularly gas and surface chemistry, composition of the products, reaction mechanism, etc.

Because of the many parameters, different results are often obtained for the same precursor, making comparison of results from different systems difficult. Coatings properties influenced by the process variables are composition, thickness, morphology, density, uniformity, adhesion to the substrate, crystallinity, stress (▶ glossary), etc.

Growth rates

The growth rate of the product layer must often be greater than $0.1 \ \mu\text{m·min}^{-1}$ for CVD processes used in microelectronics processing, and higher for large-area coating processes such as those in the glass industry.

  The typical dependence of the film growth rate on the substrate temperature is shown in Figure 3-9.

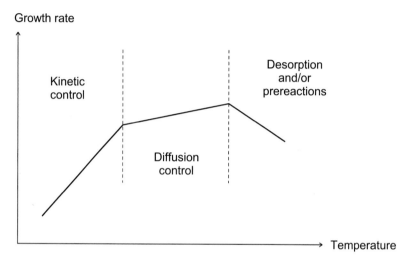

Figure 3-9. Dependence of the film growth rate on the substrate temperature.

  There are three regimes:

1. At low temperatures, the growth rate is *surface-reaction (kinetically) limited*. This is the case when the feed rate is sufficiently high and diffusion limitations do not occur.
2. In the intermediate temperature region, the growth rate is *diffusion* or *mass-transport limited*. Essentially all the reactants that reach the substrate decompose. The reaction proceeds more rapidly than the rate at which the reactant is supplied to the surface by diffusion through the boundary layer.
3. At high temperatures, the growth rate tends to decrease because of an increased desorption rate of film precursors or film components, together with depletion of reactants by reaction at the reactor walls.

If both the mass transport and diffusion are fast, the deposition rate also may be limited by the rate at which the reactant(s) are fed to the chamber (*feed-rate-limited* process). This is often encountered for precursors with low vapor pressures.

The pressure of the CVD reactor determines the relative importance of each of these regimes in the growth process. For instance, there is a significant boundary layer at atmospheric to intermediate pressures (e.g., 0.2 bar). Therefore, growth may be characterized by the intermediate and high-temperature regimes described above. However, as the pressure in the reactor is lowered (low-pressure CVD, <1 mbar) layer growth is essentially surface-reaction controlled, as in the low-temperature region.

The growth in bimolecular systems (molecules A and B) can follow two different mechanisms (Figure 3-10):

- The Elay–Rideal mechanism: Only one kind of molecule (A) is adsorbed. The adsorbed molecule reacts directly with a molecule of the other kind (B) from the gas phase. The growth rate shows a saturation behavior for high A/B ratios, i.e., the limiting growth rate is determined by the complete coverage of the surface by molecules A.
- The Langmuir–Hinshelwood mechanism: Both A and B are adsorbed onto the substrate, and reaction takes place between the adsorbed molecules. When the A/B ratio is increased, the growth rate increases to a certain limit and then drops again. The species present in excess occupies the free adsorption sites at the substrate, and therefore there are not enough sites for the other species.

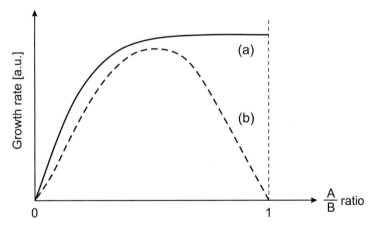

Figure 3-10. Growth rates in bimolecular systems: (a) Elay–Rideal mechanism, (b) Langmuir–Hinshelwood mechanism.

Selective deposition

In the fabrication of multi-level electronic devices, the patterning of surfaces is essential. Patterned films can be prepared by covering the whole surface with the deposited material ("blanket deposition") followed by selective-area etching. Alternatively, selective CVD deposits a material only on one substrate (the growth surface) in the presence of another substrate (the non-growth surface) (Figure 3-11). Typical examples are metals or silicon as growth surfaces and silica as the non-growth surface. The possibility of selective growth of films is unique to CVD, and not possible with other methods.

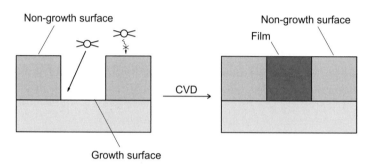

Figure 3-11. Selective deposition by CVD.

There are several strategies for selective deposition that rely on either inhibiting the adsorption and reaction of the precursor on the non-growth surface or promoting its reaction on the growth surface:

- The reaction rate of the precursor on the non-growth surface is intrinsically slower than its reaction rate on the growth surface. For example, copper CVD using $(hfac)Cu(PMe_3)$ (see Section 3.2.2) occurs on Cu, Pt and other metal surfaces, but not on SiO_2 and other non-metal surfaces.
- The growth surface acts as a coreactant and is selectively consumed by a precursor (e.g., reaction of Si with WF_6 or MoF_6, see Eqs. 3-15 and 3-16, Section 3.2.2), while the reaction rate at the non-growth surface (e.g., SiO_2 or Si_3N_4) is slower.
- A chemical reaction of a gaseous coreactant occurs on the growth surface (e.g., dissociation of H_2 on a metal), but not on the non-growth surface (SiO_2 or metal oxides).

- The rate is increased on the growth surface by radiation while the thermal reaction at the non-irradiated non-growth surface is slow.
- Selective passivation of the non-growth surface prevents adsorption and reaction of the precursor, while adsorption and reaction occur readily on the growth surface. For example, adsorption on silica can be prevented in many cases by conversion of the surface Si–OH groups in Si–OSiMe$_3$ groups.
- A free energy barrier exists for nucleation (▶ glossary) on the non-growth surface. It is assumed that the surface reaction is rapid and that the nucleation rate is only limited by the rate of formation of thermodynamically stable clusters on the growth surface.

CVD and CVD-related techniques

CVD processes resulting in the epitaxial growth (▶ glossary) of a crystalline film are called *vapor phase epitaxy* (VPE). Epitaxial films have the same crystallographic orientation as the single crystal substrate with similar lattice constants. VPE is generally used to prepare semiconductor films (▶ glossary; see Section 3.2.6) and is utilized in large-scale commercial operations.

Several prefixes are used to either specify the kind of precursors or variations of the CVD technique:

- The prefix "OM" (OMCVD, OMVPE, etc.) is used when the precursor is an organometallic compound, i.e., a compound in which one or more organic ligands are bonded to the metal by a metal–carbon bond. "MO" (MOCVD, MOVPE, etc.) is used for metal–organic precursors, i.e., metal compounds with ligands containing organic substituents that are not bonded to the metal by a metal–carbon bond, such as amides, alkoxides, etc.
- The basic CVD process is often modified, such as low-pressure CVD (LPCVD), plasma-enhanced or plasma-assisted CVD (PECVD or PACVD), laser-assisted (laser-induced) CVD (LCVD) or remote-plasma CVD (RPCVD).
- Chemical vapor infiltration (CVI) is CVD on internal surfaces of porous preforms.

There are numerous CVD-related deposition methods with widespread applications. These include atomic-layer epitaxy (▶ glossary; ALE) and chemical-beam epitaxy (CBE). The latter is also called metal-organic molecular-beam epitaxy (MOMBE). Some of the more important acronyms are summarized in Table 3-2.

Table 3-2. Survey of frequently used acronyms in CVD and related methods.

CVD	chemical vapor deposition
PVD	physical vapor deposition
VPE	vapor phase epitaxy (▶ glossary)
OMCVD	organometallic CVD
OMVPE	organometallic VPE
MOCVD	metal–organic CVD
MOVPE	metal–organic VPE
LPCVD	low-pressure CVD
PECVD	plasma-enhanced CVD
PACVD	plasma-assisted CVD
LCVD	laser-induced CVD
RPCVD	remote-plasma CVD
CVI	chemical vapor infiltration
ALE	atomic-layer epitaxy (▶ glossary)
CBE	chemical-beam epitaxy (▶ glossary)
MOMBE	metal–organic molecular-beam epitaxy

Plasma-enhanced chemical vapor deposition (PECVD). This involves generation of reactive species in a glow discharge. The working pressure is usually in the range 0.1 to 1.0 mbar. Inelastic collisions between high-energy electrons and the gaseous precursors generate excited or ionized molecules. As a result, reactions proceed at lower temperatures. This is the major advantage of PECVD over a thermal CVD, along with potential high growth rates. PECVD is the established commercial technique for low-temperature deposition of a number of important materials, especially insulating films such as Si_3N_4 and SiO_2. The low gas and substrate temperatures in PECVD processes reduce the mobility of the reactants on the substrate surface. Therefore, PECVD layers are often amorphous, and the step coverage (▶ glossary) in microelectronic devices is poorer. Other potential disadvantages are the lack of substrate selectivity and plasma-induced substrate damage. In *remote plasma-enhanced chemical vapor deposition* (RPECVD), the potential of plasma damage to the substrate and film is reduced by separating the plasma excitation region from the growth region.

Laser-assisted chemical vapor deposition (LCVD). In thermal LCVD, a laser is used to heat the substrate analogous to local heating in a cold-wall reactor. In photochemical LCVD, and photo-assisted CVD processes in general, the precursor molecule is activated by photochemical processes. This allows the deposition temperature to be significantly lower. Furthermore, deposition processes in selected areas are possible. For example, three-dimensional patterns or structures can be created in microelectronic devices without using masks by scanning of the laser over the surface.

Chemical beam epitaxy (CBE) (= MOMBE). This technique is mainly used in semiconductor technology. The molecular beams are effusive jets of the gaseous precursors. The method combines the advantages of MOCVD with those of molecular beam epitaxy (MBE; see below). In contrast to MBE, the precursor is supplied from outside the system to the deposition zone. In contrast to MOCVD, CBE film growth is performed in a ultra-high vacuum (UHV) chamber (pressure $<10^{-10}$ mbar). At these low pressures, gas transport proceeds via molecular flow, with no collisions in the gas phase. Furthermore, a hot boundary layer of gas close to the substrate is absent under these conditions, and layer growth is controlled by the substrate temperature and by the kinetics of the surface reactions. The major drawbacks of CBE is the very high cost of UHV equipment and the low deposition rates and throughput.

Atomic-layer epitaxy (ALE). This method is used for the controlled deposition of semiconductor (▶ glossary) monolayers. The film is grown one atomic layer at a time, and the deposition process is based on alternating chemisorption of the different precursors and their surface reaction. This is shown in Figure 3-12 for the formation of a ZnS film from $ZnCl_2$ and H_2S, as an example. Repetition of the growth cycle produces controlled layer-by-layer film growth.

Ideally, the precursor chemisorbs only to active sites on the substrate until all sites are occupied. When carried out within certain temperature limits ("ALE-window"), ALE is a self-limiting deposition process. It has the potential to grow extremely homogeneous crystalline films over large areas and offers unprecedented control at the monolayer level. One major disadvantage of ALE is that film growth is much slower than conventional CVD, because the growth rate is typically limited by the time required for switching from one precursor to the other, which usually requires several seconds.

ZnCl$_2$ (ad)

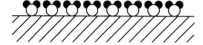

ZnCl$_2$ (ad) + H$_2$S (g) $\longrightarrow$ ZnS (ad) + 2 HCl (g)

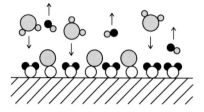

ZnS (ad)

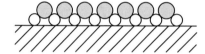

Figure 3-12. ALE reaction cycle leading to the formation of a ZnS thin film.

Non-CVD processes for the gas-phase deposition of thin films

These processes include PVD (*physical vapor deposition*) by evaporation or sputtering of a target material onto a substrate, and MBE. They are discussed briefly at this point for comparison with the CVD methods.

Today, most metallization for microelectronics is performed using PVD processes. These involve three steps:

1. vaporizing a solid;
2. transporting the gaseous compound from the source to the substrate; and
3. condensing the gaseous compound on the substrate surface, followed by nucleation and growth of a new layer.

The main difference between CVD and physical methods is that in CVD, films are formed by reaction of some precursor on the substrate (e.g., deposition of metallic copper by thermal decomposition of metal–organic precursors, see Section 3.2.2), while in PVD (and related processes) some material is volatilized and deposited as such on the substrate (e.g., elemental copper itself).

PVD by *thermal evaporation* is inherently a simple method, and is widely used for depositing metallic films for integrated circuit fabrication. An electron or laser beam is directed at the target material under high vacuum conditions (10^{-5} to 10^{-8} mbar) to vaporize the material. Alternatively, a crucible containing the material is heated by induction or by resistive heating. The rate of evaporation is controlled by the temperature of the source. The evaporated material deposits on a substrate that is maintained at a lower temperature than that of the vapor. At the low pressure, the mean free path of the evaporated species is very long (5 to 5000 m) compared to the source-to-substrate distance. This allows an essentially collisionless transport of the evaporated material to the substrate surface.

PVD by *sputtering* is a process where the surface atoms of a target material are liberated by bombardment with energetic ions that are generated in a glow discharge or plasma. The sputtered atoms are then ballistically transported to the substrate surface, where they condense. Sputter-deposited films are widely used in integrated circuits. Sputtering processes are generally applicable to all inorganic materials. The deposition of multi-element materials by sputtering is possible because the composition of the film is the same as the composition of the target. An alternative for the deposition of ceramic films is "reactive sputtering". Chemical compounds are then formed by reaction of the metal vapor with a reactive gas. Examples of reactive gases are methane, ammonia or nitrogen, and diborane to deposit carbide, nitride, and boride materials, respectively.

One of the biggest advantages of PVD methods is their lower substrate temperature which allows the deposition of films also on thermolabile substrates such as organic polymers. However, in contrast to CVD, the coating of surfaces not facing the vapor source is a serious problem. This also includes step coverage (▶ glossary) in electronic devices. In CVD processes, diffusion or convection can transport reactants to hidden areas (Figure 3-13).

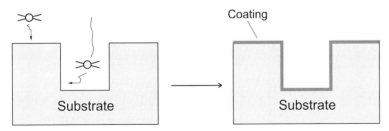

Figure 3-13. Step coverage by CVD methods.

In PVD processes, the particle stream from the evaporating or sputtering sources is directed. Furthermore, the deposited species are not sufficiently mobile on the substrate surface to migrate to areas not facing the source. Other advantages of CVD over PVD include the possibility of epitaxial growth (▶ glossary) and selective deposition, the capability of large-scale production, and the ability to produce metastable (▶ glossary) materials.

Molecular-beam epitaxy (MBE). This is a relatively simple process in which elemental sources are independently evaporated at a controlled rate, forming molecular beams that intercept at the heated substrate (Figure 3-14). MBE is usually carried out under UHV conditions (10^{-10} mbar) and at low growth rates. Films with high purity and very complex layer structures (nanostructures, for example) can be deposited with a precise control of doping of the deposited layers. The low processing temperature is also important for microelectronic production.

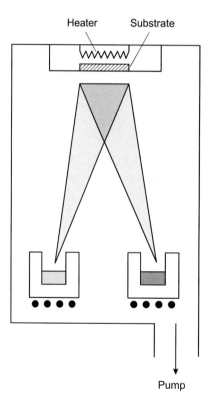

Figure 3-14. Principle of molecular-beam epitaxy (MBE).

3.2.2 Metal CVD

Metal coatings on various substrates are needed for a variety of applications. Typical examples include oxidation, corrosion or abrasion-protection films, reflective or conducting coatings, or electrodes. Particularly important applications are in microelectronics. The increasing demand for more and more sophisticated electronic devices requires the speed of the circuits to be increased and their performance to be improved. Therefore, the size of the devices must be decreased and the number of levels in complex multi-level structures must be increased. Figure 3-15 shows an example for the complexity of such a device.

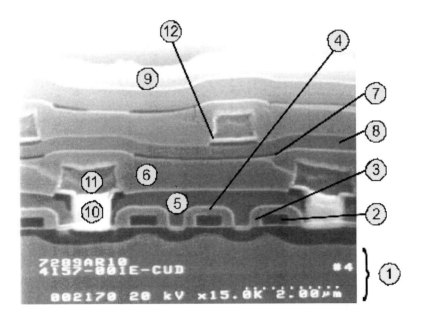

Figure 3-15. A cross-section through a multi-level structure of an electronic device. 1: silicon substrate; 2: gate, 3: spacer (CVD oxide); 4: covering oxide; 5: borate-phosphate-silicate glass (PACVD). 6: silica (PACVD from Si(OEt)$_4$); 7: glass layer (spin-on). 8: phosphate-doped silica (plasma CVD from Si(OEt)$_4$); 9: oxynitride/nitride passivation layer (PACVD); 10: tungsten plug (thermal CVD); 11: AlCu (PVD); 12: TiN barrier (reactive PVD).

The primary reason for metallization is to provide interconnections between the various circuit components. In addition to having good electrical properties, it is essential that these interconnection materials do not react with the materials which they contact. This is a problem for many metal–silicon contacts. Silicon may diffuse into the metal film to form solid solutions (▶ glossary) or metal silicides during thermal processing or by electromigration (▶ glossary). This becomes an increasingly important topic as the size of the interconnect lines decreases. Therefore, diffusion barrier layers are often introduced between the metal and the semiconductor (▶ glossary). Common materials that suppress this migration for Si/Al contacts are TiN, Ti, TiW, ZrN, or RuO_2.

Most commercial metallic films are produced by PVD methods. In this section, the CVD of Al, Cu, and W is discussed as an example, because this allows an elaboration to be made on the chemistry of different types of precursors.

Aluminum

The main uses for aluminum films are:

- Metallized polymers in food packaging (where the Al film provides a gas diffusion barrier) and as reflective films (mirrors, CDs, etc.).

- Interconnects in microelectronics. The resistivity (▶ glossary) of bulk Al (2.74 μΩ·cm) is only slightly greater than that of Cu (1.70 μΩ·cm) or Ag (1.61 μΩ·cm). An important aspect of its use in device technology is that Al as an amphoteric element is easily etched both by strong bases or acids. On the negative side, Al is prone to electromigration (▶ glossary) and is soluble in or reacts with semiconductor materials (▶ glossary). One possibility of reducing electromigration is to alloy Al with small amounts of Cu and Si.

The best investigated precursor for depositing high-quality Al films by CVD is triisobutylaluminum, Al^iBu_3, a colorless, pyrophoric liquid, which at room temperature has a vapor pressure of about 0.1 mbar. It is monomeric in the gas phase.

The first isobutyl ligand of Al^iBu_3 is degraded by β-hydride elimination (Eq. 3-11) already at temperatures above 50 °C. The diisobutylaluminum hydride formed is a trimer, and therefore has a substantially lower vapor pressure (0.01 mbar at 40 °C). The formation of $[HAl^iBu_2]_3$ in the precursor vessel and the gas tubings can be suppressed by addition of isobutene to the carrier gas, because the β-elimination process is reversible.

Typical conditions for the deposition of Al from Al^iBu_3 are temperatures of 200–300 °C in hot-wall reactors. Growth rates of 20 to 80 nm·min^{-1} are achieved. The decomposition chemistry of Al^{i}Bu$_3$ on aluminum surfaces was very well studied, and sheds light on the processes leading to film growth. The overall reaction is given in Eq. 3-12.

$$3 \text{ Al(CH}_2\text{CHMe}_2)_3 \rightleftharpoons 3 \text{ CH}_2\text{=CMe}_2 \; + \; \text{[HAl}^i\text{Bu}_2]_3$$

(Al^{i}Bu$_3$)

[HAl^{i}Bu$_2$]$_3$

(3-11)

$$\text{Al}^i\text{Bu}_3 \longrightarrow \text{Al} + 3 \text{ CH}_2\text{=CMe}_2 + {}^3\!/_2 \text{ H}_2 \qquad (3\text{-}12)$$

The three isobutyl groups become equivalent after adsorption. They diffuse over the Al surface and are no longer attached to just one aluminum atom (roughly speaking, once a layer of aluminum is formed on a particular substrate, further decomposition of Al^{i}Bu$_3$ occurs on an aluminum surface). Therefore, each isobutyl group equally participates in the rate-determining β-hydrogen elimination (Figure 3-16).

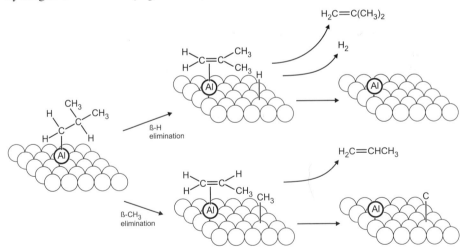

Figure 3-16. Schematic representation of the thermal decomposition of Al^{i}Bu$_3$. The upper pathway (at low temperatures) results in the clean deposition. At higher temperatures (lower pathway) carbon impurities may be incorporated into the film.

As a consequence, the hydrogen atoms created by β-elimination are also spread over the Al surface. Both H_2 and isobutene readily desorb from Al at temperatures below the decomposition temperature.

Carbon is only incorporated in the films at temperatures above 330 °C. β-Methyl elimination, leading to propene and surface-bonded methyl groups, has a higher activation energy than β-hydrogen elimination, and therefore only plays a role at higher temperatures. The surface-bonded methyl groups eventually decompose to give carbon (see Figure 3-16).

Trimethylaluminum, Al_2Me_6, is a less suitable precursor for the same reason. β-Hydrogen elimination is not possible; the decomposition path of Al_2Me_6 is complex, and depends strongly on the conditions. Al films deposited between 350 and 550 °C from Al_2Me_6 contain high levels of carbon. Thermal decomposition under N_2 even results in the formation of thermodynamically favored Al_4C_3 (Eq. 3-13). However, Al films with little carbon incorporation are formed when the atmosphere is changed to hydrogen (Eq. 3-14). Thermodynamic calculations predict that the reaction of Al_2Me_6 at 200 °C in the presence of H_2 results in the formation of AlH_3 and methane which readily desorbs.

$$2\ Al_2Me_6 \longrightarrow Al_4C_3 + 9\ CH_4 \tag{3-13}$$

$$Al_2Me_6 + 3\ H_2 \longrightarrow 2\ Al + 6\ CH_4 \tag{3-14}$$

With Al^iBu_3 as the precursor, the selective deposition of Al is possible. Film growth occurs on Si or Al, but not on silica. When Al^iBu_3 is adsorbed on silica, two of the three isobutyl groups are readily lost already at room temperature. Due to the inductive effect of the two oxygen atoms then bonded to the aluminum atom (Figure 3-17), elimination of the third isobutyl group is suppressed. The surface-bonded $^iBuAl=$ species inhibit film growth on silica by preventing further adsorption.

Al^iBu_3 can be considered as a source of AlH_3 formed by β-elimination of iso-butene. However, AlH_3 itself is a solid not suited for CVD. AlH_3 is also contained in donor–acceptor complexes of the type $L\text{-}AlH_3$, where L is a Lewis base. For CVD purposes, adducts with L = tertiary amine are mainly used, their stability and volatility being influenced by the electronic properties and the steric bulk of L. They are easily prepared and are significantly less air-sensitive than are aluminum alkyls. Trimethylamine forms a solid 1:1 and a 2:1 adduct, $Me_3N\text{–}AlH_3$ and $(Me_3N)_2AlH_3$. Liquids are obtained for L = $EtMe_2N$ or Et_3N.

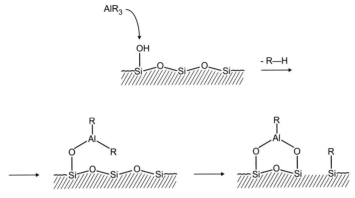

Figure 3-17. Inhibition of Al deposition from AlR_3 on a silica surface.

There are only a few problems with carbon incorporation with these precursors, because the compounds contain no Al–C bond, and the Al–N bond is easily cleaved. The decomposition mechanism of $Me_3N–AlH_3$ on Al surfaces is shown schematically in Figure 3-18. The first step after adsorption is dissociation of NMe_3 and its fast desorption. Similar to the dispersion of the butyl groups after adsorption of Al^iBu_3, the hydrogen atoms of the AlH_3 group also disperse over the Al surface.

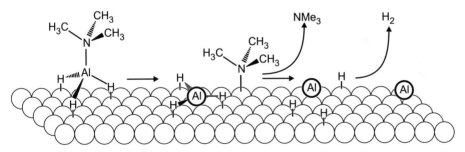

Figure 3-18. Thermal decomposition mechanism of $Me_3N–AlH_3$ on Al surfaces.

Tungsten

There are two important areas of application for tungsten films:

- Wear protection for cutting and grinding tools and corrosion protection for various materials, even at high temperatures. The hardness (▶ glossary) and chemical inertness can be enhanced by adding small amounts of carbon, oxygen, or nitrogen.
- Metallization in integrated circuits. In addition to favorable electrical, mechanical and chemical properties, tungsten has a high resistance to electromigration (▶ glossary), a low thermal expansion coefficient, and does not react with major semiconductor materials (▶ glossary) below 600 °C. A disadvantage is the low adhesion to the SiO_2 layer on Si, which necessitates the use of adhesion promoters such as TiN.

The dominant tungsten CVD precursor is tungsten hexafluoride (WF_6) despite some disadvantages (see below). This is used routinely in microelectronics industry. WF_6 has a boiling point of 17 °C, a high vapor pressure (1.3 bar at 25 °C), and is commercially available at low cost. Although thermal decomposition can be achieved above 750 °C, most methods of tungsten CVD involve reducing agents (mostly H_2, Si, or SiH_4) to lower the deposition temperatures.

The reduction of WF_6 by a silicon substrate allows the selective deposition of W. SiF_4 is formed below 400 °C (Eq. 3-15), and SiF_2 above 400 °C (Eq. 3-16). The high-temperature reaction (Eq. 3-16) consumes twice the amount of Si than does the low-temperature reaction.

$$2\ WF_6 + 3\ Si \xrightarrow{\ <400\ °C\ } 2\ W + 3\ SiF_4 \tag{3-15}$$

$$WF_6 + 3\ Si \xrightarrow{\ >400\ °C\ } W + 3\ SiF_2 \tag{3-16}$$

The reactions are very fast, but stop when a certain film thickness is reached (mostly within seconds). For film growth, either silicon has to diffuse through the W layer, or WF_6 has to diffuse to the W/Si interface. Both become increasingly difficult with increasing film thickness, particularly since tungsten is an excellent diffusion barrier for silicon. The presence of water has to be carefully avoided, because WF_6 is readily hydrolyzed and the thus generated HF attacks the silica layer on silicon and the tungsten layer.

WF_6 is reduced by hydrogen (Eq. 3-17) at temperatures between 300 and 800 °C.

$$WF_6 + 3\ H_2 \longrightarrow W + 6\ HF \tag{3-17}$$

Because of the formation of HF and the problems associated with that, tungsten is not deposited directly on silicon (Eq. 3-18). This problem is avoided when SiH_4 is used as the reductant. Furthermore, no silicon is consumed (as in the WF_6/Si combination), and the deposition rates are high (up to ~1 $\mu m \cdot min^{-1}$ in cold-wall LPCVD). In principle, two deposition processes compete with each other: reduction of WF_6 (giving W) and thermal decomposition of SiH_4 (giving Si which then reacts with elemental W). The deposition reactions are less straightforward as in the Si or H_2 reduction. Equation 3-18 gives the most probable reactions. There is no reduction of WF_6 by the produced H_2, because the reduction by SiH_4 is much faster.

$$2\ WF_6 + 3\ SiH_4 \xrightarrow{\ 250\ ^\circ C\ } 2\ W + 6\ H_2 + 3\ SiF_4$$

$$(3\text{-}18)$$

$$4\ WF_6 + 3\ SiH_4 \xrightarrow{\ T > 600\ ^\circ C\ } 4\ W + 12\ HF + 3\ SiF_4$$

The advantages of the SiH_4 reduction are somewhat impaired by the very difficult handling of SiH_4/WF_6 mixtures. The two compounds react vigorously with each other. Therefore, silane reduction is only used to initiate tungsten deposition, followed by the hydrogen reduction process.

Their volatility and thermolability makes organometallic compounds potentially interesting precursors for tungsten deposition under less corrosive conditions in the electronics industries. A variety of such compounds has been investigated, for example, tungsten hexacarbonyl $[W(CO)_6]$, or bis(cyclopentadienyl)dihydridotungsten $[(\eta^5\text{-}C_5H_5)_2WH_2]$. However, the major drawback of tungsten CVD using organometallic compounds is contamination of the films with sometimes high levels of carbon and/or oxygen.

Copper

Copper may replace Al in part of the interconnections in multilevel integrated circuits, because it is more resistant to electromigration (▶ glossary). Furthermore, it has a better thermal expansion coefficient, a higher melting point, and a lower electric resistivity (▶ glossary). Thus, Cu interconnects could allow an increase of the operation frequency of devices and higher current densities compared to Al. Disadvantages of Cu are its fast diffusion in Si and drift in SiO_2-based insulating materials (dielectrics,▶ glossary), the poor adhesion to SiO_2, and the lack of suitable etch processes.

Since copper does not form organometallic compounds suitable for CVD, and copper halides are not volatile enough, metal–organic compounds are mainly used. There are two chemically different approaches for the CVD of Cu films. In the first approach, volatile Cu(II) compounds are decomposed at the substrate surface in the presence of hydrogen. The second approach makes use of the well-known property of Cu(I) compounds to disproportionate into elemental Cu and Cu(II) compounds.

Copper(II) precursors. Attention has mostly focused on the use of Cu(II) bis(β-diketonate) complexes and, to a limited extent, on bis(β-ketoiminate) complexes. Two of the best investigated complexes are shown in Figure 3-19.

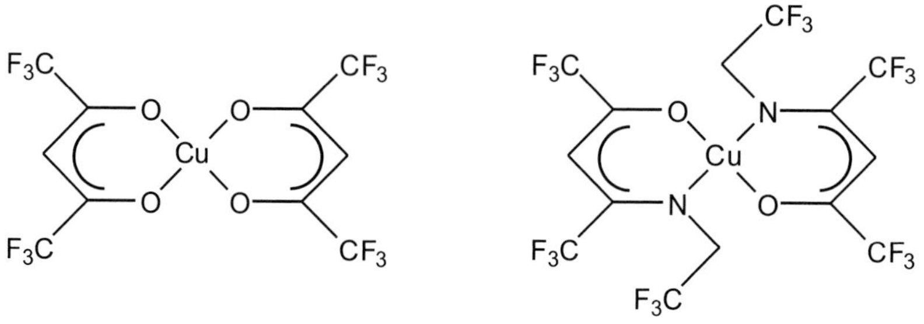

Figure 3-19. Cu(hfac)$_2$ and Cu(nona-F)$_2$ as representative Cu(II) precursors with β-diketonate (left) and β-ketoiminate ligands (right).

An important feature, also characteristic of the Cu(I) compounds discussed below, is the use of fluorinated ligands, such as 1,1,1,5,5,5-hexafluoroacetylacetonate (hfac) (Figure 3-19). The volatility of β-diketonate complexes increases with the number of fluorine atoms due to the reduction of van der Waals interactions between the molecules. The high electronegativity of the fluorine atoms may also lead to a weakening of the Cu–ligand bonds.

Equation 3-19 gives the stoichiometry of the CVD process for Cu(hfac)$_2$ in the presence of hydrogen. The ligands are liberated in their protonated forms.

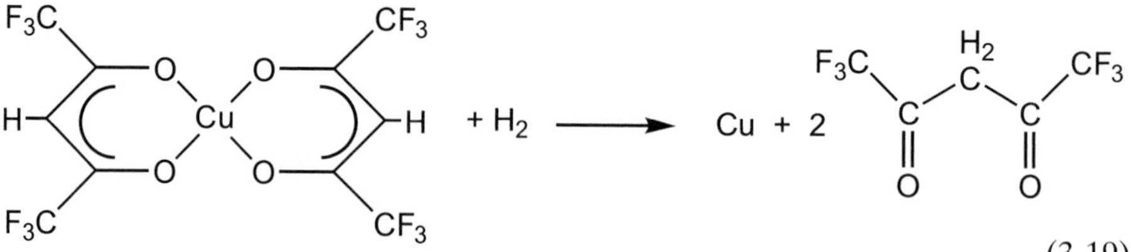

(3-19)

Cu(hfac)$_2$ adsorbs dissociatively under CVD conditions (Eq. 3-20). If the film grows on a copper surface (i.e., after the first Cu layer has been formed), the Cu atoms become part of the surface, which corresponds to an electron transfer from the surface to the incoming Cu atom. The hfac ligands spread over the Cu surface, as discussed above for the butyl ligands of Al^{i}Bu$_3$. When Cu(hfac)$_2$ is adsorbed onto non-metallic surfaces, electron-transfer is no longer facile and the intact molecule is adsorbed. The reverse reactions leading to desorption of Cu(hfac)$_2$ may occur readily under CVD conditions where bimolecular surface reactions are possible. It is an essential step in the CVD of Cu(II) compounds (see below) .

$$
\begin{aligned}
\text{Cu(hfac)}_2 \text{ (g)} &\rightleftharpoons \text{Cu(s) + 2 (hfac)}^\cdot_{\text{ads}} \\
\text{H}_2\text{(g)} &\rightleftharpoons \text{H}^\cdot_{\text{ads}} \\
\text{(hfac)}^\cdot_{\text{ads}} + \text{H}^\cdot_{\text{ads}} &\rightleftharpoons \text{H-hfac(g)}
\end{aligned}
\qquad (3\text{-}20)
$$

Dissociative adsorption of H$_2$ occurs readily and reversibly at room temperature and above on most metals. The heat of adsorption is rather low, which means that the surface will not be blocked by hydrogen under reaction conditions. There is a significant kinetic barrier for hydrogen adsorption (about 40–60 kJ·mol^{-1}). Therefore, when the coverage of the surface by hfac ligands is high and the partial pressure of H$_2$ in the gas phase is low, there may be conditions where H$_2$ adsorption becomes the rate-limiting step for the deposition of Cu films.

Kinetic measurements suggest that the rate-limiting step in the overall Cu deposition is the desorption of H-hfac formed by bimolecular surface reaction between adsorbed H and hfac. Instead of desorbing as H-hfac, the hfac ligands may undergo decomposition reactions. Therefore, the surface coverage by hfac also affects film purity.

Copper(I) precursors. Although the Cu(II) precursors have high thermal stability and high vapor pressures, deposition rates are usually low (0.1–0.5 nm·min^{-1}) and high substrate temperatures (~400 °C) are required. They only result in pure Cu films in the presence of a reducing agent, such as H$_2$. Cu(I) precursors are mostly liquids and have lower vapor pressures and lower decomposition temperatures (<200 °C). Typical deposition rates are around 0.1 to 1 µm·min^{-1}, and no reducing agent is needed.

The most widely studied family are Lewis base-stabilized Cu(I) β-diketonates. The ligand L may be phosphines (PMe$_3$), olefins (mostly 1,5-

cyclooctadiene [COD] or vinyltrimethylsilane [VTMS]), alkynes (3-hexyne), etc.

These complexes have the following advantages:

- the β-diketonate ligand provides volatility to the complex, particularly if it contains fluorinated substituents;
- the neutral ligands L do not decompose at temperatures where Cu deposition occurs; and
- both the β-diketonate ligand and the ligand L allow easy modifications to tailor the physical state of the complex and its volatility and reactivity.

$$(3\text{-}21)$$

The overall deposition process is a thermally induced disproportionation reaction according to Eq. 3-21. The proposed mechanism for deposition is shown in Eqs. 3-22 and 3-23 for (hfac)CuL. Since the ligand L is only weakly bound, the thermal decomposition of the Cu(I) complex occurs at relatively low temperatures once the precursor has been adsorbed to the surface. The neutral ligand L desorbs readily, while at the same time adsorbed Cu(hfac) disproportionates into Cu(hfac)$_2$ and elemental Cu (Eq. 3-23). Cu(hfac)$_2$ desorbs from the surface, because the deposition temperature of copper is below the decomposition temperature of Cu(hfac)$_2$.

$$[(hfac)CuL]_{ads} \longrightarrow [Cu(hfac)]_{ads} + L(g) \qquad (3\text{-}22)$$

$$[Cu(hfac)]_{ads} \longrightarrow Cu(s) + Cu(hfac)_2\,(g) \qquad (3\text{-}23)$$

3.2.3 Diamond CVD

Diamond has outstanding properties, most of which are relatively insensitive to lattice defects:

- mechanical properties: extreme hardness (~90 GPa), very high bulk modulus (1.2×10^{12} N·m^{-2}), very low compressibility (8.3×10^{-13} m^2·N^{-1})

- acoustic properties: high sound velocity (18.2 km·s^{-1})
- thermal properties: very high thermal conductivity at room temperature (2×10^3 W·m^{-1}·K^{-1}), low thermal expansion coefficient at room temperature (0.8×10^{-6} K^{-1})
- optical properties: transparency from the deep UV to the far IR
- electrical properties: good insulator (resistivity [▶ glossary] ~10^{16} Ω·cm at room temperature), semiconductor (▶ glossary; 10–10^6 Ω·cm) after doping, with a band gap of 5.4 eV
- chemical properties: resistant to chemical corrosion and radiation.

These properties result in several highly interesting applications for diamond thin films. For most applications the films can be polycrystalline (Figure 3-20). The most developed are heat sinks in electronic devices and wear-resistant hard coatings for tools. Applications close to commercialization include optical windows, radiation detectors and surface-acoustic-wave (SAW) devices. In the future, diamond films could be used as high-temperature semi-conducting devices, field emitters, or X-ray-lithography masks.

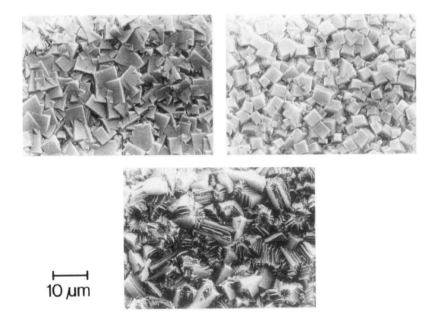

10 µm

Figure 3-20. Diamond films (on a SIALON surface) deposited by microwave activation of 1.7 % methane in hydrogen. The different morphologies originate from different gas pressures and different microwave powers.

The synthesis of polycrystalline diamond has been achieved by several CVD methods (Figure 3-21).

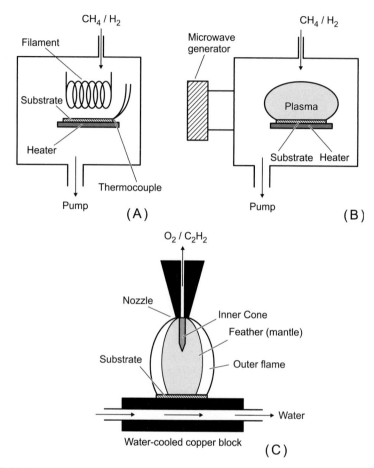

Figure 3-21. Schematic diagrams of the three most commonly used apparatus for diamond deposition. (A) Hot filament reactor. (B) Microwave plasma-enhanced CVD reactor. (C) Oxyacetylene torch.

The most common methods for diamond synthesis are the activation of methane in an excess of hydrogen (typically 1–2 vol% CH_4 in H_2) by using microwave, r.f. or d.c. discharges (Figure 3-21(B)) and the decomposition of the gas mixture by hot (~2300 °C) filaments (Figure 3-21(A)). These variations have in common that atomic hydrogen is produced from H_2 near the surface during the decomposition process. Gas-phase hydrogen-abstraction reactions lead to the formation of hydrocarbon radicals that probably are the main precursors for diamond. However, the chemistry at the growth surface is very complex. The choice of appropriate deposition parameters is important for obtaining specific diamond morphologies.

Diamond is metastable (▶ glossary) with respect to graphite at low pressure. The growth of metastable diamond is possible because of the presence of atomic hydrogen ($H^{\bullet}$) which selectively etches codeposited graphitic nuclei and leaves diamond. The dissolution rate of graphite and amorphous carbon by hydrogen is about 50 times that of diamond. Thus, the basic chemistry of diamond deposition can be written in two equations (Eq. 3-24)

$$\text{etching of graphite:} \quad y\,H^{\bullet} + x\,C_{graphite} \longrightarrow C_xH_y$$

$$\text{diamond deposition:} \quad m\,H^{\bullet} + C_xH_y \longrightarrow m + ^y/_2\,H_2 + x\,C_{diamond}$$

$$(3\text{-}24)$$

Furthermore, hydrogen prevents the formation of sp^2-hybridized carbon atoms at the growth front, and thus stabilizes the growing diamond surface. Hence, the growth surface is usually hydrogen-terminated.

Diamond films are also obtained when a hydrocarbon (typically acetylene or methane) is burnt with oxygen in a welding torch. If this flame is directed to a water-cooled substrate at temperatures of 700–1000 °C as shown in Figure 3-21(C), diamond growth occurs at the intersection between the substrate and the primary combustion zone. The method is inherently simple, but mostly not economical. The yield of diamond and the area of deposition is relatively small, the deposits are rough and have varying crystal quality, and the synthesis is not very reproducible.

Diamond nucleation (▶ glossary) is very sensitive to the substrate material itself as well as to its surface condition (surface composition and morphology, dislocations, pretreatment, etc.). Diamond nucleation (▶ glossary) on highly perfect substrate surfaces, for example silicon wafers, is sluggish and has a long initiation period. Nucleation is also difficult if the surface carbon concentration is lowered rapidly by diffusion into the substrate (e.g., Fe, Co, Ni). It only starts when the substrate is saturated. Chemical reactions occurring between the substrate and reaction gas components (for example, atomic hydrogen) can also have an important influence on the nucleation (▶ glossary) and on the adhesion of the films. To enhance the diamond nucleation density, pretreatment of the substrate surface by polishing, abrading, etc., is generally very effective.

A major problem is to obtain sufficient adhesion between the substrate and the diamond film (mismatch of the thermal expansion), particularly at the high temperatures during the operation of cutting tools.

An interesting alternative to the direct coating of tools with diamond films is the bonding of freestanding diamond sheets. Sheets in sizes of several hundreds cm^2 are commercially produced by diamond deposition on dummy substrates, which are dissolved after formation of the film.

3.2.4 CVD of Metal Oxides

Thin films of metal or semimetal oxides are employed in numerous technological applications. Although there is major competition from sol–gel techniques (Section 4.5), CVD is widely used, particularly in the microelectronics industry. A variety of transition metal oxide films has been prepared by CVD, mainly for dielectric (▶ glossary) or optical purposes. Since this book concentrates on chemical aspects rather than applications, only SiO_2 and $YBa_2Cu_3O_{7-x}$ (YBCO) will be discussed as case studies.

The precursors for metal oxide CVD are mostly the same as those for the deposition of the metals together with an oxygen source. The most common oxygen sources are O_2, N_2O, or water. Interesting alternatives are often metal alkoxides which combine a (mostly) sufficient and tailorable volatility with the advantage of being single-source precursors, and their use as precursors will be exemplarily discussed for SiO_2.

SiO_2 and silicate glasses

The formation of SiO_2 thin films is one of the most important processes in industrial microelectronic device manufacturing. SiO_2 thin films are used for gate insulation layers, surface passivation, planarization, and packaging (see Figure 3-15).

Each device within a microcircuit needs to be *isolated* from adjacent devices. This is achieved by the deposition of an insulator material. For this application, step coverage (▶ glossary) is an important issue. Borosilicate, phosphosilicate, and borophosphosilicate glasses are also frequently used. They have lower intrinsic stress, lower melting temperatures, and better dielectric properties (▶ glossary) than SiO_2 itself. *Passivation* of a semiconductor (▶ glossary) is a process by which a film is deposited to protect it from the environment and/or to provide electronic stabilization of the surface by saturating all bonds of the surface atoms. During processing of multi-level electronic devices, the surface becomes increasingly non-planar. In order to allow the subsequent deposition of conducting layers without breaking metal lines, the surface must be flat and smooth. This is achieved by the *planarization* process in which a dielectric material (▶ glossary) such as silica or silicate glass is deposited and then etched to give a smooth surface.

The silica layer covering the surface of silicon under ambient conditions is of no technological value. Thicker films can be prepared by heating the silicon in either dry oxygen or water vapor, or by plasma oxidation. However, these methods are not suitable during multi-step fabrication, and CVD is used instead.

Silane (SiH$_4$) as precursor. Silane is used widely as a precursor for SiO$_2$. It is a highly reactive gas which thermally decomposes around 1000 °C to give Si and H$_2$. Therefore, SiH$_4$ is also used for the CVD of elemental silicon. SiO$_2$ is obtained by oxidation of SiH$_4$ with O$_2$ or N$_2$O as most common oxygen sources. H$_2$O$_2$ is gaining importance as an oxidant. Growth rates up to 3 μm·min^{-1} have been achieved by rapid thermal LPCVD which also provides a highly uniform step coverage (▶ glossary). Equation 3-25 over-simplifies the reaction with O$_2$; the detailed mechanism involves complex branching chain reactions. At high partial pressures of oxygen an alternative reaction occurs, resulting in the formation of water (Eq. 3-26).

$$SiH_4 + O_2 \longrightarrow SiO_2 + 2\,H_2 \tag{3-25}$$

$$SiH_4 + 2\,O_2 \longrightarrow SiO_2 + 2\,H_2O \tag{3-26}$$

N$_2$O is an alternative source of oxygen in the reaction with SiH$_4$. The overall reaction is given in Eq. 3-27. The reaction is probably initiated by decomposition of N$_2$O generating N$_2$ and atomic oxygen and proceeds in a complex sequence of radical reactions.

$$SiH_4 + 2\,N_2O \longrightarrow SiO_2 + 2\,H_2 + 2\,N_2 \tag{3-27}$$

Silicon halides as precursor. The most widely used high-temperature LPCVD process for the growth of SiO$_2$ (~900 °C) is the N$_2$O oxidation of H$_2$SiCl$_2$ (Eq. 3-28) which is a liquid at room temperature.

$$H_2SiCl_2 + 2\,N_2O \longrightarrow SiO_2 + 2\,HCl + 2\,N_2 \tag{3-28}$$

Tetraalkoxysilane (Si(OEt)₄, TEOS) as single-source precursor. The high-temperature growth of SiO_2 (~750 °C at ambient pressure and ~600 °C under LPCVD conditions) from TEOS requires no external oxygen source. The reaction is much more complicated than the overall reaction given in Eq. 3-29.

$$Si(OEt)_4 \longrightarrow SiO_2 + 2\ C_2H_4 + 2\ EtOH \qquad (3\text{-}29)$$

Surface-bound alkoxysilyl species apparently play a very important role and are the immediate source of ethylene (Eq. 3-30). The thus created surface silanol groups (Si–OH) can react with gaseous TEOS by elimination of ethanol to form new surface-bound alkoxysilyl species (Eq. 3-31).

$$\equiv Si-OEt \longrightarrow \equiv Si-OH\ +\ C_2H_4 \qquad (3\text{-}30)$$

$$\equiv Si-OH\ +\ Si(OEt)_4 \longrightarrow \equiv Si-O-Si(OEt)_3 +\ EtOH \qquad (3\text{-}31)$$

The deposition temperature in atmospheric pressure CVD is not influenced by the addition of O_2, which, however, removes carbon contaminations from the films. The deposition temperature can be lowered to 300 °C when ozone is added. Furthermore, the quality of the films is improved. For these reasons, the TEOS/O_3 system has become widely used.

Boron-containing glass films are grown by using $B(OR)_3$ (R = Me, Et, $SiMe_3$)/TEOS mixtures, and $P(OMe)_3$ or $PO(OMe)_3$ are preferred as phosphorus sources. The use of the alkoxides allows for deposition at lower temperatures (500–650 °C) at comparable growth rates than using the hydrides.

Yttrium barium copper oxide (YBCO)

Since the discovery of high-temperature superconductors (▶ glossary), major efforts have been made to grow high-quality films by CVD. The discussion in this section is restricted to YBCO (see also Section 2.1.1), but the strategies are similar for other high-temperature superconductors.

For reasons already discussed in Section 3.2.2, β-diketonates are very useful precursors: they are available in high purity for nearly all metal ions, and their volatility can be maximized by preventing molecular association. Saturating the metal coordination sphere with sterically demanding, non-polar or fluorinated groups is an attractive strategy for that purpose. Thus, β-diketonate derivatives of Y, Ba, and Cu are nearly exclusively employed as precursors for YBCO, mostly the dipivaloylmethanate (dpm) derivatives (which is also named tetramethylheptanedionate [tmhd or thd]) (Figure 3-22). Other derivatives used for CVD include Cu(acac)$_2$ and β-diketonates with fluorinated substituents.

For the delivery of the precursors in the reaction zone, a non-reactive gas (usually argon) is passed over or through the Ba, Y, and Cu complexes heated in bubblers to temperatures at which they are sufficiently volatile for transport (for example, source temperatures for Y(dpm)$_3$ 100–170 °C, Ba(dpm)$_2$ (Figure 3-22) 230–300 °C, Cu(dpm)$_2$ 120–160 °C). The metal complex vapors are mixed, and the oxygen source is then added just before entry into the reaction chamber.

With Y(dpm)$_3$, Ba(dpm)$_2$ and Cu(dpm)$_2$ or Cu(acac)$_2$ as precursors, high-temperature decomposition (800–900 °C) in the presence of O$_2$ or N$_2$O presumably involves both pyrolysis and oxidation, while hydrolysis is probably also involved when H$_2$O/O$_2$ is used.

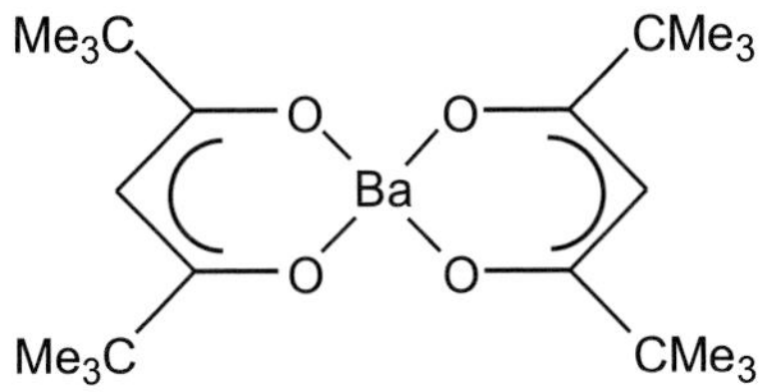

Figure 3-22. Barium dipivaloylmethanate, Ba(dpm)$_2$.

While Y and Cu precursors fulfill the requirements for CVD precursors such as stability and no degradation during delivery to the deposition zone, significant problems exist with the stability of Ba precursors. This is a consequence of the large size-to-charge ratio of the Ba^{2+} ion. Coordination numbers of up to 9 have been observed for Ba compounds. Two β-diketonate ligands balance the charge of the cation, but occupy only four coordination sites. The empty coordination sites must be blocked by Lewis bases in order to prevent intermolecular aggregation. One means of solving this problem is to add Lewis bases to the carrier gas, such as ammonia, amines, or ethers. Another strategy is the deliberate design of the ligand environment.

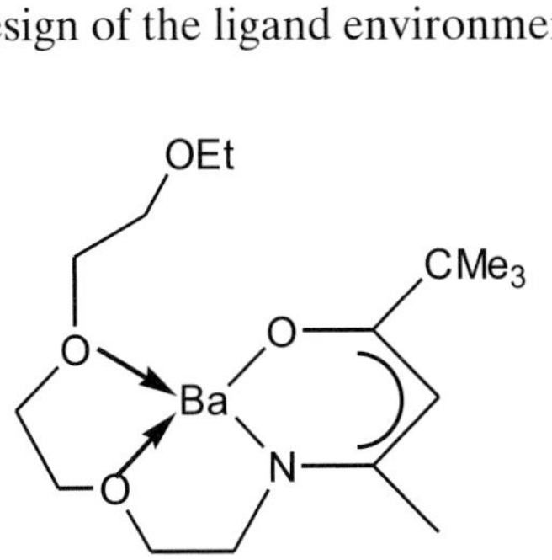

Figure 3-23. A polyether modified bis(β-ketoiminate) barium derivative (only one of the two multidentate [▶ glossary] ligands is drawn for clarity).

For example, complexes of the type $Ba(hfac)_2 \cdot L$ in which L is a polyether $(RO(CH_2CH_2O)_nR, n = 3-6)$ exhibit high vapor pressures and sometimes are even liquids. In these derivatives the metal is coordinatively saturated by the polydentate ligand L. In an extension of this approach, ligands were tailored in which the polyether moiety is appended to the chelating (▶ glossary) unit. An example is shown in Figure 3-23.

3.2.5 CVD of Metal Nitrides

Many metal and semimetal nitrides exhibit interesting properties such as high hardness (▶ glossary), high melting points, and high chemical inertness. The electrical properties vary from insulating (e.g., silicon nitride, Si_3N_4) to conducting (e.g., titanium nitride, TiN). Owing to these properties, nitride layers are applied for a variety of purposes. Some aspects of the CVD of Si_3N_4 and TiN are discussed here as case studies.

The deposition of *silicon nitride*, Si_3N_4, is a broadly applied industrial process. Si_3N_4 layers are often used in the microelectronics industry for passivation and encapsulation, because they are a very good barrier to diffusion of water, oxygen, and sodium ions. Additionally, Si_3N_4 is very hard, and chemically resistant.

Titanium nitride (TiN) has a unique combination of properties, including high hardness (▶ glossary), good electrical conductivity, a high melting point (3300 °C), and chemical inertness. Thin films of TiN have found practical uses as wear-resistant and friction-reducing coatings for machine tools. The gold-like color of TiN makes it useful for decorative purposes, for example coatings on jewelry and watches. More recently, TiN has found applications in microelectronics applications as low-resistance contact material (~22 $\mu\Omega \cdot cm$) and as a metal diffusion barrier.

Elemental nitrogen is only rarely used as the nitrogen source for the CVD of nitrides, because it is too unreactive. The most common nitrogen source for nitride films is ammonia, NH_3. This is a cheap gas and easily purified, but it is only sufficiently reactive at high temperatures. Approaches to lower the reaction temperatures in nitride CVD include the use of hydrazine (N_2H_4) as a more reactive nitrogen source and precursors containing nitrogen–element bonds.

An inherent disadvantage of the often-used metal halide/ammonia combinations is the formation of ammonium halides (NH_4X) by reaction of ammonia (which is often used in large excess) and the hydrogen halide (HX) generated as a by-product (Eq. 3-32). This secondary reaction is not explicitly stated in the following examples. Although NH_4X is sublimable, it can lead to chlorine contamination of the deposited films. Since the equilibrium in Eq. 3-32 is shifted to the left side at higher temperatures, and HX are gaseous compounds, the level of chlorine contaminations is lowered at higher temperatures.

$$NH_3 + HX \rightleftharpoons NH_4X \qquad (3\text{-}32)$$

As in the CVD of SiO_2, the most often employed silicon sources for the CVD of Si_3N_4 are SiH_4 or H_2SiCl_2 (Eq. 3-33), and, more recently, Si_2Cl_6. In commercial systems, a large excess of NH_3 is used to obtain films with a stoichiometric composition.

$$3\ H_2SiCl_2 + 4\ NH_3 \longrightarrow Si_3N_4 + 6\ HCl + 6\ H_2 \qquad (3\text{-}33)$$

Only Ti(IV) compounds are sufficiently volatile to be suitable as CVD precursors for TiN. The metal atom in TiN is in the formal oxidation state +III, Thus, other than in the analogous reactions leading to Si_3N_4 (Eq. 3-33, for example), the overall process of TiN deposition also involves the reduction of the metal atom from +IV to +III.

While the combination $TiCl_4/N_2/H_2$ (Eq. 3-34) needs temperatures >700 °C for high-quality films, the use of NH_3 (Eq. 3-35) reduces the necessary temperature to between 320 and 700 °C. In the latter reaction, ammonia has a dual role: it delivers the nitrogen for TiN, and it acts as a reductant. Therefore, nitrogen is formed as a by-product.

$$2\ TiCl_4 + 4\ H_2 + N_2 \longrightarrow 2\ TiN + 8\ HCl \qquad (3\text{-}34)$$

$$6\ TiCl_4 + 8\ NH_3 \longrightarrow 6\ TiN + 24\ HCl + N_2 \qquad (3\text{-}35)$$

Thermodynamic calculations suggest that $TiCl_3$ may be the actual precursor in the $TiCl_4/N_2/H_2$ system. $TiCl_3$ is probably formed by gas-phase reaction between $TiCl_4$ and H_2, and is then adsorbed to the substrate surface where it reacts with chemisorbed hydrogen and nitrogen.

A very important development with respect to the nitride deposition are transamination reactions between metal dialkylamides and ammonia. Metal dialkylamides are relatively easy to prepare and to handle. Transamination reactions provide a means of producing the corresponding amides (with M–NH₂ groups) *in situ*, which readily undergo condensation reactions to give M–N–M linkages (Eq. 3-36, see also Section 5.6). Examples are given in Eqs. 3-37 and 3-38.

$Si(NMe_2)_4$ (b.p. 196 °C), $HSi(NMe_2)_3$ (b.p. 142 °C) or $H_2Si(NMe_2)_2$ (b.p. 93 °C) are good precursors for transamination reactions with ammonia (Eq. 3-37). In TiN CVD, $Ti(NMe_2)_4$ and $Ti(NEt_2)_4$ are primarily employed because of their commercial availability. $Ti(NMe_2)_4$ is a liquid at room temperature with a vapor pressure of 1 mbar at 60 °C, and $Ti(NEt_2)_4$ has a vapor pressure of about 0.1 mbar at 100 °C. TiN films have been deposited from both precursors by transamination reactions with NH_3 (Eq. 3-38) at temperatures as low as 450 °C with low impurity levels.

$$\equiv M\!-\!NR_2 + NH_3 \longrightarrow \equiv M\!-\!NH_2 + HNR_2$$

$$3 \equiv M\!-\!NH_2 \longrightarrow \underset{\underset{M}{|}}{M\!-\!N\!-\!M} + 2\,NH_3 \tag{3-36}$$

$$3\,Si(NMe_2)_4 + 4\,NH_3 \longrightarrow Si_3N_4 + 12\,HNMe_2 \tag{3-37}$$

$$6\,Ti(NR_2)_4 + 8\,NH_3 \longrightarrow 6\,TiN + 24\,HNR_2 + N_2 \tag{3-38}$$

3.2.6 CVD of Compound Semiconductors

The majority of important semiconducting compounds (▶ glossary) is iso-electronic with elemental silicon, and they have related structures. These include combinations of groups III and V such as GaAs or InP ("III–V compounds"), or combinations of groups II and VI such as CdS or ZnSe ("II–VI compounds"). The band gaps of common semiconductors are shown in Figure 3-24. They are in "useful" regions of the electromagnetic spectrum. For example, those in the visible region can be used for displays and solar cells, and those in the infrared region for thermal imaging technologies.

Technical applications of III–V materials are very well developed. They are mainly in optoelectronics, and utilize the red to near-IR range of the electromagnetic spectrum. Examples for current commercial applications are

- solar cells: GaAs;
- light-emitting diodes (LED): GaAs, GaP, GaAsP, GaN, AlGaAs, InGaP;
- solid-state laser: InGaP, AlGaInP (red laser pointer);
- electronic devices: electrons can move with a higher velocity through GaAs than through Si, which allows the manufacturing of faster electronic devices. However, the electronics industry is still dominated by silicon technology.

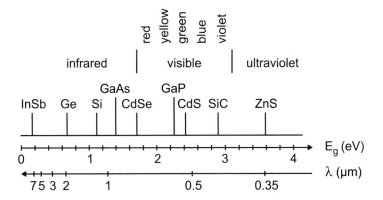

Figure 3-24. The band gaps of common semiconductors.

The energy gap (and the lattice parameters) can be engineered by simple variation of the composition ("semiconductor alloys";▶ glossary). For example, while the emission maximum in Si-doped GaAs LEDs is between 0.91 and 1.02 μm, alloys with aluminum ($Al_{0.4}Ga_{0.6}As$, Zn-doped) are used to form highly efficient photodiodes with an emission at 0.65 μm (red). GaN is the basis for blue LEDs.

The band-gap energies of II–VI compounds are larger than those of the III–V compounds. Thus, these materials are potentially important for the emission, detection or modulation of light in the green to near-UV. For example, ZnS and ZnSe are promising materials for green/blue laser diode materials. Progress in II–VI semiconductor applications has been hampered by inherent disadvantages, which include the difficulty to grow high-quality single crystals, problems associated with polytypism (cubic and hexagonal modification), and the reproducible control of the conductivity. However, new technologies are being developed and may open new applications. Thin-film electroluminescent displays based on ZnS are already commercially available. Applications for Cd chalcogenides (CdS, CdSe, CdTe) include solar cells, photoconductors, sensors, and transducers.

A particular problem in semiconductor (▶ glossary) CVD is the purity of the precursors. Since the semiconducting properties of a material are strongly influenced by impurities, very pure precursors must be applied. Special purification techniques have been developed for this purpose. These include not only classical techniques such as distillation, sublimation, or chromatography, but also chemical techniques such as "adduct-purification". Group II and III organometallic compounds are Lewis acids, and therefore form adducts with Lewis bases such as ethers, amines, or phosphines. The donor–acceptor bond is usually the most labile in the molecule. Although the adducts can be purified intact by distillation or recrystallization, the donor–acceptor bond can be thermally cleaved at moderate temperatures, liberating the purified organo-metallic precursor. For example, the bis-adducts between the trialkyl com-pounds of Al, In or Ga (MR_3) and 4,4'-methylene-bis(N,N'-dimethylaniline) or 1,2-bis(diphenylphosphino)ethane (diphos) (Figure 3-25) are already formed at room temperature. Traces of Si, Sn, Zn, or Mg compounds do not form such adducts and are thus removed by purification of the adducts. The diphos adducts liberate the pure MR_3 on heating to $T > 80\,°C$ at 10^{-2} mbar.

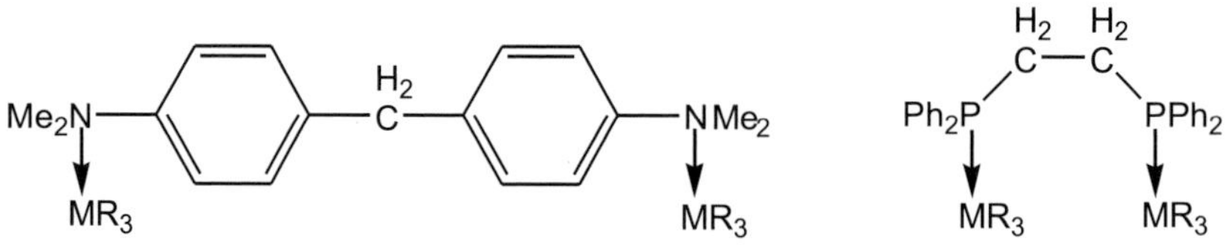

Figure 3-25. 4,4'-Methylene-bis(N,N'-dimethylaniline) (left) and 1,2-bis(diphenyl-phosphino)ethane (right) as Lewis bases in donor–acceptor complexes.

The chemistry of compound semiconductor CVD is exemplarily discussed for GaAs. The principles of the precursor and deposition chemistry for other III–V and II–VI combinations (the nitrides have been discussed in Section 3.2.5) is very similar. Semiconductor alloys can be obtained by adding a third precursor to the gas mixture.

The most commonly employed precursors for GaAs OMVPE or CBE are the commercially available trimethylgallium ($GaMe_3$) and arsine AsH_3. Typical growth temperatures are 600–750 °C. The deposited GaAs layers have remarkably low levels of carbon, considering that organometallic com-pounds are used.

A mechanism has been proposed that involves both gas-phase and surface reactions. Pyrolysis reactions occur in the boundary layer which primarily lead to $GaMe_x$ species (x <3, mainly x = 1) by abstraction of methyl radicals. The methyl radicals can react with AsH_3 in the gas phase to give methane and AsH_2 (Eq. 3-39). The formation of these species lowers the decomposition temperature of AsH_3. The gas phase formation of the adducts Me_3Ga–AsH_3 has no significant contribution to the GaAs growth mechanism due to the weakness of the Ga–As interaction. After adsorption of AsH_2 and the $GaMe_x$ species, the remaining methyl group(s) are removed by surface reactions. Hydrogen is transferred from the AsH_x species to surface-adsorbed methyl groups, and the formed methane is desorbed. This clean removal of carbon-containing groups from the surface leads to layers with very little carbon incorporation. Similarly, high-purity InP can readily be grown from InMe$_3$ and PH$_3$.

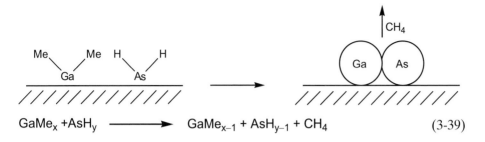

$$\text{GaMe}_x + \text{AsH}_y \longrightarrow \text{GaMe}_{x-1} + \text{AsH}_{y-1} + \text{CH}_4 \qquad (3\text{-}39)$$

GaEt$_3$ is a less convenient precursor than GaMe$_3$, because of its lower vapor pressure (4 mbar at 20 °C compared with 243 mbar for GaMe$_3$). However, it has been used successfully in combination with AsH$_3$ at low pressures. The pyrolysis chemistry of GaEt$_3$ is different to that of GaMe$_3$. As expected, the major decomposition route is by β-elimination which yields C$_2$H$_4$ together with Et$_2$GaH and EtGaH$_2$. These species are thermally unstable and, when formed in the boundary layer, can lead to the premature formation of Ga and GaAs. Therefore, GaAs layers deposited by using GaEt$_3$ are generally less uniform than those obtained from GaMe$_3$. However, they contain even less residual carbon due to the "cleaner" decomposition mechanism (see also discussion in Section 3.2.2, the related chemistry of Al compounds).

Efforts are being made to replace the highly toxic and gaseous AsH$_3$ with less hazardous liquid precursors, which may also be more easily purified and pyrolyze at lower temperatures. Since hydrogen is needed to remove the methyl radicals from the growth surface, the most promising layer properties were obtained with precursors containing both alkyl and hydrogen substituents. The most successful alternative to AsH$_3$ to date is *tert*.butylarsine (^{t}BuAsH$_2$). This is a liquid with a vapor pressure of 108 mbar at 10 °C, and it is considerably less hazardous than AsH$_3$. Furthermore, the pyrolysis of ^{t}BuAsH$_2$ creates more active AsH$_2$ or AsH species near the surface. Dominant decomposition reactions of ^{t}BuAsH$_2$ in the gas phase are dissociation in Bu$^\bullet$ and AsH$_2$$^\bullet$ radicals, and elimination of butane with the concomitant formation of AsH species.

A wide variety of III–V semiconducting compounds have been deposited from single-source precursors. Potential advantages include a higher air and moisture stability, a lower toxicity and less pre-reactions in the gas phase. The advantages are somewhat balanced by a lower volatility, a more difficult control of the stoichiometry, particularly for semiconductor alloys (▶ glossary;

for example, $Al_xGa_{1-x}As$), and a lower surface mobility of the polynuclear decomposition fragments. The most widely investigated III–V single source precursors are dimeric compounds $[R_2GaAs^tBu_2]_2$ (R = Et, ^tBu) as in Eq. 3-10. An important feature is that the 1:1 stoichiometry is retained during pyrolysis.

3.3 Aerosol Processes

CVD processes (Section 3.2) are controlled in a way that the gaseous precursors are adsorbed to the substrate surface and there form the solid product by thermal reactions. Gas-phase reactions and particle formation in the gas phase are usually undesirable. Contrary to that, in gas-phase powder syntheses – also called aerosol (▶ glossary) processes – particles are produced in the gas phase by chemical or physical processes. The advantages of aerosol processes, compared to other processes, are that:

- they do not involve the large volumes of liquids as in wet processes;
- their time scales are much shorter than those for solid–solid reactions;
- they can produce materials of high purity at high yields and with a high throughput; and
- multicomponent or nanophase materials (see Chapter 7) can be produced.

The example we will start with is the Aerosil® process for the production of fine powders of inorganic compounds. We will then treat some general aspects of the two major routes for aerosol processes: the gas-to-particle conversion route, and spray pyrolysis. After a short excursion to reactor types we will then turn to products that can be produced by aerosol processes.

The use of gas-phase reactions to produce titania, silica, or carbon black (soot) with nanometer-size particles is well established. Several million tons of products per year are produced worldwide. A typical example is highly dispersed amorphous silica. The so-called Aerosil® process was patented in 1942 by Degussa. The chemical reactions (Eq. 3-40) explain why this process is also called *flame hydrolysis*.

$$2\,H_2 + O_2 \longrightarrow 2\,H_2O$$

$$SiCl_4 + 2\,H_2O \longrightarrow SiO_2 + 4\,HCl$$

$$\text{overall:} \quad SiCl_4 + 2\,H_2 + O_2 \longrightarrow SiO_2 + 4\,HCl \qquad (3\text{-}40)$$

Silicon tetrachloride (obtained by reaction of elemental silicon with HCl) is volatilized and fed into an oxygen–hydrogen flame (Figure 3-26). The water formed by reaction of hydrogen and oxygen serves for the very fast and quantitative hydrolysis of $SiCl_4$ at about 1000 °C.

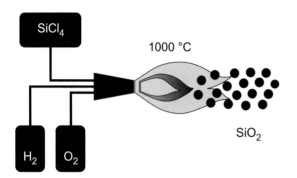

Figure 3-26. Formation of silica by the Aerosil process.

The only by-product is HCl, which is separated and recycled. The formed fumed silica is cooled, collected by conventional means (e.g., by cyclones, electrostatic precipitators, or baghouse filters) and de-acidified to remove adsorbed residual HCl. The surface properties of the silica can be modified by post-treatment with various silanes.

Highly dispersed titania, which is mainly used as pigments, is prepared by the same procedure. Fumed titania and silica are major chemical products. They are very light powders – a 200-liter bag filled with fumed silica can be easily lifted, because it weighs only about 10 kg! Its low weight is paralleled only by aerogels (see Section 6.3).

Fumed silica consists of agglomerated spherical, amorphous primary particles of only 7–40 nm diameter (Figures 3-27 and 3-28). The primary particle size can be influenced by the reaction parameters. Owing to the small primary particles, the specific surface area (S) is high (50 to 400 $m^2 \cdot g^{-1}$). For spherical, unagglomerated particles of diameter d and density ϱ, the specific surface area is given by

$$S = 6/(d \cdot \rho) \qquad (3\text{-}41)$$

A material composed of discrete small particles will possess a higher specific surface area than the same material in a coarser form (see Figure 2-5), because diameter and surface area are inversely related. In contrast to aerogels, the high surface area of pyrogenic silica originates only from the outer surface, i.e., it is not caused by the presence of pores.

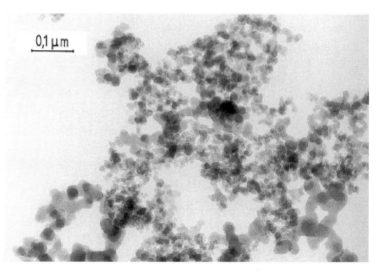

Figure 3-27. Transition electron micrograph of fumed silica with a primary particle size of 16 nm and a specific surface area of 130 ± 25 m$^2 \cdot$g^{-1} (Aerosil® 130 of Degussa–Huels).

There are many uses for pyrogenic silica. It was originally developed as "white soot" as a reinforcing filler (▶ glossary) for rubber, but now has found manifold use for various applications. For example, all types of liquids are thickened by the addition of pyrogenic silica. Many of these formulations show thixotropic behavior (▶ glossary). This behavior is explained by the formation of a three-dimensional network by hydrogen-bonding between the surface OH groups of the silica particles, either directly or via molecules of the liquid. This causes an increase of viscosity. When mechanical stress (▶ glossary) is applied, the network is degraded and the viscosity falls. Influencing the rheology (▶ glossary) of a system by addition of pyrogenic silica is used, for example, in drilling fluids, lacquers, paints, plastics, adhesives, greases, creams, ointments, or toothpastes. An additional effect is that the addition of pyrogenic silica prevents or at least slows down the sedimentation of solids in dispersions, such as pigments in lacquers. Pyrogenic silica is also used as a filler (▶ glossary) in silicones (see Section 5.2) and other elastomers, as adsorbent or support, to improve the storage stability and free flow of powders (fire-extinguisher powders, tablets, cosmetic powders, toners, table salt, etc.), or to control the triboelectric properties (▶ glossary) of powders. It has excellent heat insulation properties (see also Section 6.3), and is therefore used for thermal insulation of ovens, furnaces, pipelines, heating elements, or turbines.

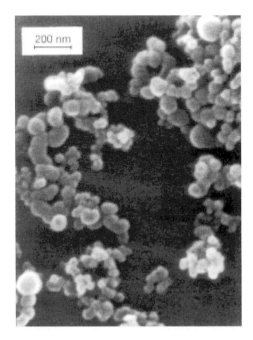

Figure 3-28. Scanning electron micrograph of fumed silica with a primary particle size of 40 nm and a specific surface area of 50±15 $m^2 \cdot g^{-1}$ (Aerosil® OX 50 of Degussa–Huels).

Aerosol process routes

There are two major routes for aerosol processes:

1. *Gas-to-particle conversion route*: Powders are made at high temperatures by reacting gases or vapors.
2. *Spray pyrolysis*: Precursor particles or droplets are converted to the desired product powder either by reaction with gaseous species or by pyrolysis. Such processes are also called particle-to-particle conversion, evaporative decomposition, spray roasting, or aerosol decomposition. The characteristics of the product powders made by this route are distinctly different to those made by gas-to-particle conversion. Although the solid powders are actually not formed from the gas phase, this route is nevertheless dealt with in this section, because of the similarity of the

physical processes and the reactor operation with that of the gas-to-particle conversion route. The *chemical* processes converting the droplets or particles to the products are dealt with in the relevant sections.

The precursors used for aerosol processes are often the same as in CVD and PVD processes. Examples will be given below. Since mainly commodities are produced, the precursors for aerosol synthesis must be inexpensive and convenient to use. In CVD processes, relatively low partial pressures can yield adequate film deposition rates, whereas powder synthesis reactors require substantial precursor partial pressures to be economical.

Some technical terms associated with aerosol processes include:

Coagulation: attachment of two particles when they collide.
Coalescence: fusion (sintering, condensation) of two particles.
Agglomerates: assemblies of primary particles physically held together by weak interactions. They are also called "soft agglomerates", because they are easier to break up.
Aggregates: assemblies of primary particles connected by the stronger chemical bonds. They are also called "hard agglomerates", because they make powder sintering and consolidation more difficult.

Gas-to-particle conversion. In this route (Figure 3-29), mixtures of gaseous precursors are fed into the reactor. They react at high temperatures to form molecular clusters and eventually ultrafine particles of the product. The particles then form aggregates and agglomerates of solid powder. Having passed the reaction zone, the powders are collected. Additional post-processing is sometimes required to produce high-purity powders.

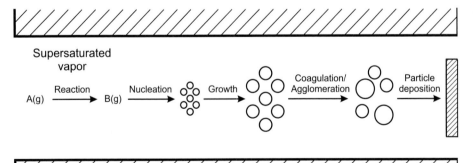

Figure 3-29. Particle generation by gas-to-particle conversion.

The formation of the powders proceeds in several steps (see also Sections 4.1.2 and 4.2 for analogous processes in melts and solutions):

1. *Homogeneous gas phase reactions* between precursor gases result in the formation of molecular or cluster (oligomeric) compounds.
2. *Nucleation.* A supersaturated vapor is inherently unstable and tends to form the condensed phase. The initially formed molecules and clusters will form particles by *homogeneous nucleation* (▶ glossary). If the clusters are thermodynamically stable, nuclei are formed by coagulation. In the case of unstable clusters, the nuclei are formed by balanced condensation and evaporation of molecules or atoms to and from the clusters until a thermodynamically stable particle is formed.
3. *Particle growth.* The nuclei grow by several mechanisms, including condensation, surface reactions, and coagulation. These processes strongly influence the properties of the product particles. If the rate of particle collision is faster than that of their coalescence, then non-spherical (agglomerate, aggregate) particles are formed. If the sintering rate is faster than the particle collision rate, then regularly shaped, monolithic (dense) particles are formed. The sintering rate depends strongly on the material, primary particle size, and temperature.

The particles can also grow by condensation of precursor molecules on the particle surface. The involved processes are similar to CVD processes, i.e., the precursor must first be adsorbed to the surface, and the by-products must be desorbed after formation of the product, etc. Ceramic powders are frequently made by reacting precursor particles with gases or vapors. If the rate of reaction is faster than the rate of vapor transport to the particle surface, the latter controls the rate of particle growth (similar to diffusion or mass-transport limited processes in CVD). The reverse situation (rate of chemical reaction is slower than that of vapor transport) is similar to reaction-limited processes in CVD.

Powders made by the gas-to-particle conversion route are usually aggregates or agglomerates of fine, non-porous primary particles. Temperature, reactor residence time and chemical additives affect the particle sizes and size distributions, extent of agglomeration or aggregation and, consequently, powder morphology. Short reactor residence times (higher flow rates) lead to smaller primary particles, since growth takes place for a shorter time in the reaction zone. Higher reaction temperatures tend to have the same effect, perhaps by increasing the nucleation rate, and hence decreasing the amount of reactant available for each particle to grow. Typically, the number of primary particles forming an agglomerate particle range from a few up to several thousand, varying in size from 1 to 500 nm.

The difference between soft and hard agglomerates is frequently due to the temperature at which the particles aggregate. If the aggregation takes place at the high temperatures during the synthesis, interparticle diffusion will lead to strong bonds between the primary particles. If aggregation can be suppressed until the temperature has decreased sufficiently, then readily dispersed agglomerates may be formed. Other mechanisms such as vapor deposition into the necks between aggregated primary particles may also contribute to the formation of hard agglomerates. This process also takes place primarily in the high-temperature region of the reactor.

The gas-to-particle conversion route is most suitable for the synthesis of single-component, high-purity powders of small particle size, high specific surface area, and controlled particle size distribution. Its major disadvantage is that it results in aggregates or agglomerates that can lead to problems in consolidation and sintering during fabrication of large ceramic parts.

Multicomponent powders are more difficult to synthesize by the gas-to-particle conversion route because of differences in vapor pressure, nucleation (▶ glossary) and growth rates of the various components that can lead to non-uniform product composition. For this reason, coated particles or particles with varying composition from particle to particle are often obtained. As in CVD, single-source precursors have been used. The advantages and disadvantages of such precursors were discussed in Section 3.2.

Spray pyrolysis. In this route (Figure 3-30), a solution or slurry is atomized, or solid precursor powders are suspended in a carrier gas. The aerosol (▶ glossary) is passed through a heated region. Inside the furnace the solvent evaporates and the particles are pyrolyzed or reacted with a gas to yield the product powder, which is then removed from the process stream by conventional means. The size of the product particles is proportional to that of the aerosol droplets or particles.

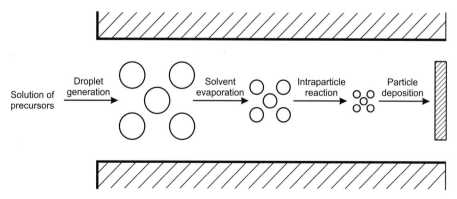

Figure 3-30. Particle generation by spray pyrolysis.

Most work has been done for the aerosol synthesis of oxide powders. With improved syntheses of polymeric precursors for ceramic materials (Section 5.5), spray pyrolysis (in the presence or absence of a reactant gas) could become a viable means for production of non-oxide powders. Multicomponent powders are easier made by spray pyrolysis than by the gas-to-particle conversion route. Each droplet contains precursors in the same stoichiometry as in the product.

The first step in spray pyrolysis is the suspension or atomization process. This can be carried out for liquid precursors using a variety of atomizers depending on the rheological characteristics of the liquid or the required size of the atomized droplets. In general, atomization results in broad droplet size distributions. There are limited options for suspension of solids without assistance of a liquid phase carrier (atomization of slurries).

Particles obtained by this route have a high purity, and they are generally amorphous, unagglomerated, and have a monolithic, spherical morphology. A major advantage of spray pyrolysis is that it can be easily scaled up.

A limitation of this route is that hollow and porous material is easily formed. It is possible to control the porosity of powders by changing the precursor concentration in the droplets and the reactor temperature profile.

Hollow particles can be formed when a solute concentration gradient is created during evaporation (Figure 3-31). The solute precipitates first at the more highly supersaturated surface if there is not enough time for solute diffusion in the droplet. If the crust thus formed is impermeable to solvent, exploded particles may result when the pressure within the particle builds upon further heating.

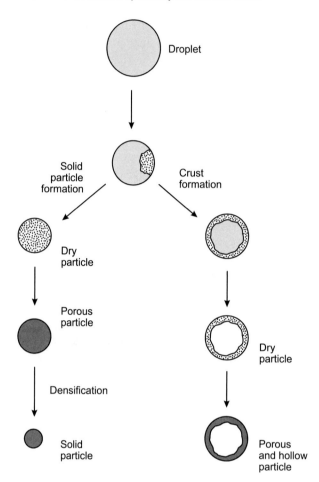

Figure 3-31. Formation of dense (left) and hollow particles (right) by spray pyrolysis.

Reactors

One of the most important aspects of gas-phase powder synthesis is the design and operation of the units that provide the high temperatures needed for gas- or particle-phase reactions.

Flame reactors. These employ the combustion of hydrocarbons or hydrogen (see Aerosil process above), or the reaction of hydrogen and chlorine.

They are attractive for particle generation as they are easy to construct and operate. They only need to bring fuel and oxidant into contact, and make the maximum use of the heating energy. However, the product powder may be contaminated because of its direct contact with the combustion reactants and flue gases. A unique characteristic of flame reactors is the short residence time in the high-temperature region. As a result, the flame-produced powders are typically aggregates of fine, non-porous primary particles with a relatively narrow size distribution. A disadvantage of powders made in flames is that the agglomerates are often hard to break up and have a broad size distribution. The temperature profile of the flame, additives (to modify powder phase composition, morphology and size) and residence time determine the crystallinity, phase composition and extent of aggregation.

Furnace reactors. These are externally heated metallic or ceramic tubes through which the precursor gases are flowing. Such reactors can produce high-purity materials at the expense of energy and material.

The closed environment offers significant advantages, because these reactors provide excellent control over temperatures and residence times, and hence over the product particle characteristics. Two of the key problems associated with externally heated reactors are the loss of product powder by deposition on the reactor walls and the formation of hard agglomerates. To prevent extensive agglomeration and further growth of the agglomerates, the submicron powders formed in the reaction zone must be removed from the hot and possibly still reacting gases and must be effectively quenched. One method is to expose the product particles to a large volume of inert gas, or to cool the gases containing the particles by sudden expansion. Another effective technique is to limit the length of the reaction zone and rapidly withdraw the submicron particles from the reactor.

Laser reactors.   These utilize the energy of lasers. The aerosol synthesis of powders using high-temperature furnaces involves relatively slow heating of the reactant gases by convection or radiation. Lasers improve the efficiency of the energy transfer to gaseous reactants. Nucleation at the walls is avoided, because the pre-mixed gases react in a small, well-defined zone where the focused laser beam intersects the reactant jet and induces a rapid increase in temperature in the gas stream. Reactants are dilute, and hence give fine, loosely agglomerated powders. In contrast to the thermal method, the steep temperature gradients at the reaction zone allow precise control of particle nucleation and growth rates. It is necessary for one of the reactants to absorb the laser radiation strongly. For example, SiH_4 or BCl_3 strongly absorbs CO_2 laser radiation. If none of the reactants is absorbing, a sensitizer may have to be added to the gas mixture, but these are a potential source of impurities in the product material. Laser-heated reactors have been used primarily for non-oxide ceramics. Powders produced via laser reactions tend to be of high purity, controlled stoichiometry, and uniform particle size.

Plasma reactors. The energy of highly ionized gases is transferred to molecules participating in chemical reactions. In plasmas, a significant proportion of the particles are ionized, although overall positive and negative charges compensate each other. The plasmas are generated by ionizing a flowing gas such as argon, either using d.c. or low-frequency a.c., with the electrodes in direct contact with the gas, or by inductively coupling a radio-frequency (r.f.) source to it. The r.f. plasma technique avoids the use of electrodes, which may be sources of product contamination. Solid reactants are used in most plasma syntheses. They are injected into the discharge zone, where vaporization occurs. Since the reactants are heated quickly to high temperatures and quenched rapidly, plasma methods produce very finely divided materials with high specific surface areas.

Products

Large-scale industrial aerosol processes are mainly used for the preparation of highly dispersed silica, titania, and alumina powders from the corresponding chlorides ($SiCl_4$, $TiCl_4$, $AlCl_3$) by flame synthesis, as described earlier in this section. On a smaller scale, other oxides are also produced by the Aerosil process, such as Bi_2O_3 from $BiCl_3$, Cr_2O_3 from CrO_2Cl_2, Fe_2O_3 from $FeCl_3$ or $Fe(CO)_5$, GeO_2 from $GeCl_4$, NiO from $Ni(CO)_4$, MoO_2 from $MoCl_5$, SnO_2 from $SnCl_4$ or $SnMe_4$, V_2O_5 from $VOCl_3$, WO_3 from WCl_6 or $WOCl_4$, ZrO_2 from $ZrCl_4$, $AlBO_3$ from $AlCl_3$ and BCl_3, Al_2TiO_5 from $AlCl_3$ and $TiCl_4$, and $AlPO_4$ from $AlCl_3$ and PCl_5. Other reactor types – and particularly also the spray pyrolysis method – have been used, particularly for metal oxides containing two or more metallic species (spinels (▶ glossary), high-temperature superconductors (▶ glossary), magnetic oxides).

The increased demand for high-purity, non-oxide powders has led to the development of novel routes for their manufacture. Aerosol methods have therefore been employed in the synthesis of various non-oxides such as nitrides, carbides, borides, or silicides (Table 3-3). Even metal powders can be prepared, when the gaseous halides are reacted with the vapors of reactive metals (Eqs. 3-42 and 3-43). Many of these processes are variations of similar ones developed for the synthesis of oxide particles. However, oxygen contamination must be prevented in all process stages.

$$SiCl_4(g) + 4\ Na(g) \longrightarrow Si + 4\ NaCl \tag{3-42}$$

$$2\ NbCl_5(g) + 5\ Mg(g) \longrightarrow 2\ Nb + 5\ MgCl_2 \tag{3-43}$$

Table 3-3. Examples of non-oxide powders prepared by aerosol methods.

Product	Method	Reactants
Carbides		
B_4C	Plasma/Laser	$BCl_3 + CH_4$
B_4C	Thermal	$B_2O_3 + C$
SiC	Laser/Plasma/Thermal	$SiH_4 + CH_4$
SiC	Plasma	$SiCl_4$ or SiO_2 or $SiO + CH_4$
SiC	Plasma/Thermal	$SiMe_4$
SiC	Laser/Plasma	$SiO_2 + C$
SiC	Laser	$H_2SiCl_2 + C_2H_4$
TiC	Plasma/Thermal	$TiCl_4 + CH_4$
Mo_2C	Thermal	$MoCl_5$ or $MoO_3 + CH_4$
WC, W_2C	Plasma/Thermal	WCl_6 or $W + CH_4$
Nitrides		
BN	Laser/Thermal	$BCl_3 + NH_3 + N_2$
AlN	Plasma/Thermal	$Al + N_2 [+ NH_3]$
AlN	Thermal	$AlCl_3 + NH_3 + H_2$
AlN	Thermal	$Al_2Et_6 + NH_3$
Si_3N_4	Laser/Plasma/Thermal	$SiH_4 + NH_3$
Si_3N_4	Laser/Plasma/Thermal	$SiCl_4 + NH_3 + H_2$
Si_3N_4	Plasma	$Si + NH_3$ or N_2
Si_3N_4	Thermal	polysilazanes $+ NH_3 + N_2 + H_2$
SiAlON	Plasma	$Si + Al + NH_3 + O_2$
TiN	Plasma	$Ti + N_2 [+ H_2]$
TiN, ZrN	Thermal	$TiCl_4/ZrCl_4 + NH_3 + N_2 + H_2$
VN_x	Thermal	$VCl_5 + NH_3 + N_2 + H_2$
Borides and Silicides		
B_4Si	Plasma	$B_2H_6 + SiH_4$
$TiSi_2$	Laser	$TiCl_4 + SiH_4$
TiB_2	Laser	$TiCl_4 + B_2H_6$
TiB_2	Thermal	$TiCl_4 + BCl_3 + Na$ or H_2
WSi_2	Plasma	$WF_6 + SiH_4$

Nitrides are usually prepared using ammonia or ammonia/nitrogen mixtures as the nitrogen source, and carbides using methane (or other hydrocarbons) as the carbon source (see Table 3-3). The presence of hydrogen (not explicitly mentioned in Table 3-3) helps in reducing the quantity of excess carbon in the final product, converting it to methane.

The chemistry of the gas-phase reactions is very complex, and mostly not fully understood. For example, 119 separate reactions have been identified in the gas-phase synthesis of SiC from SiH_4 and propane, and the mechanism probably is still incomplete. The underlying chemistry is often related to what has been discussed in Section 3.2 for CVD processes, but other processes may also occur due to the high temperatures involved.

For example, laser syntheses involving SiH_4 are thought to be initiated by the decomposition of SiH_4 to elemental Si. The concentration of Si in the vapor is supersaturated and, therefore, nucleation (▶ glossary) of Si particles occurs. Collision of these nuclei results in the growth of larger particles. The addition of a hydrocarbon or ammonia then results in the carburization or nitridation of the elemental silicon particles.

Film generation

Although aerosol routes are mainly used for powder synthesis, they also offer a variety of approaches for film generation which are summarized in Figure 3-32. These processes have been used to deposit many materials at high rates, including ceramic superconductors (▶ glossary), simple metal oxides, non-oxide ceramics, metals, and composites.

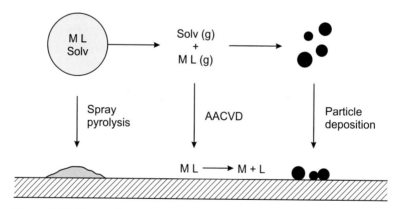

Figure 3-32. Comparison of processes for film formation during aerosol syntheses.

In *spray pyrolysis (droplet deposition)* for film generation, an aerosol (▶ glossary) of a solution containing the reactants is formed and then deposited onto a heated surface where solvent evaporation and chemical reactions take place resulting in a film. This process is usually carried out at atmospheric pressure, often without any enclosure. Some of the advantages of this process include: simplicity and low cost, many choices for the precursors, and high deposition rates (0.1–1 μm·min^{-1}). Film thicknesses of 1–10 μm are common. The primary disadvantages are that porous films are sometimes formed, that purity is low in some cases, and that the method is limited to oxidation-resistant materials. Industrial applications of the technique are for solar cells, oxide superconductor films (▶ glossary), photochemical cell electrodes, gas sensing elements, antireflection coatings, and thermal coatings.

In *aerosol-assisted chemical vapor deposition* (AACVD), a solution containing volatile precursors is carried in the form of an aerosol close to the substrate that is to be coated. Before or near the substrate, the precursors evaporate into the gas phase. This is followed by CVD of gaseous molecular precursors on the heated surface to deposit a thin film. This approach has several advantages for the reproducible generation of multicomponent films (high-temperature superconductors, for example) with thermally sensitive precursors. Deposition rates as high as 5 μm·min^{-1} have been obtained while producing high-quality films. One disadvantage is that a solvent is present in many cases and may be incompatible with the precursors or the chemistry of film formation.

In *particle deposition*, films consisting of particles can be formed by a variety of deposition mechanisms including impaction, diffusion, sedimentation, thermophoresis, and electrophoresis. This method has been used to form ceramic filters, sensors, and hard coatings. Typically, a carrier gas stream is directed at a surface. Particle deposition takes place by impaction, in the absence of temperature gradients, and porous polycrystalline films are formed. Typically, post-processing is required to densify the porous films. Under suitable conditions, high deposition rates can be obtained. The primary strength of this method is for generation of thick ceramic films. Deposition rates as high as 1–5 μm·min^{-1} have been obtained.

3.4 Further Reading

A. R. Barron, "CVD of SiO$_2$ and related materials: an overview", *Adv. Mater. Optics and Electronics* **6** (1996) 101–114.

T. M. Besman, B. M. Gallois (Eds.), *Chemical Vapor Deposition of Refractory Metals and Ceramics* (*Mater. Res. Soc. Symp. Proc.* **168** (1990)).

T. M. Besman, B. M. Gallois, J. Warren (Eds.), *Chemical Vapor Deposition of Refractory Metals and Ceramics II* (*Mater. Res. Soc. Symp. Proc.* **250** (1992)).

S. B. Bhaduri, S. Bhaduri, "Combustion synthesis" in *Non-Equilibrium Processing of Materials*, Pergamon, Amsterdam 1999, Ed. C. Suryanarayana, pp. 289–309.

R. W. Chorley, P. W. Lednor, "Synthetic routes to high surface area non-oxide materials", *Adv. Mater.* **3** (1991) 474–485.

J. S. Colligon, "Physical vapor deposition" in *Non-Equilibrium Processing of Materials*, Pergamon, Amsterdam 1999, C. Suryanarayana (Ed.), pp. 225–253.

G. J. Davies, J. S. Foord, W. T. Tsang (Eds.), *Chemical Beam Epitaxy and Related Techniques*, Wiley, Chichester, 1996.

M. M. Factor, I. Garrett, *Growth of Crystals from the Vapor*, Chapman & Hall, London, 1974.

B. M. Gallois, W. Lee, M. Pickering (Eds.), *Chemical Vapor Deposition of Refractory Metals and Ceramics III* (*Mater. Res. Soc. Symp. Proc.* **363** (1994)).

A. V. Gelatos, A. Jain, R. Marsh, C. J. Mogab, "Chemical vapor deposition of copper for advanced on-chip interconnects", *Mater. Res. Soc. Bull.* **19**(8) (1994) 49–54.

D. M. Gruen, I. Buckley-Golder (Eds.), "Diamond films: recent developments", *Mater. Res. Soc. Bull.* **23** (1994), No. 9.

R. Gruen, R. Glaum, "New results of chemical transport as a method for the preparation and thermochemical investigation of solids", *Angew. Chem. Int. Ed. Engl.* **39** (2000) 692–716.

A. Gurav, T. Kodas, T. Pluym, Y. Xiong, "Aerosol processing of materials", *Aerosol Sci. Technol.* **19** (1993) 411–452.

M. Hampden-Smith, T. T. Kodas, "Chemical vapor deposition of metals", *Chem. Vap. Depos.* **1** (1995) 8–23, 39–48.

M. Hampden-Smith, T. T. Kodas, A. Ludviksson, "Chemical vapor deposition" in *Chemistry of Advanced Materials – an Overview*, Wiley-VCH, New York, 1998, Eds. L. V. Interrante, M. Hampden-Smith, pp. 143–206.

M. L. Hitchman, K. F. Jensen (Eds.), *Chemical Vapour Deposition: Principles and Applications*, Academic Press, New York, 1993.

A. C. Jones, "Developments in metalorganic precursors for semiconductor growth from the vapor phase", *Chem. Soc. Rev.* (1997) 101–110.

A. C. Jones, P. O'Brian, *CVD of Compound Semiconductors*, VCH, Weinheim 1997.

A. E. Kaloyeros. M. A. Fury, "Chemical vapor deposition of copper for multi-level metallization", *Mater. Res. Soc. Bull.* **18**(6) (1993) 22–29.

T. Kodas, M. Hampden-Smith (Eds.), *The Chemistry of Metal CVD*, VCH, Weinheim, 1994.

T. Kodas, M. Hampden-Smith (Eds.), *Aerosol Processing of Materials*, VCH, Weinheim, 1997.

I. Konyashin, J. Bill, F. Aldinger, "Plasma-assisted CVD of cubic boron nitride", *Chem. Vap. Depos.* **3** (1997) 239–255.

G. W. Kriechbaum, P. Kleinschmit, "Superfine oxide powders – flame hydrolysis and hydrothermal synthesis", *Angew. Chem. Adv. Mater.* **101**(1989) 1446–1453.

M. Lenz, R. Gruehn, "Developments in measuring and calculating chemical vapor transport phenomena demonstrated on Cr, Mo, W, and their compounds", *Chem. Rev.* **97** (1997) 2967–2994.

A. H. Lettington, J. W. Steeds (Eds.), *Thin Film Diamond*, Chapman & Hall, London, 1994.

B. Lux, R. Haubner, P. Renard, "Diamond for tooling and abrasives", *Diamond Relat. Mater.* **1** (1992) 1035–1047.

F. Maury, "Trends in precursor selection for MOCVD", *Chem. Vap. Depos.* **2** (1996) 113–116.

C. E. Morosanu, *Thin Films by Chemical Vapour Deposition*, Elsevier, New York, 1990.

D. A. Neumayer, J. G. Ekerdt, "Growth of group III nitrides. A review of precursors and techniques", *Chem. Mater.* **8** (1996) 9–25.

L. Niinistö, "Atomic layer epitaxy", *Curr. Opin. Solid State Mater. Sci.* **3** (1998) 147–158.

L. Niinistö, M. Ritala, M. Leskelä, "Synthesis of oxide thin films and overlayers by atomic layer epitaxy for advanced applications", *Mater. Sci. Eng. B*, **6** (1996) 23–29.

S. E. Pratsinis, "Flame aerosol synthesis of ceramic powders", *Prog. Energy Combust. Sci.* **28** (1998) 197–219.

A. Rabenau, "Zur Chemie der Glühlampe", *Angew. Chem.* **79** (1967) 43–49.

W. S. Rees (Ed.), *CVD of Nonmetals*, VCH, Weinheim, 1996.

H. Schäfer, *Chemical Transport Reactions*, Academic Press, New York, 1964.

K. E. Spear, G. W. Cullen (Eds.), *Chemical Vapor Deposition*, The Electrochemical Society, Pennington, New Jersey, 1990.

G. P. Stringfellow, *Organometallic Vapor Phase Epitaxy; Theory and Practice*, Academic Press, 1998.

T. Suntola, M. Simpson (Eds.), *Atomic Layer Epitaxy*, Blackie, Glasgow, 1990.

F. Teyssandier, A. Dollet, "Chemical vapor deposition" in *Non-equilibrium Processing of Materials*, Pergamon, Amsterdam 1999, Ed. C. Suryanarayana, pp. 257–285.

E. Wagner, H. Brünner, "Aerosil – Herstellung, Eigenschaften und Verhalten in organischen Flüssigkeiten", *Angew. Chem.* **72** (1960) 744–750.

A. W. Weimer (Ed.), *Carbide, Nitride and Boride Materials – Synthesis and Processing*, Chapman & Hall, London, 1997.

W. A. Yarbrough, "Vapor-phase-deposited diamond – problems and potential", *J. Am. Ceram. Soc.* **75** (1992) 3179–3199.

P. Zanella, G. Rossetto, N. Brianese, F. Ossola, M. Porchia, J. O. Williams "Organometallic precursors in the growth of epitaxial thin films of group III–V semiconductors by metal-organic chemical vapor deposition", *Chem. Mater.* **3** (1991) 225–242.

4 Formation of Solids from Solutions and Melts

This section deals with reactions and processes, in which solid products are obtained from a liquid phase. The simplest case, from a chemical point of view, is that the liquid has the same composition as the resulting solid, i.e., the solid is formed from its melt, without a chemical transformation, just by change of the physical state. The involved crystallization and glass-forming processes are discussed in Section 4.1. The intention of this book is not to become involved with physical–chemical details, and to keep the number of mathematical formulas as few as possible, without mathematical or physical derivations. Rather, we focus on a broader picture in order to allow an understanding of the chemical processes involved. Crystallization and precipitation processes from solutions are treated in Section 4.2. The physical–chemical fundamentals are very much related to those relevant to melts, but the processes involved are more complex due to the different chemical composition of the solid and liquid phases. In Section 4.3 we will examine how nature controls crystallization processes in biological systems. In solvothermal processes (Section 4.4), the dissolution and re-crystallization of compounds is speeded up by increases in temperatures and pressures. In the final section (4.5), we will discuss sol–gel processes which allow the obtainment of solid products by gelation rather than crystallization or precipitation.

4.1 Glass

When talking about glass, it is window or bottle glass, or perhaps the glass front of computer or TV screens that first comes to one's mind. However, glasses are also used in "high-tech" applications such as communications technologies or as biomaterials. At the end of this section we will see that even metals are able to form glasses, and we will discuss methods of producing such metallic glasses. We will start with this section by defining what a "glass" is. After touching structural issues, we will find out under which conditions a melt will form a glass upon cooling instead of crystallizing. This is followed by a short introduction into the glass-making process.

Cooling a liquid compound below its melting temperature (T_m) normally results in its crystallization. Crystals are characterized by a long-range, periodic arrangement of atoms or molecules. When a compound crystallizes, its structure rearranges discontinuously from the disordered liquid structure to the ordered crystal structure. Concomitant with that, the enthalpy decreases abruptly from the value for the liquid to that for the crystal (Figure 4-1; the volume dependence on the temperature is very similar). Continued cooling below T_m results in a further enthalpy decrease due to the heat capacity (▶ glossary) of the crystal.

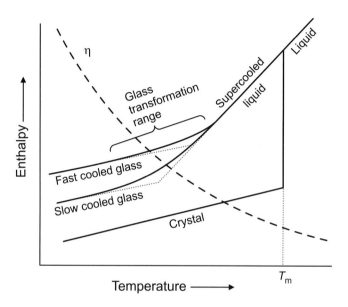

Figure 4-1. Effect of temperature on the enthalpy of a glass-forming melt. η is the viscosity of the melt.

A supercooled liquid is obtained if the liquid can be cooled below the melting temperature (T_m) without crystallization. If the supercooled liquid could be cooled indefinitely slowly, the atoms would rearrange into the equilibrium structure of the liquid (which depends on the temperature) without the abrupt decrease in enthalpy observed upon crystallization. However, as the liquid is cooled, its viscosity (η) increases and eventually becomes so great that the atoms can no longer completely rearrange to the equilibrium structure in a finite period of time. The enthalpy thus begins to deviate from the straight equilibrium line until it eventually becomes determined by the heat capacity (▶ glossary) of the frozen liquid (Figure 4-1). The viscosity of a frozen liquid is so great that the structure becomes fixed and is no longer temperature-dependent. The frozen liquid is now a glass. The temperature region lying between the limits where the enthalpy is that of the equilibrium liquid and that of the frozen liquid is the *glass transformation region*.

A glass can thus be defined as an amorphous (non-crystalline) solid without a long-range, periodic structure that exhibits a region of glass transformation behavior. Any inorganic, organic, or metallic material which exhibits glass transformation behavior is a glass.

Figure 4-1 shows that the glass transformation behavior is a time-dependent phenomenon. When the supercooled liquid is cooled very slowly, the enthalpy begins to deviate from the equilibrium line at a lower temperature, i.e., the glass transformation region is shifted to lower temperatures. Owing to the higher viscosity at lower temperatures, more time is required to reach the equilibrium structure. A slower cooling rate thus allows the supercooled liquid to adjust its equilibrium structure to a lower temperature. The glass obtained with a lower cooling rate has a lower enthalpy than that obtained with a faster cooling rate.

Although the glass transformation actually occurs over a temperature range, it is convenient to use a single temperature as an indicator for the transition between a melt and a glassy solid. This temperature is called either the glass-transformation temperature or the glass-transition temperature (T_g). Since T_g is a function of both the heating (or cooling) rate and the experimental method used for the measurement, it cannot be considered a true property of the glass. Standardized conditions have to be used to make the T_g of different samples comparable.

4.1.1 The Structural Theory of Glass Formation

Early theories of glass formation were centered around the question why some materials form glasses while others do not. These theories are often called *structural theories of glass formation*. It was assumed that the ability to form *three-dimensional networks* by linking some basic building blocks is the ultimate condition for glass formation. Highly ionic materials do not form network structures. The basic building blocks consist of a central electropositive element (called "cation" in the following discussion, although the networks are not ionic) surrounded by a certain number of electronegative elements (called "anion" in the following discussion).

Silicates, for example, readily form glasses instead of recrystallizing after melting and cooling (Figure 4-2). They have network rather than close-packed structures. In contrast to the corresponding crystalline compounds, the vitreous ("glass-like") networks are not periodic and exhibit no symmetry. Their average behavior in all directions is the same, and therefore the properties of glasses are isotropic.

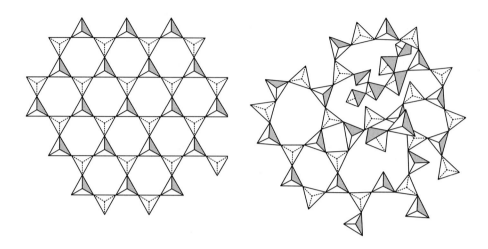

Figure 4-2. Schematic structure of crystalline (left) and amorphous silica (right). In the crystalline forms of silica the [SiO$_4$] tetrahedra (the silicon atoms are located in the center of the tetrahedra, and the oxygen atoms at the vertices) are regularly arranged. There is no long-range order in the amorphous form.

The most commonly used models for glass structures are based on the original ideas of Zachariasen, and are called "random network theory". The rules for simple chalcogenide or halide glasses are:

1. Each anion is linked to no more than two cations (higher coordination numbers for the anion than two prevent variations in bond angles necessary to form a non-periodic, random network).
2. The coordination polyhedra share only corners and not edges or faces (this is a consequence of rule 1).
3. The coordination number of the network-forming cations ("network former") is small (higher coordination polyhedra, such as octahedra, tend to share edges or faces instead of corners).
4. At least three corners of each polyhedron must be shared in order to form three-dimensional networks (only then can the network be three-dimensional. Sharing only two corners results in polymeric structures instead [as in silicones, see Section 5.2]).

Figure 4-3 provides a schematic drawing of the structure of an alkali silicate glass to illustrate these rules. Silicon (the "cation") in silicates and silicate glasses is always tetrahedrally surrounded by four oxygen atoms (the "anions"), i.e., the coordination number of silicon is four (rule 3). The network is formed from connected [SiO$_4$] tetrahedra. The tetrahedra only share corners, i.e., two adjacent silicon atoms are connected by only one oxygen atom (rule 2). In vitreous silica, each oxygen atom bridges two silicon atoms (rule 1). In silicate glasses, there is a certain portion of non-bridging oxygen atoms (see Eq. 4-6 below). The negative charge of each non-bridging oxygen atom must be compensated by a nearby cation to maintain local charge neutrality. However, rule 4 is obeyed.

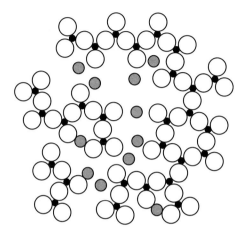

Figure 4-3. Schematic two-dimensional drawing of the structure of a silicate glass. The fourth oxygen atom of each $[SiO_4]$ tetrahedron positioned above or below the silicon atom is not drawn for clarity.

The structure of fluoroberyllate glasses (derived from BeF_2 as the parent compound) is the same. This is not surprising because SiO_2 and BeF_2 are iso-electronic and have the same structures.

Borate glasses exhibit much more complicated structures, owing to a larger number of building blocks. Some structural elements of borate glasses are shown in Figure 4-4. Note that there are both trigonal planar $[BO_3]$ and tetrahedral $[BO_4]$ units.

The Zachariasen rules were modified for complex glasses that *some* anions are linked only to two network cations, there must be a *high percentage* of network cations which are tetrahedrally or trigonally planar surrounded by the anions, and the tetrahedra or triangles share only corners.

In general, the structure of a glass is determined by:

- Coordination number of the network-forming cations. The coordination polyhedra of these cations are the building blocks of the glass structure. There is the possibility that the blocks may be connected to slightly larger units, such as rings or clusters, that have a more ordered arrangement than predicted by a random connection (see Figure 4-2).

- Network connectivity; i.e., the average number of bridging bonds per network-forming cation.
- Bond angle distributions. Bonds angle and dihedral angle distributions introduce randomness into the structure and are therefore inherent to amorphous materials.
- Dimensionality of the network. A network does not need to be three-dimensional in order to form a glass. For example, long-chain polymers that have a one-dimensional network may form glasses by three-dimensional entanglement of the polymer chains.

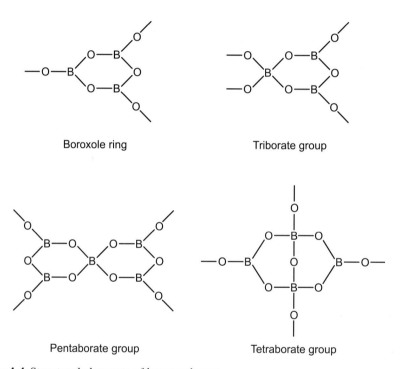

Boroxole ring   Triborate group

Pentaborate group   Tetraborate group

Figure 4-4. Structural elements of borate glasses.

The structural theory of glass formation considers only the relative ease of glass formation. Any compound or mixture which forms a glass during cooling from the melt at a moderate cooling rate is considered to be a good glass-former, while materials which require a more rapid cooling rate are considered to be poor glass-formers. Melts which cannot be cooled to form glasses without extreme cooling rates are considered not to be glass-formers.

It is now recognized that virtually any material will form a glass if cooled so rapidly that insufficient time is provided to allow the reorganization of the structure into a periodic crystal lattice. Therefore, the question is not whether a material will form a glass, but rather how fast it must be cooled to avoid detectable crystallization. This leads to *kinetic theories of glass formation* processes (see below).

4.1.2 Crystallization versus Glass Formation

Crystallization actually includes two processes: the formation of a crystalline nucleus (nucleation, ▶ glossary); and subsequent growth of the crystal. The nucleus may be either homogeneous, i.e., formed spontaneously within the melt, or heterogeneous, i.e., formed at a pre-existing surface (impurity, crucible wall, etc.). If no nuclei are present, crystal growth cannot occur and the material will form a glass. Thus, melts which exhibit a large barrier to nucleation also exhibit good glass-forming behavior. A melt which is free of potential heterogeneous nuclei can be cooled more easily to form a glass than a melt which contains a large concentration of such nuclei. On the other hand, even if some nuclei are present, but no crystal growth occurs, the solid is still a glass.

Melts composed of many different elements inhibit rearrangement of the melt into the ordered crystalline structure because the redistribution of ions to the appropriate sites on the growing crystals is more difficult. This approach is used routinely in commercial glass technology, and partially explains the complex compositions of many common glasses.

In the process of homogeneous nucleation (▶ glossary), nuclei are formed with equal probability throughout the liquid or melt. In the classical nucleation theory, the nucleation rate I (nuclei per unit volume per second) is given by Eq. 4-1,

$$I \propto e^{\left[\frac{-(\Delta G_N + \Delta G_D)}{kT}\right]} \tag{4-1}$$

where ΔG_N is the free energy change in a system when a crystalline nucleus is formed (*thermodynamic barrier* to nucleation) and ΔG_D is the *kinetic barrier* for diffusion across the liquid–nucleus interface.

For spherical nuclei, the thermodynamic barrier (ΔG_N) is expressed by Eq. 4-2.

$$\Delta G_N = {}^4/_3 \pi r^3 \Delta G_v + 4\pi r^2 \gamma \tag{4-2}$$

where γ is the crystal–melt interfacial energy and ΔG_v is the change in volume free energy per unit volume. Note that $^4/_3\pi r^3$ is the volume and $4\pi r^2$ the surface area of a sphere.

- The first term represents the change of the volume free energy (ΔG_v). ΔG_v is negative, since the crystalline state has a lower free energy than the melt.
- The second term represents the increase in surface energy ($\Delta G_s = 4\pi r^2 \gamma$) due to the formation of a new interface, the interface between the solid phase (nucleus) and the melt.

Since nuclei are small, the surface energy term will dominate at very low radii r. The energy of the system (ΔG_N) will first increase with increasing radius (Figure 4-5), and the nucleus will dissolve or melt. If, however, the nucleus survives (by a statistical event) to grow to a large enough size, the first term of Eq. 4-2 will become larger than the second (surface energy), and the energy of the system will begin to decrease with increasing nucleus size. The nucleus will become stable.

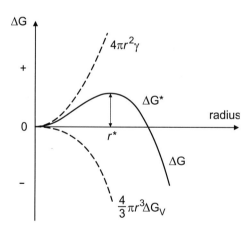

Figure 4-5. Changes in the thermodynamic barrier (ΔG_N) upon formation of a crystalline nucleus.

The value of the radius where the nucleus just becomes stable, is known as the critical radius, r^* (Eq. 4-3),

$$r^* = \frac{-2\gamma}{\Delta G_v} \tag{4-3}$$

(the radius r^* is positive because $\Delta G_v < 0$).

If the temperature is only a little below T_m, the absolute value of the volume free energy (ΔG_v) is very small. It follows that the critical radius, r^* for a stable nucleus (Eq. 4-3) is very large. Since the probability of a nucleus reaching such a large size is extremely low, the melt will remain effectively free of nuclei, even though the temperature is below T_m. As the temperature decreases further, ΔG_v will increase, thus decreasing the value of r^*. Eventually, r^* will become so small (often only a few tenths of a nanometer) that the probability of formation of a nucleus will become significant. The degree of undercooling may be as little as a small fraction of a degree or as much as a few hundred degrees.

At further lowering the temperature, the thermodynamic barrier to nucleation further decreases, allowing nuclei to form at an ever-increasing rate. However, the viscosity is also highly temperature-dependent, so that the kinetic barrier to nucleation will increase rapidly with decreasing temperature. As the kinetic barrier increases, the nucleation rate (▶ glossary) begins to decrease and eventually approaches zero. The opposed effects due to changes in the thermodynamic and kinetic barriers thus results in a maximum in the temperature dependence of the nucleation rate.

Heterogeneous nucleation occurs when the energy of the solid–solid interface between the pre-existing surface and the growing crystal is small. Then the energy balance in Eq. 4-2 tips in favor of the reduction of the volume free energy (ΔG_v); the volume of the crystal can form with less energy consumed by the solid–liquid interface.

The temperature dependence of the crystal growth rate is very similar to that for the nucleation rate. The principal difference is that crystals can grow at any temperature below T_m so long as a (homogeneous or heterogeneous) nucleus is available. As for nucleation, if the viscosity is low, the growth rate will be determined by the thermodynamic values, and will tend to be large. As the temperature decreases, the rapid increase in viscosity will eventually stop crystal growth. The resulting curve of the crystal growth rate versus temperature thus also exhibits a maximum.

In reality, nucleation (▶ glossary) and crystal growth occur simultaneously during cooling of a melt, with rates which change continuously as the temperature decreases. The dependence of the nucleation rate and crystal growth on the temperature is shown in Figure 4-6 for two cases: crystallizing melts (upper diagram); and glass-forming melts (lower diagram). Upon cooling a melt below T_m, crystals will grow if there is a sufficient number of nuclei. If the maximum of nucleation is at a similar temperature than the maximum of crystal growth (i.e., if ΔT is small), a large number of nuclei will be produced at a temperature at which crystal growth is optimal. This means that crystallization of the melt will occur easily (upper part of Figure 4-6). On the other hand, if the maximum of nucleation is at a much lower temperature than the maximum of crystal growth (i.e., if ΔT is large), nuclei will be formed, but they cannot grow because the kinetic barrier at this temperature (higher viscosity) inhibits crystal growth. This means that the melt will form a glass (lower part of Figure 4-6).

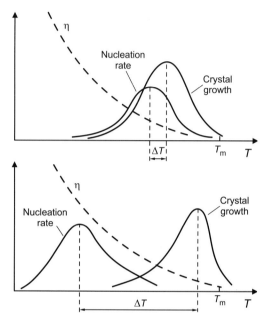

Figure 4-6. The diagrams show the different relation between nucleation rate and crystal growth in a crystallizing system (upper diagram) and glass-forming system (lower diagram). η is the viscosity of the melt.

Formation of a glass requires cooling of the melt in such a manner that significant crystal formation is prevented. If the nucleation rate (I) and the linear crystal growth rate (U) as a function of the temperature are known, then the volume fraction of crystals in a sample, V_x/V, is given by Eq. 4-4 under isothermal conditions.

$$\frac{V_x}{V} = 1 - e^{\frac{-\pi I U^3 t^4}{3}} \tag{4-4}$$

where V_x is the volume of crystals, V is the sample volume, and t is the time the sample has been held at the experimental temperature.

From Eq. 4-4 one can calculate the time required to form a particular volume fraction of crystals at a given temperature. At another temperature, the time required for the formation of the same volume fraction of crystals is different, because of the different temperature dependence of the nucleation (▶ glossary) and crystal growth rate. One can thus calculate the curve in time/temperature space that corresponds to a particular value V_x/V (Figure 4-7), the so-called time–temperature-transformation (TTT) curve. This curve has the general shape shown in Figure 4-7, and is therefore also called the "nose curve". From the TTT curve, the critical conditions for cooling can be obtained. Since I and U approach zero as the temperature approaches T_m, the time required to form the specified volume fraction of crystals will approach infinity. At very low temperatures, the values of I and U also approach zero due to the very high viscosity of the melt, and the time to reach the specified value of V_x/V also approaches infinity. The least favorable conditions for glass formation occur at the temperature T_n, corresponding to the "nose" of the curve, where the time required for crystal formation, t_n, is minimal.

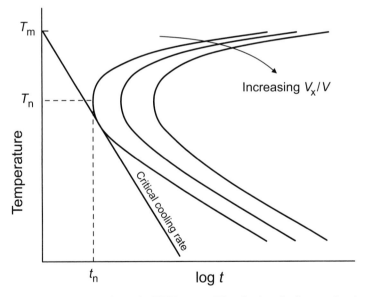

Figure 4-7. Schematic illustration of a TTT curve. The abscissa indicates the time required at each temperature to produce a volume fraction V_x/V of crystals.

To what volume fraction of crystals a sample is considered a glass is a question of definition; a typical value is 1 ppm. The crystal content acceptable for a window glass will be quite different from that acceptable for an optical fiber or a lens, for example. The critical cooling rate, $(dT/dt)_c$, i.e., the minimum cooling rate required to yield a glass (with an acceptable crystal content), can be obtained from the slope of the tangent to the curve, with the initial conditions defined as T_m at time zero. It is given by the expression

$$\left(\frac{dT}{dt}\right)_c \approx \frac{(T_m - T_n)}{t_n} \tag{4-5}$$

A typical value for the critical cooling rate is 9×10^{-6} K·s^{-1} for SiO$_2$ glass. At the opposite end, the formation of metallic glasses (see below) requires cooling rates of 10^6 to 10^{10} K·s^{-1}. The melt with the smaller critical cooling rate has the better glass-forming ability.

One of the best "rules of thumb" for predicting the glass-forming ability of any liquid is the "Turnbull" criterion. When the "reduced glass-transition temperature" $T_{rg} = T_g/T_m$ is close to a value of $^2/_3$, homogeneous nucleation in undercooled melts becomes very sluggish, compared with lower values.

Unintended crystallization of glass during the production process results in a locally inhomogeneous concentration of crystals of different size. This has to be avoided. However, controlled crystallization by re-heating of a glass (Figure 4-8) first to the temperature of maximum nucleation rate (T_1) and then to the temperature of maximum crystal growth (T_2) leads to a class of materials called glass ceramics with many interesting properties and uses. The name reflects the fact that these materials are *not* obtained by sintering (see Section 2.1.4) as the usual ceramic materials. Glass ceramics consist of small (usually some µm) and uniform crystallites irregularly distributed in an amorphous matrix.

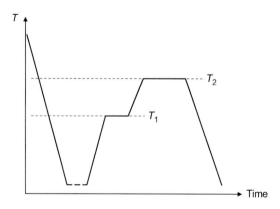

Figure 4-8.
Temperature profile during the production of glass ceramics.

Since the thermal expansion coefficient of a sample is a volume averaged function of the contribution of each of the phases present, formation of crystals with thermal expansion coefficients very different from the initial glass, can radically alter the overall thermal expansion coefficient. The initial glass used for the production of transparent cookware, for example, has a thermal expansion coefficient of about 4 ppm·K^{-1}. After processing as glass ceramics, the thermal expansion coefficient is only about one-tenth of this value.

4.1.3 Glass Melting

Although glasses can be made by a wide variety of methods, the vast majority are still produced by melting of batch components at an elevated temperature. The production steps are shown in Figure 4-9.

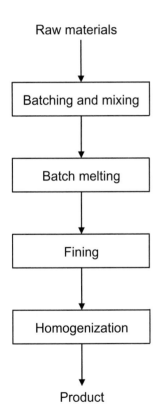

Figure 4-9. Steps in the fabrication of glass.

Examples for the composition of technical and optical glasses are given in Table 4-1.

Table 4-1. Composition of some typical glasses

Window glass	72 % SiO_2, 1.5 % Al_2O_3, 3.5 % MgO, 8.5 % CaO, 14.5 % Na_2O
Laboratory glass	80 % SiO_2, 10 % B_2O_3, 3 % Al_2O_3, 1 % MgO, 1 % CaO, 5 % Na_2O
Fluoride fiber glass	53 % ZrF_4, 20 % BaF_2, 20 % NaF, 2 % LaF_3, 3 % AlF_3, 2 % LnF_3

Raw materials of glass production

The *raw materials* can be divided into four categories based on their role in the process: glass-formers, network-modifiers, colorants, and fining agents. The same compound may be used for different purposes. Alumina, for example, serves as a glass-former in aluminate glasses, but is a modifier in most silicate glasses. Arsenic oxide may be either a glass-former or a fining agent.

Glass-formers (*network formers*). The primary glass-formers in commercial oxide glasses are silica (SiO_2), boric oxide (B_2O_3), and phosphoric oxide (P_2O_5), which all readily form single component glasses. A large number of other compounds may act as glass-formers when mixed with other oxides, including GeO_2, Bi_2O_3, As_2O_3, Sb_2O_3, TeO_2, Al_2O_3, Ga_2O_3, and V_2O_5. As_2S_3, As_2Se_3 and GeS_2 are important glass-formers in chalcogenide glasses, the three most common halide glass-formers are BeF_2, AlF_3 and ZrF_4.

Network modifiers. While silica itself forms an excellent glass, with a wide range of applications, the use of pure silica glass for bottles, windows, and other bulk commercial applications would be too expensive due to the high melting temperature (>2000 °C). The processing temperature is lowered by addition of alkali or alkaline earth oxides which break Si–O–Si bonds (Eq. 4-6, for example) and thus lower the melting temperature. The use of PbO is becoming much more limited due to concerns regarding its toxicity. PbO is especially useful in dissolving any refractory (▶ glossary) or other impurity particles which might otherwise result in flaws in the final glass. A combination of different network modifiers is often necessary to modify the properties of the glass.

$$\equiv Si-O-Si\equiv \; + \; Na_2O \longrightarrow \; 2 \;\equiv Si-O^-Na^+ \qquad (4\text{-}6)$$

Colorants. These are used in small quantities to control the color of the final glass. In most cases, colorants are oxides of either the 3d transition metals or the 4f rare earths. Gold and silver are also used to produce colors by formation of colloids (▶ glossary) in glasses. Iron oxides, which are common impurities in the sands used to produce commercial silicate glasses, act as unintentional colorants in many products. When colorants are used to counteract the effect of other colorants to produce a slightly gray glass, they are referred to as *decolorants.*

Fining agents. These are added to glass-forming batches to promote the removal of bubbles from the melt. Fining agents for oxide glasses include the arsenic and antimony oxides, potassium and sodium nitrates, NaCl, fluorides, and a number of sulfates. These materials are usually present in very small quantities (<1 wt%) and have only minor effects on the properties of the final glasses.

Batch melting

Batch melting involves the melting and decomposition of the raw materials to form the initial melt. Initial heating usually results in the release of some moisture, which may have been adsorbed onto the raw materials or may be formed by dehydration of hydroxides. Far more gas is released during the decomposition of carbonates, sulfates, and nitrates. The gases released result in considerable mixing and stirring, which helps to homogenize the melt. However, the gas bubbles must be removed from the melt before processing is completed (see below).

The time required to dissolve the original batch completely is called the *batch-free time* (Figure 4-10). Factors that influence the batch-free time are temperature, overall glass composition, the raw materials used, batch homogeneity, grain size (▶ glossary) of batch components, and the amount of scrap glass (*cullet*) added to the batch. The use of cullet not only reduces waste, but also reduces the batch-free time both by reducing the amount of refractory material (▶ glossary) in the batch and by providing additional liquid throughout the melting process.

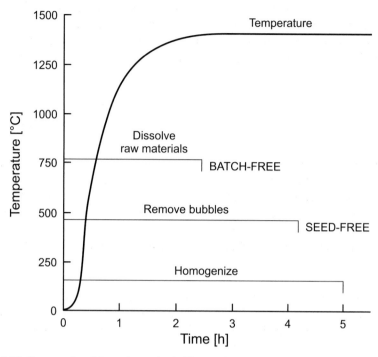

Figure 4-10. Stages of melting of a typical silicate glass.

A large number of the components of glasses are quite volatile at elevated temperatures. Loss of these components can significantly alter the composition of the glass obtained after prolonged melting. Volatilization losses are particularly significant for alkali, lead, boron and phosphorus oxides, halides, and other components which have high vapor pressures at high temperatures. The rate of loss of alkali increases rapidly in the order of Li < Na < K < Rb < Cs. These losses can usually be reduced dramatically by lowering the melt temperature. Covering the melt increases the partial pressure of the volatile components directly above the melt, establishes an equilibrium between the dissolved and vaporized species, and prevents significant loss of those components. However, covering the melt is usually not possible in large commercial melting processes. It may be necessary to allow for losses by providing excess concentrations of components known to vaporize from a given melt.

Many glass-forming melts require special techniques. For example, the components of chalcogenide glasses are toxic and highly volatile. Furthermore, contamination by very small quantities of oxygen will destroy the infrared transmission of the glass. These melts are usually prepared by heating the mixture of powders in a sealed vitreous silica tube under a vacuum and then quenching the melt to form a glass. Heavy metal halide glasses also require melting in an oxygen-free atmosphere to preserve their optical properties. These glasses are often melted under a reactive atmosphere, such as CCl_4 or SF_6. The atmosphere acts as both a getter for oxygen and as a source of halide to replace volatilization losses and maintain the stoichiometry of the melt.

Fining of melts

The terms fining (or refining) refer to the removal of bubbles from the melt. Bubbles are undesirable in most commercial glasses, and are considered as flaws. Fining of a melt begins during the melting process, but typically extends beyond the batch-free times (Figure 4-10).

Bubbles can be formed by the physical trapping of gases of the furnace atmosphere during the initial phase of batch melting, or by the decomposition of batch components. The gases in the interstices between batch particles may be trapped as the particles begin to soften and form a viscous liquid.

Decomposition of batch materials can produce extremely large quantities of gases such as CO_2, SO_3, NO_x, H_2O, etc. Reactions of materials in contact with the melt can also generate gases. Since many gases have a large enthalpy of solution in glass-forming melts, their solubility in these melts is a strong function of temperature and the melt composition. Bubbles can therefore also be formed whenever supersaturation occurs for a specific gas.

Bubbles can be removed from melts either by their rising physically to the surface or by chemical dissolution of the gas into the melt. If bubbles do not rise sufficiently quickly, the melt can be moved by convection or stirring to carry the bubbles to the surface. The velocity of rise of a bubble is inversely proportional to the viscosity of a melt and directly proportional to the density of the melt and to the square of the bubble radius. Procedures which increase bubble size will therefore rapidly accelerate fining. Very small bubbles do not rise fast enough, and other processes are needed.

Bubbles can be removed chemically by the addition of fining agents. These release large quantities of gases, which form large bubbles that rise rapidly to the surface of the melt and also carry the smaller bubbles to the surface.

Arsenic and antimony oxides are the most efficient chemical fining agents. They are usually added to the batch in 0.1-1 wt% quantities, and are especially useful when combined with alkali nitrates in the batch. Their fining action results from a series of chemical reactions that occur at different stages of the melting process. One possible reaction might be as given in Eq. 4-7.

$$4\,KNO_3 + 2\,As_2O_3 \longrightarrow 2\,K_2O + 2\,As_2O_5 + 4\,NO\uparrow + O_2\uparrow \qquad (4\text{-}7)$$

After batch melting is complete, melts are usually heated to higher temperatures and held until completely fined. Since the trioxide is more stable than the pentoxide at these temperatures, the pentoxide produced via reaction with nitrates decomposes by release of oxygen (Eq. 4-8).

$$E_2O_5 \rightleftharpoons E_2O_3 + O_2$$

$$E = As, Sb \tag{4-8}$$

The released oxygen can either form new bubbles or diffuse into nearby smaller bubbles, thus increasing their size and rise rate. Any bubbles remaining in the melt at this point will be highly enriched in oxygen relative to the bubbles previously present. The equilibrium expressed by Eq. 4-8 is very temperature-dependent. Lowering the temperature will result in a shift toward the pentoxide, which requires absorption of oxygen from the melt. This eventually leads to the complete consumption of the bubbles.

Although other fining agents are usually less efficient, the toxicity of arsenic and antimony oxides often necessitates alternative fining agents. Sodium sulfate, for example, also releases considerable gas volumes during batch melting, and also provides a portion of the sodium for soda-lime-silicate melts (Eq. 4-9).

$$Na_2SO_4 + n\ SiO_2 \longrightarrow Na_2O \cdot n\ SiO_2 + SO_2 + \tfrac{1}{2}O_2 \tag{4-9}$$

Sulfate fining is strongly affected by the reactions with furnace gases or other sources of carbon. Reactions between SO_3 and either C or CO produce SO_2 and CO_2, releasing large quantities of gas.

Halides are most useful as fining agents because they efficiently lower the viscosity of melts. They are particularly effective in fining of high alumina content melts.

Homogenization of melt

The initially formed melt is very heterogeneous. This heterogeneity is gradually reduced by the stirring action of rising bubbles during fining. Production of an acceptably homogeneous glass, however, usually requires additional time for diffusion processes to improve the homogeneity of the melt.

Gross defects such as bubbles and "stones" (particles of undissolved material) are often clearly visible and cannot be tolerated in commercial glasses. The terms "striae" and "cord" are used to describe variation in local composition within a glass. Striae are two-dimensional regions of a composition different from the bulk, while cords are similar regions which are effectively one-dimensional veins in the glass. These regions cause a "wavy" appearance due to local variations in refractive index. In colored glasses, regions of inhomogeneity can often be detected by variations in color intensity.

Poor homogeneity frequently results from poor mixing of the original batch materials. Striae and cords can be formed by reactions with refractories and at the melt–atmosphere boundary owing to volatilization of batch components, especially alkali, boron, or lead. Decreasing the grain size (▶ glossary) of the batch improves homogeneity by reducing the scale of the inhomogeneity of the initial melt. Stirring by mechanical stirrers, creation of convection flow in the melt, or bubbling of a gas through a melt can improve homogeneity.

4.1.4 Metallic Glasses

Thermodynamically speaking, amorphous materials (glasses) are solids in a non-equilibrium state. Accordingly, the preparation of glasses requires techniques in which a non-equilibrium state is frozen. The first metal alloy (▶ glossary) vitrified by rapidly quenching the melt ("rapid solidification") had the composition Au_3Si, and was reported in 1960. Since this time, a number of methods have been developed that allow very high cooling rates. Although solids can be formed from a liquid (melt) or a gas (vapor), the term "rapid solidification" is normally applied to forming solids from a melt. It should be pointed out, however, that metallic glasses can also be obtained by physical vapor deposition (PVD) methods (Section 3.2.1).

Metallic glasses, mostly alloys, can have unique properties. They can exhibit high flow stresses (▶ glossary) and hardness (▶ glossary) with fracture stress approaching the theoretical limit. This can be combined with ductile (▶ glossary) behavior in bending, shear, and compression. The increased scattering of electrons by the disordered structure gives rise to an increase in electric resistivity (▶ glossary) that is two to three times higher than that of the same composition in crystalline form. In addition, metallic glasses based on ferromagnetics (▶ glossary) exhibit excellent soft ferromagnetic behavior.

As discussed above, poor glass-formers (such as metals) require a more rapid cooling rate than the good glass-formers discussed so far (see TTT curves in Figure 4-7). The very high cooling rates during rapid solidification of a melt are achieved by generating thin layers, filaments or droplets of the melt in good contact with an efficient heat sink.

Droplet method. These are variants of the long-established technology of making lead shots by pouring molten lead through a perforated pre-heated steel dish. The principle behind is that a pending drop or a free-falling melt stream breaks up into droplets as a result of surface tension. There are several methods to intensify this process. The standard method is *spray-atomization (spray-forming)*, where high-velocity gas or water jets rapidly fragment a volume of melt into large numbers of small droplets (Figure 4-11). The jets form, propel, and cool the droplets that may freeze completely to form a powder or may form individual splats ("splat-quenching") or a thick deposit by impact on a suitable substrate. For example, in the splat-quenching method, the cooling rate can be as high as 10^4 K·s^{-1}. The materials need to be consolidated for typical engineering applications.

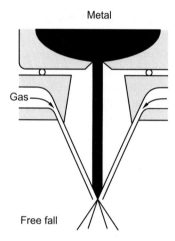

Figure 4-11. Principle of spray-atomization by impingement of a high-velocity gas jet on a free-falling melt stream.

Spinning methods. A single melt stream is stabilized by surface film formation before it can break up into droplets ("melt-spinning" or "melt-extrusion"). In the standard chill-block melt-spinning method (Figure 4-12), the melt-stream is flinged on a chilled roll rotating at a surface speed of some tens of meters per second. A single ribbon is formed that is typically 10–50 μm thick and a few mm wide.

Surface melting methods. In its simplest form, a heat source, such as an electron or laser beam, is scanned over the surface of a material to be melted. The molten pool cools very rapidly behind the beam, because the unmelted bulk acts as the heat sink (Figure 4-13). This process can also be used to alloy the surface. An alloying material pre-placed on the substrate is fused to and mixed with the surface material. This results in an alloyed layer after solidification.

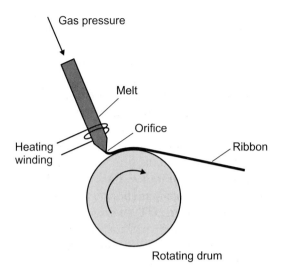

Figure 4-12. Schematic drawing of chill-block melt-spinning for the formation of glass ribbons.

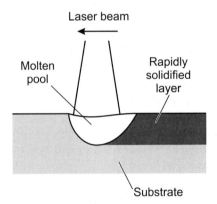

Figure 4-13. Laser surface melting and quenching.

Ion-mixing. This is a completely different approach. Multilayer films are first prepared by depositing alternatively the metals A and B on a substrate (Figure 4-14). The as-deposited films are then irradiated by a scanning ion beam (such as Xe ions) to achieve a uniform mixing between metal layers A and B, resulting in forming an A_xB_y alloy (▶ glossary). The effective cooling rates can be as high as 10^{14} K·s^{-1}. The supporting substrate can be an inert compound such as silica or NaCl, or it can be metal A as a matrix for forming a surface alloy.

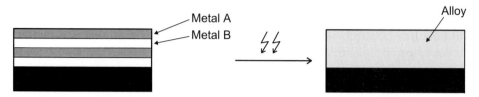

Figure 4-14. Ion-mixing of multilayered films consisting of metal layers A and B.

Bulk metallic glasses

Recently, certain high-order metallic alloys (▶ glossary) were synthesized that have exceptional glass-forming abilities, and do not need rapid quenching to form glasses. Such alloy systems include La-Al-Cu-Ni or Zr-Ti-Cu-Ni-Be (for example with the composition $Zr_{41.2}Ti_{13.8}Cu_{12.5}Ni_{10}Be_{22.5}$ or $Zr_{46.75}Ti_{8.25}Cu_{7.5}Ni_{10}Be_{27.5}$). They have high values of the reduced glass-transition temperature (T_g/T_m) of above 0.6 and form glasses at cooling rates lower than 10^2 K·s^{-1}. This allows the casting of glassy rods or bars with maximum sample thickness up to a few centimeters by conventional metallurgical methods. They are called "bulk metallic glasses". Three empirical rules were proposed for achieving high glass-forming abilities for metallic alloys:

- The alloy system should contain more than three elements.
- There should be a significant difference in atomic size ratios (above about 12 %) among the main constituent elements.
- The component elements should exhibit negative heats of mixing among themselves.

These factors induce a higher dense random packing in the supercooled liquid which leads to a high liquid–solid interfacial energy (see Eq. 4-2) and renders atomic rearrangement more difficult. The energetic advantages of forming a periodically ordered structure become progressively more marginal as the systems become more complex.

4.2 Precipitation

Every first-year science student knows how to precipitate sparingly soluble salts from solution. However, simply adding a sufficiently concentrated solution of one constituent ion to the solution of the other does not give control over particle size, particle morphology, etc.

For example, mixing aqueous solutions of silver nitrate and sodium bromide results in the immediate precipitation of silver bromide (AgBr), but the thus obtained precipitate would be useless for photographic emulsions, because it contains a mixture of crystals of different size and shape. The properties of photographic materials are strongly influenced by the size, shape and composition of the light-sensitive microcrystals in the photographic emulsions. The precipitation process of silver halides can be controlled in a way that crystals uniform in size and shape can be obtained. A selection of silver halide crystals used for photographic emulsions is shown in Figure 4-15. The methods by which the size and shape can be controlled will be discussed later in this section. In general, aqueous solutions of silver nitrate ($AgNO_3$) and an alkali halide are added to an aqueous gelatine solution (about 2–5 wt%) with vigorous stirring at 30–80 °C. The gelatine serves to prevent the coagulation of the microcrystals (see Section 4.5 and Chapter 7). The soluble alkali salts generated during precipitation and the excess alkali halides must then be removed before the emulsions can be further processed.

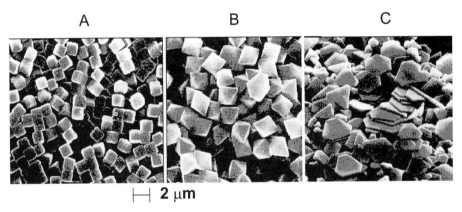

Figure 4-15. Shapes of silver halide crystals of photographic emulsions. A, cubes; B, octahedra; C: plates.

Precipitation is a rather complex phenomenon that can involve the following processes:

- nucleation (▶ glossary);
- crystal growth;

- Ostwald ripening (▶ glossary);
- recrystallization;
- coagulation; and
- agglomeration.

Each of these steps must be controlled in order to create monodisperse particles with a well-defined and reproducible morphology. Together with the chemical composition, the size, size distribution and habit of the crystalline particles contribute to the actual properties and performance of the desired material. Uniform particle size facilitates, inter alia, the preparation of stable dispersions, uniform ceramic powders, pigments with a reproducible color or catalysts with an optimized activity.

In technical precipitation processes, water is mostly the solvent. The following general considerations can, of course, be applied to any other solvent. A very special "solvent" is a molten salt. For example, the perovskite (▶ glossary) $BaTiO_3$ was prepared from $Ba(NO_3)_2$ and partially hydrolyzed $TiCl_4$ in a molten mixture of $NaNO_3$ and KNO_3 (the eutectic mixture [▶ glossary] melts at $225\,°C$) at $500\,°C$. Note that preparation of $BaTiO_3$ by the ceramic method (Section 2.1.1) would require temperatures $>1000\,°C$.

Some theoretical considerations concerning nucleation and crystal growth were already made in Section 4.1 for the crystallization from melts. Formation of crystalline nuclei in melts occurs by lowering the temperature. For precipitation (crystallization) from solution, the concentration of the precipitating solute species must be increased until nuclei are formed. This can be achieved by different ways, for example:

- direct reaction of ions (e.g., addition of bromide ions to a solution containing silver ions to give AgBr);
- redox reactions (e.g., reduction of $HAuCl_4$ with formaldehyde to give colloidal [▶ glossary] gold);
- precipitation by poor solvents (e.g., addition of water to ethanolic solutions of sulfur to precipitate sulfur);
- decomposition of compounds (e.g., addition of an acid to an aqueous thiosulfate solution to obtain elemental sulfur); and
- hydrolysis reactions (see below).

Further examples will be given below.

The initial formation of particles from solution proceeds as indicated in Figure 4-16 (LaMer model):

- The concentration of the solute is continuously increased up to the minimum concentration for nucleation, c_0, by one of the processes mentioned before. No precipitation takes place.
- When c_0 is reached, nucleation sets in. The concentration of the solute keeps climbing up to the maximum concentration for nucleation, c_N, and then decreases again due to the consumption of the solute by nucleation and precipitation of particles. The range between c_s and c_N corresponds to a supersaturation. At the critical concentration, c_N, nucleation is very rapid.
- Once the minimum concentration for nucleation, c_0, is reached again, new nuclei are no longer formed. Crystal growth reduces the concentration until the equilibrium solubility, c_s, is reached.

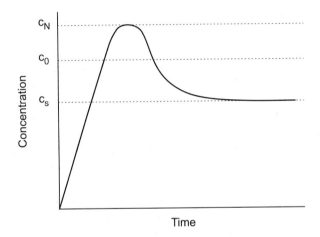

Figure 4-16. Concentration of the solute before and after nucleation.

If new nuclei are formed during the growth period, then a range of particle sizes results. Thus, separation between the nucleation and the growth stage is the primary requirement for obtaining uniform particles. This means that after nucleation, the particle-forming solute species must continue to be formed at a rate that allows their removal from solution by reaction with the existing particles so that no secondary nucleation can take place. The rate of growth of the particles may be controlled by diffusion of dissolved species to the particle or by the rate of the condensation reactions between the particle and dissolved species.

An alternative supply of particle-forming species is by dissolution of small particles. As discussed in Section 2.1.4, smaller particles are more soluble than larger particles, due to curvature effects. Thus larger particles grow at the expense of smaller ones. Larger, more stable particles are thus formed – a process called Ostwald ripening (▶ glossary).

Particles formed by these mechanisms should be crystalline, although amorphous, porous particles are frequently obtained. Crystal growth is only possible if the attachment of the molecular species is weak enough to allow the species, if necessary, to redissolve and attach to the surface in a position required for a crystalline structure. This is more likely in dilute solutions. Amorphous particles are obtained if the molecular species attach irreversibly to the particle and are unable to reorient. Note the analogy to CVD processes (Section 3.2), for example, where the probability that a precursor molecule reacts at the first point of contact with the surface should be significantly smaller than unity to obtain smooth films.

The growth mechanism discussed so far does not account for all the available experimental results. For example, monodisperse spheres were obtained under conditions where the concentration of the particle-forming species in solution is above c_0 (Figure 4-16) and new nuclei should be created continuously. This should result in broad particle size distributions according to the classical theory. Electron micrographs of such materials show that the particles consist of a large number of tiny, aggregated primary particles. A *nucleation–aggregation model* was proposed for such cases that assumes the initial small primary particles to be too unstable owing to their small size. They aggregate to secondary particles, and freshly formed primary particles attach to the aggregates. Monodispersity (▶ glossary) of the final precipitate is achieved through size-dependent aggregation rates. The aggregate structure apparently causes a collective reduction in surface area. The concomitant reduction in surface energy is the driving force for the ordered aggregation. Rather large particles (up to 250 nm) with this internal structure can be obtained for various compositions. This excludes an alternative explanation of the observed behavior based on Ostwald ripening (▶ glossary), because the difference in solubility and hence Ostwald ripening disappears when particles exceed about 5 nm in diameter.

As indicated above, there are several methods to increase the concentration of the precipitating species in a controlled way until supersaturation is reached and to supply new species for crystal growth.

The method of *forced hydrolysis* for the preparation of metal oxides or hydroxides proceeds by deprotonation of hydrated metal ions which then undergo polycondensation reactions (▶ glossary). The best results are obtained under mild conditions and low concentrations. When a metal salt is dissolved in water, the metal ions are solvated by water molecules. The coordinated water molecules can be deprotonated to give hydroxide or even oxide species. The deprotonation processes are equilibria (Eq. 4-10) which depend on the charge of the metal and the pH.

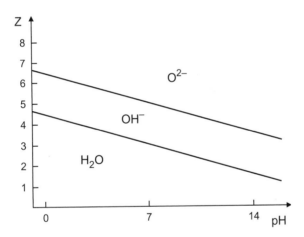

$$(4\text{-}10)$$

Figure 4-17 shows that the formation of oxides is favored for highly charged metal ions and/or high pH. This diagram also explains qualitatively why the hydrolysis of low-valent cations (Z <4) yields aquo, hydroxo, or aquo-hydroxo complexes over the complete pH scale, while high-valent cations (Z >5) form oxo or oxo-hydroxo compounds.

Figure 4-17. Charge (Z) versus pH diagram indicating the domains of aquo, hydroxo, and oxo species.

There are several possibilities to shift the equilibria in Eq. 4-10 to the right, i.e., to obtain oxides from hydrated metal salts. For instance, a compound such as formamide, which thermally decomposes to yield ammonia, can be used to raise the pH gradually, i.e., to remove the protons from the equilibrium. This corresponds to moving from left to right in Figure 4-17 at a given charge (Z). Alternatively, the solution can simply be aged at elevated temperatures. A higher temperature promotes dissociation of protons from the hydrated metal ions. The mechanisms of the polycondensation (▶ glossary) of the hydroxo compounds eventually leading to metal oxide (or oxide/hydroxide) precipitates will be discussed in Section 4.5.

The forced hydrolysis method is very sensitive to factors such as salt concentration, pH, nature of the counterion, and temperature. In most cases, the solids are formed by interaction of various solute complexes. The composition and concentration of these particle-forming species vary from system to system and strongly depend on the experimental conditions. Changing the experimental conditions influences the equilibria of different monomeric and oligomeric complexes, with a different degree of hydroxylation, which undergo condensation at different rates and sometimes different mechanisms.

The precipitation of iron oxide/hydroxide by aging acidic solutions of Fe(III) salts is a particularly well-investigated example. Depending on the reaction conditions, the chemical composition may be FeOOH or Fe_2O_3, and it is possible to produce systems consisting of cubic, ellipsoidal, pyramidal, rodlike, or spherical particles. The color may range from yellow to red, and from brown to black, depending on the particle size and shape.

The counter-anion is particularly important in determining the properties and morphology of the precipitate. One possible reason is that ions such as phosphate or sulfate may promote polycondensation (▶ glossary) by forming polynuclear complexes (see also Section 4.5.3). For example, chromium hydroxide grows as spherical amorphous particles when chromium sulfate or phosphate solutions are aged, but no particles grow from chromium chloride, nitrate, or acetate solutions. Consequently, it is necessary in each case to elucidate quantitatively all species in a given solution to control kinetically the reaction and thus to obtain reproducible results.

Another widely applicable method of preparation of monospheres is *thermal decomposition of complexes* formed by metal ions with chelating agents (▶ glossary), such as triethanolamine, ethylenediamine tetraacetic acid (EDTA), nitrilotriacetic acid, etc. When the chelated metal complexes are dissolved in strongly basic solutions, the chelating ligand (▶ glossary) dissociates, and the released metal ions react with water at a rate determined by the stability constant of the metal complex. Since different chelating agents (▶ glossary) result in different stability constants of the resulting metal complex, the reaction rate can be controlled by the choice of the chelating ligand. This approach allows a wider range of experimental conditions than the forced hydrolysis method, including the addition of oxidizing or reducing agents. In addition to a variety of single-component oxides, this method can be used to make composites (metal oxide crystals coated by another oxide) and also some non-oxide materials. For example, CdS particles can be grown by decomposing thioacetamide (a source of sulfide ions) in a dilute solution of $Cd(NO_3)_2$.

The drawback of these methods is that they require rather dilute solutions, resulting in small quantities of product in a given time. A method for the synthesis of large quantities of well-defined dispersions is the *controlled double jet precipitation* (CDJP) process. The technique was developed for the photographic industry to make precisely defined silver halide crystals, but it is now used to produce various other sparingly soluble salts. Particle sizes from the nanometer to the micrometer range can be obtained. The method is based on the simultaneous introduction of the reactant solutions through separate input lines into a stirred reactor under precisely defined conditions (Figure 4-18).

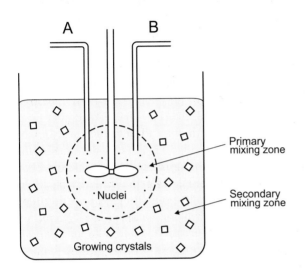

Figure 4-18. Model of a crystallizer for the controlled double jet precipitation (CDJP) process.

The idea of the CDJP technique is to achieve a single burst of stable nuclei and to use the newly added material for their growth without increasing their number. In the region where the highly concentrated solutions of the precursors are introduced, the primary mixing zone (Figure 4-18), a very high supersaturation is reached, typically $10^5 - 10^8$ times the solubility. In this zone a large number of unstable nuclei is formed during addition of the reactants. The unstable nuclei are transported in the well-mixed bulk of the vessel, the secondary mixing zone, where they redissolve if the supersaturation in this zone is low enough (otherwise new nuclei will survive and lead to broad particle size distributions). In the early stage, the supersaturation in the secondary mixing zone increases by dissolution of the unstable nuclei. This results in an increasing number of slightly grown stable nuclei (see Figure 4-16). Once a sufficient number of stable nuclei has been built up in the bulk phase, they become able to absorb all species generated by the constant dissolution of the unstable nuclei. From this moment, the number of growing crystals does not increase any longer, and the unstable nuclei generated in the primary mixing zone act as a monomer source for crystal growth (Ostwald ripening; ▶ glossary). Thus, after a short initiation stage, the only growth of stable crystals occurs in the secondary mixing zone.

Alternatively, monodisperse particles can also be formed by nucleation–aggregation. First, the primary particles begin to form by growth of the unstable nuclei. Once a certain number of primary particles has formed, they start to coagulate to form clusters, which are the nuclei of the secondary particles. During the CDJP, the secondary particles incorporate newly formed primary particles and thus grow in size.

The crystal shape of silver halides for photographic emulsions (Figure 4-15) is controlled by holding the Ag^+ concentration (i.e., the pAg), and thus the bromide concentration at a constant value. This process is called the pAg-controlled double-jet method. AgBr forms cubic crystals only when precipitated at pAg values <7.5. For pAg >8.5, the grains (▶ glossary) become octahedral. The reason for this behavior is the difference in the strength of bromide adsorption to the cubic (100) and the octahedral (111) crystal faces. At pAg <7.5, more bromide is adsorbed to the (100) than to the (111) faces, so that growth is inhibited at the (100) faces relative to the (111) faces. The octahedral faces thus grow faster and eventually disappear, leaving only the cubic faces. In contrast, at pAg >8.5, the (111) faces are more strongly inhibited by bromide ions. Tabular crystals are obtained by controlling both the pAg and the inlet rate into the reactor.

4.3 Biomaterials

Inorganic chemistry and live sciences combined into one discipline appear to be a contradiction at first sight. However, biomineralization processes and bioinspired inorganic materials chemistry are attracting increasing attention, and a vast amount of research is now devoted to them. Two different classes of materials will be discussed in this section. On the one hand, solid biomaterials produced by living organisms such as bones, teeth, spines, and shells (Section 4.3.1) in which nature shows us a huge variety of fascinating morphologies, beauty and complexity with specific mechanical, structural, and optimally adapted functionalities. On the other hand, substances will be discussed in Section 4.3.2 which are prepared by biomimetic approaches, or which will be used in prostheses or in medical devices designed for contact with the living body. The biomimetic synthesis of soft materials such as chemically responsive hydrogels (artificial muscles, skin) will not be highlighted. Instead, we will focus on biomimetic approaches to inorganic-based materials, nanophases, and composites (▶ glossary). Some of the more important aspects from a biological point of view cannot be covered within this textbook, for example the activity of the cells that govern all the processes, the hormones and other molecules that are involved in communication between the organism and mineralized cells as well as all the complicated uptake and transport phenomena of ions to the reaction center. Further information regarding those topics can be found in the literature (see Section 4.6).

4.3.1 Biogenic Materials and Biomineralization

Biomineralization refers to the process by which living organisms form inorganic solids. One of the major questions is: Why is biomineralization different from other mineralization/crystallization processes (see Section 4.2)? The major component of, for example, seashells (Figure 4-19) that can be found along a beach is crystalline calcium carbonate. However, just try to compare these beautiful biological objects and their complicated architectures and morphologies with their "man-made" synthetic analogues! It is a fascinating comparison.

Figure 4-19. Different types of seashells.

First of all, the "test-tube" calcium carbonate is a mixture of crystals of all kinds of shapes and morphologies, while nature exerts a remarkable control over the crystal's size, shape, and orientation, as well as the properties of the resulting materials such as high strength, resistance to fracture, and aesthetic value. Second, abiogenic crystals are formed corresponding to the thermodynamic/kinetic conditions during the synthesis, while the structure of the biogenic material is species-specific and also depends strongly on the local environment in which the crystals are formed.

Biogenic minerals can be found everywhere: Oyster shells, corals, ivory, sea urchin spines, magnetic crystals, etc. They are formed on a huge scale in the biosphere, have a major impact on ocean chemistry, and are important components of marine sediments and ultimately of many sedimentary rocks as well. A major function of biogenic minerals is to provide mechanical strength to skeletal hard parts and teeth. However, biominerals cannot be considered as static systems, but display an active demineralization/regeneration behavior which makes them useful as storage media, for example for iron or calcium. Not all biomaterials are desirable – the biomineral calcium oxalate monohydrate, $CaC_2O_4 \cdot H_2O$, is a major component of urinary stones. An understanding of the processes that lead to calcium oxalate precipitation in the urinary tract could lead to the prevention or treatment of urinary stone disease.

Common constituents of biological minerals are carbonates, phosphates, halides, sulfates and oxalates of the earth alkaline metals, particularly calcium, and the oxides of silicon and some transition metals, mainly iron (Table 4-2).

The predominance of calcium-containing minerals over other alkaline earth metals (group IIA) can be explained by the low solubility products of the carbonate, phosphate, pyrophosphate, sulfate, and oxalate and the relatively high levels of Ca^{2+} in extracellular fluids (10^{-3} M). Magnesium salts, for example, are generally more soluble, and no simple Mg biominerals are known. Most biominerals are ionic salts; an important exception is silica, due to the stability of Si–O–Si units in water.

A variety of different types of biomaterials is known such as *amorphous* materials, *ordered mesoscopic crystal aggregates*, e.g., in the form of macroscopic functional materials such as bones or teeth, and *nanocrystalline* materials. Typical examples for all three types of biominerals are given in the following paragraphs.

Amorphous: Diatoms

Diatoms are microscopic unicellular algae which are important components of the phytoplankton. Diatoms possess unique exoskeletons (shells or frustules), which are composed of biogenic amorphous silica (Figure 4-20). The living part is encased inside. When diatoms die, the silica shells collect on the ocean floor. These deposits are used commercially as components in products such as shoe polish and cosmetic articles.

Table 4-2. Important biominerals, their chemical composition, and their function.

Chemical composition	Mineral	Function and examples
Calcium carbonate $CaCO_3$	calcite aragonite vaterite amorphous	exoskeletons (e.g., egg shells, corals, molluscs, sponge spicules)
Calcium phosphates $Ca_{10}(OH)_2(PO_4)_6$ $Ca_{10-x}(HPO_4)_x(PO_4)_{6-x}(OH)_{2-x}$ $Ca_{10}F_2(PO_4)_6$ $Ca_2(HPO_4)_2 \cdot 2\,H_2O$ $Ca_2(HPO_4)_2$ $Ca_8(HPO_4)_2(PO_4)_4 \cdot H_2O$ $Ca_3(PO_4)_2$	hydroxylapatite defect apatites fluoroapatite	endoskeletons (bones and teeth) calcium storage
Calcium oxalate $Ca_2C_2O_4 \cdot (1 \text{ or } 2)\,H_2O$	whewellite wheddelite	calcium storage and passive deposits in plants, calculi of excretory tracts
Metal sulfates $CaSO_4 \cdot 2\,H_2O$ $SrSO_4$ $BaSO_4$	gypsum celestite baryte	gravity sensors exoskeletons gravity sensors
Amorphous silica $SiO_n(OH)_{4-2n}$	amorphous (opal)	defense in plants, diatom valves, sponge spicules, and radiolarian tests
Iron oxides Fe_3O_4 $\alpha, \gamma\text{-FeOOH}$ $5\,Fe_2O_3 \cdot 9\,H_2O$	magnetite	chiton teeth, magnetic sensors

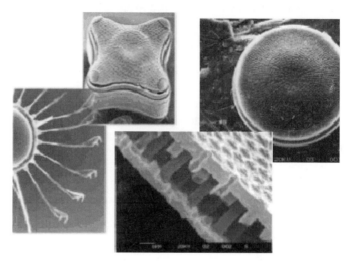

Figure 4-20. Diatom impressions – the fascinating world of diatoms in the micron range.

Although silica exhibits no long-range crystallographic order, microscopic morphological order is often observed. Such order may arise during either the nucleation (▶ glossary) or the growth processes. Energy considerations support the formation of a silica aggregate with a densely packed, covalently bonded core and a highly hydrated surface. The surface is likely to interact with organic substrates in a biological environment in a manner analogous to crystal interactions, thereby lowering the free energy of aggregate formation and controlling aggregate morphology on the microscopic scale (Figure 4-21).

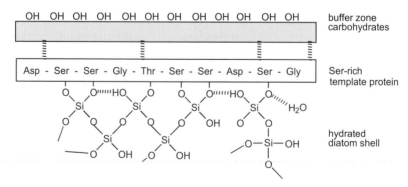

Figure 4-21. Model of a diatom shell.

Figure 4-22 shows a model of the silicon metabolism of unicellular diatoms. The formation of diatom shells is connected to the vegetative cycle during which two daughter cells with complete exoskeletons are formed by division of a mother cell.

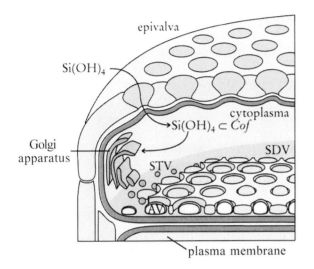

Figure 4-22. Model of the silicon metabolism of diatoms.

This synthesis of the diatom shell occurs in a stepwise manner by transport of monomeric silicic acid ($Si(OH)_4$) from the environment into the interior of the cell by active transport mechanisms. To avoid an uncontrolled polycondensation (▶ glossary) of the silicic acid with increasing concentration in the cell it is bonded to a cofactor ($Si(OH)_4 \subset Cof$) the chemical nature of which is still unknown. The Golgi apparatus of the cell probably serves as reservoir for the masked silicic acid. From the silicic acid depots of the Golgi apparatus small transport vesicles (▶ glossary), silica transport vesicles (STV) are formed which fuse with the silica deposition vesicles (SDV), the mineralization organelles of the cell. Condensation of the silicic acid takes place inside the SDVs, which are located on the bottom part of the new cell wall, and grow rapidly in all directions. Areolae vesicles are also located on this cell wall which are usually one layer of close-packed bubbles serving as negative pattern for the SDVs. This often leads to a hexagonal structure of the diatoms shells. This mold–prepattern hypothesis describes the formation of diatom shells, but does not explain it. The form and appearance of the silica shell is governed by genetic factors.

From a materials science point of view, morphogenesis of the diatom shell is an accomplishment which is unparalleled by all synthetic approaches towards other silica-based porous materials such as zeolites, MCMs etc. (see Sections 6.3 and 6.4).

Ordered mesoscopic crystalline aggregates: Bone

Bone is a fascinating material with very unusual mechanical properties. It serves two essential functions: on one hand as a structural material which is able to support its own weight, withstand acute forces, bend without shattering etc., on the other hand as an ion reservoir for both cations and anions. Both functions depend to a significant extent on the exact size, shape, chemical composition, and crystal structure of the mineral crystallites, and their arrangement within the organic matrix. Bone mineral can be generally represented by the formula $Ca_{8.3}(PO_4)_{4.3}(CO_3)_x(HPO_4)_y(OH)_{0.3}$, in which y decreases and x increases with increasing age, whereas x + y remains constant and equal to 1.7.

The structure of the bone as a composite material (▶ glossary) can be understood in terms of the different levels of organization that exist within the material:

- The lowest level of organization describes the crystals, the organic framework (mostly collagen fibrils), and the relationship between the framework and the crystals.
- The next level of organization (tens of microns) describes the longer range order of collagen and associated crystals.
- The highest level of organization describes the macroscopic build-up of bones. They are usually composed of a relatively dense outer layer (the cortical bone) surrounding a less dense, porous tissue (cancellous bone), which is filled with a gel-like tissue known as bone marrow.

Crystalline materials

Of the many transition metals that display a rich biocoordination chemistry, only iron and, to a smaller extent, manganese, have extensive roles in biomineralization. The bioinorganic solid-state chemistry of these elements is dominated by the redox chemistry as an energy source for biological activities, an affinity for O, S, and OH ligands, and ease of hydrolysis in aqueous solution. Like the calcium-containing biominerals, biological iron oxides are used to strengthen soft tissues and as storage depots (Fe^{3+}, OH^-, and HPO_4^{2-}). Furthermore, the magnetic properties of mixed-valence phases are utilized by bacteria of several types for navigation in the ambient geomagnetic field. Most magnetotactic bacteria synthesize intracellular magnetite (Fe_3O_4); species inhabiting sulfide-rich environments deposit the isomorphic mineral greigite (Fe_3S_4). The size and morphology of the crystals is controlled by organic membranes which are species-dependent. In both systems, the crystals (magnetosomes) must be aligned in chains to impart the bacterium with a magnetic dipole moment and must have dimensions compatible with those of a single magnetic domain (ca. 40–80 nm) (Figure 4-23). Particles smaller than this would show superparamagnetic behavior (▶ glossary; see Chapter 7); particles larger than this would have several domains which would be unable to function efficiently as biomagnetic compass.

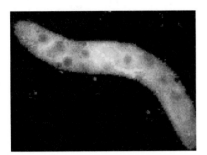 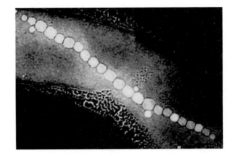

Figure 4-23. Transmission electron micrograph of a magnetospirillum bacterium (left). The chain of magnetite crystals (magnetosomes) can be seen in the picture to the right.

Mineralization processes

Much is known about the ultrastructures of biominerals and how they vary in different organisms, but precise details about the molecular interactions governing their formation are still unknown. The precipitation of many of the minerals listed in Table 4-2 from aqueous solutions is a relatively straightforward laboratory procedure, but controlling the size, shape, orientation, and assembly of these crystals, as typical of many biominerals, is a much more complex task. The fundamental physico-chemical principles are the same as already discussed in Section 4.2: supersaturation, nucleation (▶ glossary), and crystal growth. In biomineralization processes these steps depend critically not only on the ion concentration of the medium but also on the nature of interfaces (mineral–organic matrix and mineral–environment) present in the system.

Mineralization processes take place in an open system (cell with selective permeable cell membranes) far away from the thermodynamic equilibrium.

The cell is in a permanent exchange of energy and material with the environment. In this well-defined spatially delineated site regulation of the mineralization process is possible. Localized compartments (e.g., vesicles, ▶ glossary) which are surrounded by lipid membranes are very common. Furthermore, the exact regulation of the physico-chemical processes in these compartments allows the control of structure in biominerals. To achieve supersaturation (Figure 4-16), the compartments in which the mineral is formed must allow passive diffusion of ions and/or the active accumulation of ions against concentration gradients. Ion-specific pumps and channels are therefore necessary components of the machinery required for biomineralization. The site must be activated at specific times in the life of the organism, constrained in size and shape, and highly regulated with respect to the chemistry of the mineralization process.

The process of biomineralization can be divided into four stages (supramolecular preorganization, controlled nucleation, controlled crystal growth, and cellular processing) as follows.

Supramolecular preorganization. As already mentioned, one of the pre-requisites for the controlled deposition of biogenic inorganic materials in living organisms is the presence of a supramolecularly organized reaction compartment in which the mineralization zone is isolated from the cellular environment. These compartments can be located:

- on or in the membrane wall of bacterial cells (epicellularly);
- outside the cell, such as the extracellular facilitated construction of extended protein–polymer networks as in the collagen matrix for bone formation. Many shells and teeth are constructed within frameworks that may be lamellar, columnar or reticular;
- intracellularly by self-assembly of enclosed protein cages or lipid vesicles (▶ glossary) in which the molecular construction of the compartment is mainly based on the balancing of hydrophobic–hydrophilic interactions that exist for amphiphilic (▶ glossary) molecules in aqueous environments.

Controlled nucleation by interfacial molecular recognition. Controlled nucleation (▶ glossary) of inorganic clusters from aqueous solution into the framework built in the first stage by supramolecular preorganization is one of the major points in biomineralization processes. The underlying concept is that these preorganized organic architectures consist of somehow functionalized surfaces that serve as blueprints for site-directed inorganic nucleation. Electrostatic, structural and stereochemical recognition processes at the inorganic–organic interface are required (Figure 4-24).

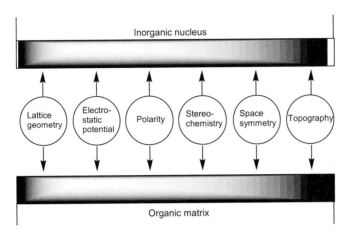

Figure 4-24. Possible modes of molecular complementarity at inorganic–organic interfaces.

It can be assumed that several of these factors act cooperatively in biological systems. The most fundamental aspect of recognition involves the matching of charge and polarity distributions. Additionally, the curvature of molecular cavities, which can be concave, convex, or planar, provides dimensional control over nucleation (▶ glossary) and a limit on the size of the nucleation site.

Most favorable for a controlled nucleation are probably concave surfaces, since here the local concentration of functional groups is very high. Convex surfaces are less active, since the binding sites are not in close proximity to each other. The same is true for planar surfaces. However, those allow for long-range structural matches which is called "biological epitaxy" (Figure 4-25). Epitaxy in biomineralization is distinct from inorganic epitaxy (▶ glossary), since the organic substrate does not show the typical smoothness or rigidity but exhibits surface stereochemistry due to exposed functional groups, etc. The role of the organic surface involved in inorganic crystallization is primarily to lower the activation energy of nucleation (see Section 4.1, heterogeneous nucleation).

It is not only short-range interactions that are important; sometimes larger periodic structures can control inorganic nucleation (▶ glossary) by the secondary, tertiary, or quaternary conformation of macromolecules which can act as blueprints for a controlled nucleation along specific crystallographic axes. A good example for such macromolecules is collagen in bone formation. Here, the bone crystals are nucleated in the interstices of a crystalline assembly of collagen fibrils.

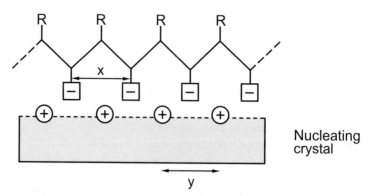

Figure 4-25. The concept of epitaxy as applied to biomineralization. Geometric matching must exist at the interface between a structured organic surface and nuclei of the inorganic crystal. Cation–cation distances in one specific crystal face are commensurate with the spacing of periodic binding sites on an organic surface (i.e., $x \approx y$).

Controlled crystal growth. By simple nucleation of an inorganic phase in a supramolecular host, followed by crystal growth according to the laws of crystallization, the resulting particles would be constrained in size, but would exhibit normal morphologies. Biological constraints such as genetic mechanisms have to be considered to explain the complexity of shapes in biominerals. They control the cellular environment and therefore the shape of organic compartments for nucleation (▶ glossary) and crystal growth. This genetically based morphogenesis cause the mineralized architectures to be unique and species-specific.

The chemistry within localized biological environments determines crystal growth, aggregation, and texture. This allows that in the same biomineralized system different compounds or polymorphs can be deposited such as $Fe_2O_3 \cdot n$ H_2O, γ-$FeOOH$, and Fe_3O_4 in molluscan teeth, or aragonite and calcite in some shells.

In some systems it is also possible that a spatial location of ion pumps within the enclosed organic compartment allows vectorial shaping of the crystal. If ions flow into the localized compartments only at specific ports of entry, then these sites will be the initial regions of mineral growth. If these sites are now turned off and other pumps along the membrane are switched on, vectorial flow of the ion stream will cause the mineral to develop along a preferred direction.

Figure 4-26 illustrates some generalized processes (a–g) for controlling the supersaturation conditions of solutions encapsulated within preformed supramolecular assemblies.

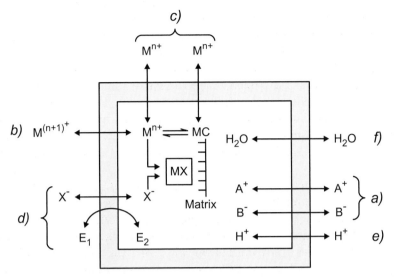

Figure 4-26. Control mechanisms involved in biomineralization processes (MX = biomineral; for an explanation of (a-f) see text).

These mechanisms include:

(a) Energy-consuming build-up of a concentration gradient in the membrane by specific ion pumps (A and B = extraneous ions).
(b) Redox processes at the membrane surface followed by selective transport of the oxidized or reduced species into the cell (e.g., Fe^{3+} is transported through the bacterial cell wall after reduction to Fe^{2+}).
(c) Selective complexation of metal ions (M^{n+}) – usually at the inner membrane surface – followed, in a later stage, by the controlled decomposition of this complex (MC) to release metal cations in aqueous solution.
(d) Enzyme (E)-mediated transport processes to increase anion (X^-) concentrations.
(e) Variation of the pH.
(f) In mineralization reactions in which water is produced by chemical reactions, control over the osmotic pressure governs nucleation, e.g., the condensation reactions of Si–OH groups to Si–O–Si units.

In brief, regulation can be achieved by facilitated ion flux, complexation–decomplexation switches, local redox and pH modifications, and changes in local ion activities.

Cellular processing. Biomineralization does not stop with the formation of small particles, but proceeds with the construction of higher-order architectures with elaborate structural properties. An example of such an organized ultra-structure is the relatively rigid and spatially fixed linear assembly of a chain of membrane-bound magnetite crystals in magnetotactic bacteria (see Figure 4-23). Another example is the macroscopic organized architectures found in the nacreous layer of shells with sheet-like organic assemblies (Figure 4-27). These are secreted periodically beyond the mineralization front, leading to progressive infilling that results in the construction of a highly organized lamellar architecture.

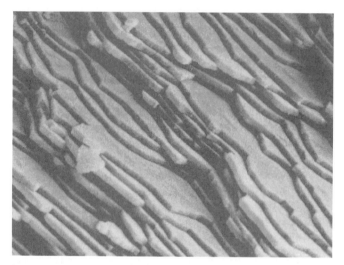

Figure 4-27. Scanning electron microscope image of the fracture of red abalone shell (*Halitois rufescens*) showing multiple tiling of $\approx$200–500 nm-thick aragonite ($CaCO_3$) crystals. Organic layers between the tiles (<10 nm-thick) are not discerned at this resolution.

The details of the recognition and organizational processes involved in the construction of these higher-order biomineralized architectures are currently unknown.

4.3.2 Synthetic Biomaterials

An exact replication of the biological architecture and formation process would be desirable for the development of implants or prostheses. However, the limit so far is the ability (or better the non-ability) to command and direct living cells in a way to tailor a material deliberately. Therefore, ways must be found to design synthetic materials which can substitute biological materials.

Biomaterials are substances that are used in prostheses or medical devices designed for contact with the living body. Almost all types of materials are represented. Polymers are used in ophthalmology, for skin wound treatments, and as soft tissue implants, for example. Metals have a large range of applications, including devices for fracture fixation, partial and total joint replacement, as well as dental amalgams. Pyrolytic carbon is used as coatings, for example, on prosthetic heart valves, and ceramics and glasses as bioactive components for a better adhesion of implants to natural tissue or bone, and as drug delivery carriers. Bioceramics are specially designed and fabricated ceramics for the repair and reconstruction of diseased, damaged, and "worn-out" parts of the human body. A schematic of some of the clinical uses of biomaterials is shown in Figure 4-28.

Biomaterials for medical applications need optimized mechanical, chemical and biological properties. In many cases composite materials (▶ glossary) and surface-modified materials are in use, because a single material cannot fulfill all the requirements. The form of the biomaterial depends on its intended function in the body. Load-bearing implants are usually made from bulk, non-porous materials, but coatings or composite structures may also be used to achieve improved mechanical and interfacial chemical properties (see Figure 4-31 for an example). Implants that serve only to fill space or augment existing bone tissue are used in form of powders, particulates, or porous materials.

Different types of interaction between tissue and biomaterials can be distinguished:

- *Bioinert materials* show minimal interactions with the neighboring tissue. They do not release any compounds to the environment, and there is no damage to the tissue. For example, implants made of metals or non-porous alumina attach by bone growth into surface irregularities, by cementing the device into the tissues, or by press fitting or screwing into a defect (called morphological fixation).
- *Biocompatible materials* do positively interact with the neighboring tissue. Due to this interaction, the mechanical stability of the implant is enhanced. For example, hydroxylapatite implants are mechanically attached by ingrowth (biological fixation).
- *Bioactive materials* increase recovering (healing) and growth of the tissue. This is the ideal case for a biomaterial. Resorbable bioactive materials for implants and prostheses are designed to be slowly replaced by bone. Bioactive, dense, non-porous surface-reactive ceramics, glasses and glass-ceramics attach directly by chemical bonding with the bone (bioactive fixation).

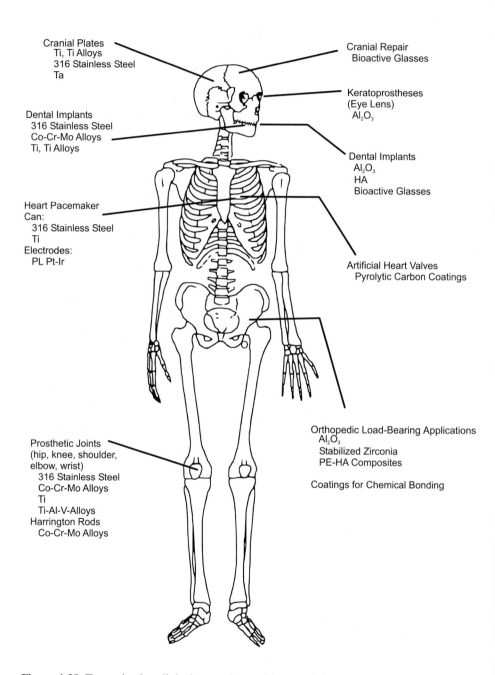

Figure 4-28. Examples for clinical uses of some biomaterials.

Bioactive ceramics and glasses

Bioactive ceramics and glasses are materials with a high potential for medical applications (Figure 4-28) since they were found to produce a reaction layer on their surface and thereby form a bond with bone tissue. Current clinical applications for these materials are mainly as coatings on metallic prostheses or as bone graft substitutes. There is also growing interest in the clinical use of calcium phosphate materials and ceramic composites as bone cements and settable bone-grafting materials.

Typical chemical components in bioactive materials are Na_2O, K_2O, MgO, CaO, Al_2O_3, SiO_2, P_2O_5, and CaF_2 in different combinations and ratios. A low silica content and the presence of calcium and phosphate ions in the glass result in rapid ion exchange in physiological solutions and in rapid nucleation (▶ glossary) and crystallization of hydroxycarbonate apatite bone mineral on the surface. This growing bone-mineral layer bonds to collagen, which is grown by the bone cells, and a strong interfacial bond is formed between the inorganic implant and the living tissues. Depending on the bulk composition the behavior with respect to interactions to the neighboring tissue varies. The sequence of reactions that occur on the surface of bioactive glass as a bond with tissue is formed is summarized in Figure 4-29. Stages 1–5 are well understood, while the overall understanding of stages 6 to 11 is sparse.

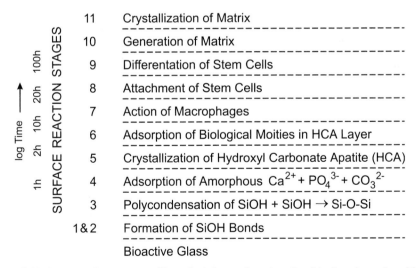

Figure 4-29. Proposed sequence of interfacial reactions involved in forming a bond between tissue and bioactive glass.

Bone substitutes

Materials that can be used as bone substitutes are also very important. The complicated hierarchical structure of bone is not easily mimicked by material scientists. Different approaches have been taken to substitute bone material. Modified biological samples such as sterilized and calcined bones from animals can be used. Alternatively, corals and algae can be chemically treated by hydrothermal treatment (see Section 4.4) to transform calcium carbonate to calcium phosphate (Eq. 4-11).

$$5\ CaCO_3 + 3\ (NH_4)_2HPO_4 + H_2O \longrightarrow Ca_5(PO_4)_3OH + 3\ (NH_4)_2CO_3 + 2\ H_2CO_3$$

$$(4\text{-}11)$$

Using this approach, it is possible to retain the porous structure in the calcium phosphate material, which is important in allowing the newly built bone to grow into the pores.

Other methods use chemically modified collagen on biologically degradable polymers, on metallic implants, on "bioglasses" and on combinations of those materials. However, despite much research an optimal bone substitute with good mechanical stability and biocompatibility is not yet available. New approaches are based on calcium phosphate (Eq. 4-12).

$$Ca(H_2PO_4)_2 \cdot H_2O(s) + \alpha\text{-}Ca_3(PO_4)_2(s) + CaCO_3(s) + Na_2HPO_4(aq)$$

$$\longrightarrow Ca_{8.8}(HPO_4)_{0.7}(PO_4)_{4.5}(CO_3)_{0.7}(OH)_{1.3}$$

$$(4\text{-}12)$$

The solid components are mixed with sodium phosphate solution to give an injectable paste which cures *in situ* after only 5 minutes. The carbonate-containing hydroxyl apatite formed during curing is very similar to the bone mineral with its very small crystals (ca. 20 nm).

Joint prostheses

The discussion of joint prostheses is intended to show that the development of biomaterials for medical applications is a complex task in which mechanical, chemical, and biological issues play a major role.

The first artificial hip prostheses were used in 1938, but today it is a standard technique in clinical medicine, with almost 500 000 such prostheses implanted annually throughout the world. The eroded hip joint is replaced by a two-part device (Figure 4-30): the shank of the ball portion of the artificial ball-and-socket hip joint is inserted into the thigh bone, and the artificial socket is fixed to the pelvic bone.

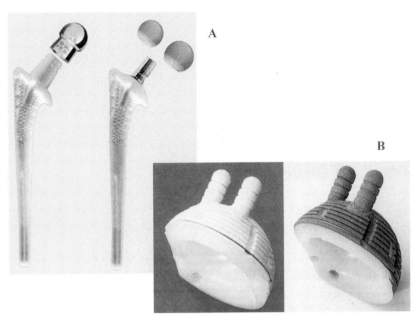

Figure 4-30. Artificial hip joint (two-part device). A. The ball portion with the shank. B. The socket portion with two pins for fixation and holes for screws. For a better biocompatibility with the tissue, the prosthesis is coated with either hydroxylapatite (B; left) or titanium powder (B; right).

One of the dominant problems is the high mechanical load on the prostheses, and the instabilities of the interface between the implant and its host tissue. Therefore, anchoring or chemical bonding of the implant to the bone is critical. Since the 1960s, major progress has been made with the development of the so-called "bone cement". This is a rapidly self-curing poly(methylmethacrylate) (PMMA) which provides a stable mechanical anchor for a metallic prosthesis in a bone bed. The mechanical properties of PMMA with respect to the connection of bone and implant are very good. However, because of the exothermic curing process, temperatures up to 100 °C can be reached, and this may damage the neighboring tissue. Additionally, toxic monomers or oligomers can be released into the body. Therefore, alternatives have had to be developed.

Today, only half of the artificial hip bones are cemented into the bone; the other half is implanted by accurate fitting. In those cases the implant is coated with a hydroxylapatite (HA), ceramic (calcium phosphate) or titanium powder to improve the bioactivity (Fig. 4-31). A biological bonding can also be achieved by coatings containing peptide ligands which bind selectively to the receptors of bone-forming cells (osteoblasts).

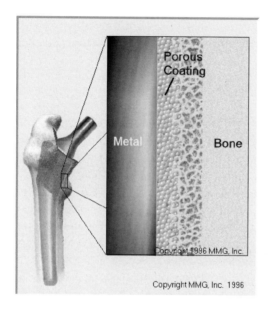

Copyright MMG, Inc. 1996

Figure 4-31. Scheme of the artificial hip joint coated with a porous coating.

The choice of the material used for the implant is crucial, since the life-time of these artificial hip joints is supposed to be very long (10–20 years). During this time they must go through millions of moving cycles. As a consequence, friction is very high and migration of small particles can occur and cause defects in the neighboring tissue. Therefore, a combination of a soft and a hard material is usually applied. For the socket part, polyethylene with a high molecular weight ($M_r > 10^6$ g·mol^{-1}) is used, and this is inserted into a titanium or steel capsule (Figure 4-30). Typical materials for the ball part are alumina or steel. Metallic titanium is mostly used to form the shank, as this generally shows good biocompatibility and corrosion resistance due to the formation of a titanium oxide (TiO_2) layer on the surface.

However, despite continuously ongoing developments in this area the optimal material that would combine high mechanical stability, corrosion resistance and good bioactivity has yet to be found.

4.3.3 Biomimetic Materials Chemistry

Bioinspired materials processing (which means exploiting the basic principles of biomineralization) is also an emerging discipline in materials science. Materials such as bones, teeth, and shells are complex composites (▶ glossary) and the organization and interfacial chemistry of the components are optimized for functional use. Mimicking such structures would be a significant step towards so-called "smart" materials. For materials scientists, biomineralization provides a unique opportunity to study solutions to key problems in materials design. Despite some successes, no system has yet been devised that approaches the integrated molecular controls in natural biomineralization, and this rudimentary development is reflected in the relatively simple materials architectures so far formed.

An example of a synthetic processing technique inspired by the compartmental strategy of, for example, magnetite crystal (Fe_3O_4) formation (see above) is the use of microemulsions, phospholipid vesicles (▶ glossary), proteins, and reversed micelles (▶ glossary) formed by surfactant–water mixtures to produce inorganic nanoparticles with precisely controlled sizes and shapes. Nanoscale particles are of special interest because they exhibit quantum-size effects in their electronic, optical, magnetic and chemical properties and have a high portion of surface atoms (see Chapter 7). The organic supramolecular cages formed by the lipids or surfactants contain a microenvironment in which controlled precipitation occurs. These micelles (▶ glossary) or vesicles (▶ glossary) can be considered to mimic the supramolecular compartments in which biomineralization processes occur. They are very versatile systems because the reaction environment can have variable diameters (from 1 to 500 nm), and the surface functional groups can be modified. Many different nanoparticles have been synthesized in this way. As each particle is surrounded by an organic membrane, particle–particle interactions are negligible, and reaction rates can be diffusion controlled (Figure 4-32; see also Chapter 7).

As seen, surfaces play an important role in biomineralization processes. A second example for biomimetic approaches in materials science is the use of synthetic surfaces to initiate crystal nucleation (▶ glossary) and growth. Other techniques, e.g., chemical vapor deposition methods (see also Section 3.2) also use oriented inorganic substrates such as gold and silicon for epitaxial growth

(▶ glossary). The use of solutions means a significant step forward, since they can be applied to complex shapes and a wide variety of surfaces. Biomimetic approaches towards a controlled nucleation and growth of inorganic materials involve the use of surfactant monolayers (see Chapter 7) or surfaces which have the advantage that their functionality and packing can be modified in a way that they exactly serve as molecular blueprint.

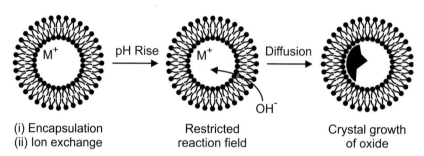

(i) Encapsulation   Restricted   Crystal growth
(ii) Ion exchange   reaction field   of oxide

Figure 4-32. Membrane-mediated precipitation of metal oxides in phospholipid vesicles. The chemical reaction is as previously discussed (Eq. 4-10).

Surfactant monolayers, when spread on surfaces of aqueous solutions, can be considered to model the surface of biological membranes, and will be discussed in more detail in Chapter 7. The potential of monolayers formed at gas–liquid interfaces for promoting crystallization was identified for the first time, when amphiphilic (▶ glossary) films of chiral amino acids were found to induce enantioselective (▶ glossary) crystallization of organic crystals (α-glycine).

Size-quantized microcrystals and ultrathin, particulate *films* of semiconductor sulfides (▶ glossary) were also synthesized by using surfactant monolayers (Figure 4-33). A surfactant monolayer is spread on an aqueous solution of the metal salt precursor. Hydrogen sulfide gas infuses through the monolayer, nucleation (▶ glossary) of the metal sulfide takes place at different sites at the monolayer/water interface (upper part of Figure 4-33), and well-separated nanocrystalline metal sulfide particles grow. The particles coalescence, forming a "first layer" of a porous sulfide semiconductor (▶ glossary) particulate film. Fresh metal species diffuse to the monolayer head group area and form a second layer of the sulfide film. These steps are repeated subsequently to build up layer-by-layer the sulfide semiconductor film up to a plateau thickness, which depends on the chemical composition (CdS ~30 nm and ZnS ~350 nm). The presence of the surfactant monolayer is absolutely essential to the formation of the semiconductor film or the nanoparticles. This can be seen in an experiment in which hydrogen sulfide gas is infused over an aqueous metal-ion solution without surfactant. This experiment results in the formation of large, irregular, and polydispersed metal–sulfide particles.

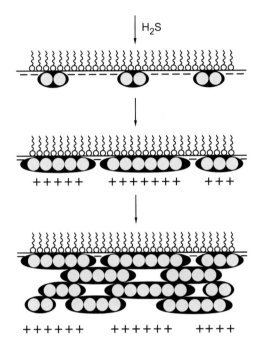

| H_2S

Figure 4-33. Schematics of the growth of nanoparticulate metal sulfide films under monolayers. The time of hydrogen sulfide treatment increases from top to bottom.

Functionality and packing of the supramolecular surface can be modified to provide complementarity between the surface chemistry and structure of a film and the crystal faces of the nucleus. An example is the nucleation and growth of crystals on template (▶ glossary) surfaces. For example, barium sulfate (baryte, $BaSO_4$) was precipitated from supersaturated solutions in the presence of a monolayer of n-eicosyl sulfate, $C_{20}H_{21}OSO_3^-$, an amphiphilic (▶ glossary) aliphatic long-chain sulfate. Crystallization of barium sulfate in the absence of monolayers results in the precipitation of rectangular tablets. Under monolayers of n-eicosyl sulfate, barium sulfate crystals nucleate with the (100) face parallel to the plane of the monolayer (Figure 4-34). The arrangement of the three oxygen atoms of the sulfate head groups at the interface mirrors a similar arrangement of sulfate anions on the (100) face of barium sulfate. Binding of Ba^{2+} cations to the monolayer may simulate the (100) face and, thus, initiate oriented nucleation from the monolayer. When a monolayer of eicosanoic acid was used instead, in which only minimal structural recognition appeared to take place, because the carboxylate end group was only bidentate (▶ glossary), the oriented growth of $BaSO_4$ was not observed.

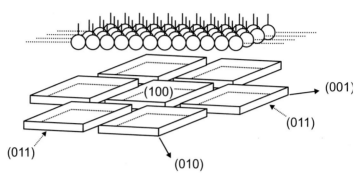

Figure 4-34. Scheme of barium sulfate precipitation in the presence of a monolayer of *n*-eicosyl sulfate.

4.4 Solvothermal Processes

The growth of many crystals is not possible by the processes discussed earlier in this book. Furthermore, many inorganic phases are unstable under traditional high-temperature, solid-state synthesis conditions. Nature serves as model for solvothermal methods, since many geological important minerals are grown under such conditions. Solvothermal processes take place in a liquid medium above the boiling point and 1 bar. Besides water (hydrothermal synthesis), ammonia (ammonothermal synthesis) is the most important solvothermal reaction medium. There are three different types of technique:

1. hydrothermal crystallization allows for the preparation of large crystals and gemstones;
2. hydrothermal syntheses can be used for the production of fine-particle oxides such as zeolites; and
3. hydrothermal leaching is used, for example, for the treatment of ores to recover metals.

   In this section, a detailed discussion of the fundamental aspects of hydrothermal crystallization is given, followed by a short description of the hydrothermal synthesis and the hydrothermal leaching process.

4.4.1 Hydrothermal Synthesis of Single Crystals

As early as 1845 Schafhäutl mentioned the aqueous synthesis of small quartz crystals in a "Papinscher Topf" (Papin's pot). Based on work of Spezia, who synthesized crystals up to 1 cm^3 under isothermal conditions, suitable ways for an industrial application were developed, such as the temperature gradient method after World War II. In 1985, the world production was about 1500 tons (Figure 4-35), and today the industrial growth of quartz crystals is performed in large autoclaves of ≥ 100 liter capacity.

Figure 4-35. Synthetic quartz crystal (left) and rock crystal (right).

Since the discovery of the piezoelectric properties (▶ glossary) of quartz in 1880 by Pierre Curie, quartz has become a significant factor in the growth of the electronics industry. The pressure resulting from a voltage being applied is displayed in the form of oscillations at a particular resonant frequency. This frequency is a function of the thickness of the crystal. By carefully preparing a crystal, it can be made to oscillate at any frequency. This piezoelectric quartz is widely used from electronic watches and electronic instruments to processors in automobiles to satellite communications. For this electronic application the crystal must be free of electrical and optical twinning, voids, and liquid or solid inclusions. Other applications for quartz crystals are in the area of optical materials such as prisms, eyeglasses, or windows of high-power lasers. Figure 4-36 illustrates some typical application of quartz crystals, including crystal filters which are widely used in communication systems with frequencies ranging from 10 MHz to 200 MHz, as clock oscillators, and as crystal wafers or blanks.

Syntheses of gemstones are also an interesting feature of hydrothermal crystallization. Quartz variants such as amethyst, citrine, smoky quartz, ruby, sapphire, and emerald can be synthesized by incorporation of ions into the quartz during hydrothermal treatment.

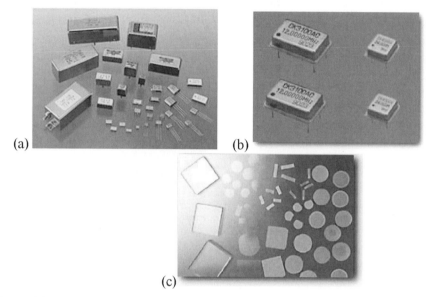

(a)

(b)

(c)

Figure 4-36. Applications of quartz crystals. (a) as filters; (b) as oscillators; and (c) as crystal wafers or blanks.

General aspects of solvothermal processes

Crystallization by solvothermal synthesis is, in the simplest case, a straightforward (isothermal) equilibrium reaction. Substances that are insoluble under ambient conditions are dissolved, and the crystalline phase, which is stable under the hydrothermal conditions, is formed. However, for the growth of single crystals temperature gradients in the reaction vessel often have to be applied. Therefore, the hydrothermal synthesis can be often treated as a special case of a transport reaction (Section 3.1).

Most of the solvothermal processes are carried out in water, and thus are termed hydrothermal. Water serves two different functions: it is the solvent, and also the pressure transmitting medium. Therefore, the pVT-data (p = pressure, V = volume, T = temperature) and the physico-chemical properties of water at high temperatures and pressures are of significant importance. For example, the ion product of water increases with increasing pressure and temperature, the viscosity decreases with higher temperatures (facilitating diffusion processes), and the polarity, i.e., the dielectric (▶ glossary) constant, decreases with higher temperatures, but increases with higher pressures.

Since hydrothermal reactions must be carried out in closed vessels, the pressure–temperature relations of water at constant volume are very important. Figure 4-37 shows the V–T diagram of water in which the volume usually is replaced by the density.

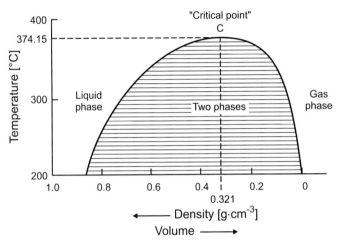

Figure 4-37. Volume (density)/temperature dependence of water.

This diagram shows a closed, *two-phase region*, in which the liquid and the gaseous phase are in equilibrium, that means they coexist. The density of the two phases at a given temperature is given by the two intersections of a horizontal line through the two-phase region with the *equilibrium curve*. For example, at 300 °C, the density of the liquid water phase is approximately 0.75 g·cm^{-3}, and that of the gas phase 0.05 g·cm^{-3}. With increasing temperature, the density of the liquid phase decreases, while the density of the corresponding gas phase increases. At the critical point C (with a *critical temperature* of 374.15 °C and *a critical density* of 0.321 g·cm^{-1}) the densities of both phases are the same, as are all other properties. That means, the difference between liquid and gaseous phases has disappeared. Above the critical temperature only one phase exists, which is called the *supercritical* or *fluid* phase.

For practical applications the corresponding P–T diagram (Figure 4-38) is more important. The curve between the triple point (Tr) (the point at which gas, liquid and solid phase coexist) and the critical point (C) represents the saturated steam curve at which the liquid and gaseous phase coexist. At pressures below this curve liquid water is absent and the vapor phase is not saturated, above this curve liquid water is under compression and the vapor phase is absent. The solid lines in Figure 4-38 (isochores) may be used to calculate the pressure that is developed inside a vessel after it has been partially filled with water, after closing and heating to a certain temperature.

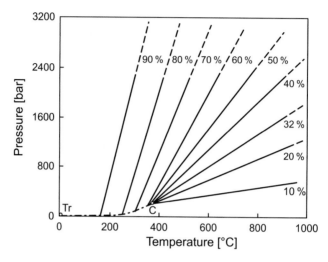

Figure 4-38. Pressure/temperature dependence of water for different degrees of filling of the reaction vessel.

When an autoclave is filled to more than 32 %, the meniscus between the liquid and the gas phase rises when the temperature is increased, and at some temperature below the critical temperature (373 °C) the whole autoclave is filled with liquid. For example, if the percentage of filling of the autoclave is 80 %, the autoclave is totally filled by the liquid phase at 245 °C, because the corresponding isochore separates from the saturated steam curve at this temperature. The higher the degree of filling of the autoclave in the beginning the lower is the temperature at which it is filled completely. If the degree of filling at room temperature is below 32 %, the liquid–gas meniscus drops, and at some temperature below 373 °C the autoclave is filled only with gas. The lower the starting percentage of filling, the lower is the temperature at which this "dry boiling" occurs. For an autoclave filled to exactly 32 %, the meniscus does not move with increasing temperature, and it disappears at the critical temperature of 373 °C. Above this temperature the medium exists only as a supercritical fluid. The density and solvent properties of this fluid, however, are very similar to those of a conventional solvent. Usually hydrothermal crystallization is performed with degrees of filling of 32 %, and, with even more success, above 65 % and pressures of 200–3000 bar. These data refer to pure water. Assuming that the solubility of the substances is small and that only a very small portion is dissolved, these data may also be used for the solutions.

The solvent properties of pure water are often – even at high temperatures – not sufficient to dissolve substances for crystallization. Therefore, it is often necessary to add a *mineralizer*. A mineralizer is any compound which added to the aqueous solution speeds up crystallization, and it usually operates by increasing the solubility of the solute through the formation of soluble species that would not normally be present in the water.

Typical mineralizers are the hydroxides of the alkali metals, especially for the growth of amphoteric and acidic oxides, of silicates, germanates or elemental metals. In addition, alkali salts of weak acids, e.g., Na_2CO_3, Na_3BO_3, Na_2S, or the chlorides of alkali metals as well as acids are applied.

For instance, the solubility of quartz in water at 400 °C and 2 kbar is too small to permit the recrystallization of quartz, in a temperature gradient, within a reasonable time. Quartz crystals are grown in an alkaline medium in a temperature gradient of 400 to 380 °C and 1 kbar. Usually polycrystalline quartz serves as precursor material which is termed "Lasca". Typical mineralizers for SiO_2 are NaOH, KOH, $NaCO_3$ or NaF. Eq. 4-13 gives an example of how bases act as mineralizers for silica.

$$SiO_{2(solid\ \alpha\text{-quartz})} + 2\ OH^- \rightleftharpoons SiO_3^{2-} + H_2O \qquad (4\text{-}13)$$

Figure 4-39 shows the solubilities of some substances under hydrothermal conditions as a function of the OH^- concentration.

Typical saturation concentrations of the substances for crystallization are in the region of 2–5 wt%. With this concentration, growth rates of around 1 mm per day can be achieved. As can be seen in Figure 4-39, this saturation concentration is reached, e.g., for SiO_2 (quartz) in solutions of 0.5 M NaOH or for ZnO in solutions of 6 M NaOH or 6 M KOH.

A typical hydrothermal procedure is as follows: The precursor substance (nutrient) is put at the bottom of the autoclave which is filled to the desired degree with the solvent (water and mineralizer). The autoclave contains a baffle – a perforated metal disc (with defined openings) in the middle – to separate the growth zone of the single crystals from the zone in which the precursor dissolves. Seed crystals, cut with defined crystallographic orientation, are hung into the growth zone to facilitate nucleation (▶ glossary; Figure 4-40).

High pressures and temperatures during the hydrothermal treatment require a special design for the reaction vessel. Up to 300 °C and 10 bar, thick-walled glass tubes can be used. These have the advantage that the crystalliza-

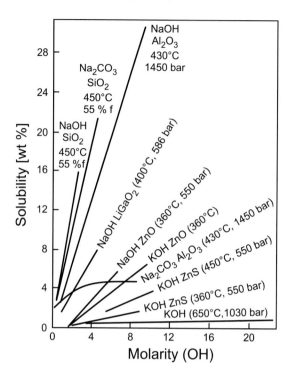

Figure 4-39. Solubility as a function of the hydroxide concentration for some hydrothermal systems.

tion reactions can be followed easily. However, a steel tube which is closed at one end is usually used.

The autoclave is put into an oven which heats the nutrient section more than the growth zone ($T_1 < T_2$). Therefore, a temperature gradient exists in the reaction vessel. With increasing temperature, the substance at the bottom part of the autoclave dissolves and is transported into the cooler growth zone by convection. Here, the solution is supersaturated due to the lower temperature, and crystal growth occurs at the single crystal nuclei. For crystallizations by the temperature gradient method, solubilities (s) of some weight percents are required. Additionally, the temperature coefficient of the solubility, ds/dT, should be around 0.01–0.1 wt% per 10 °C. Because of the convection processes, supersaturation conditions are maintained during the synthesis. The baffle hinders total mixing of the two zones and localizes the temperature gradient, mainly at the metal disc. This creates very uniform conditions within the different zones, which facilitates uniform crystal growth. The duration of growth runs varies between days and month, depending on the growth conditions.

More rarely, the case of *retrograde solubility* can be found, that means transport from cold to hot regions. For example, the solubility of SiO_2 in pure water

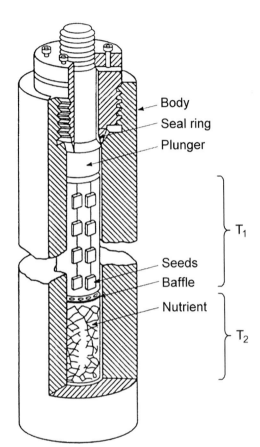

— Body
— Seal ring
— Plunger

T_1

— Seeds
— Baffle

— Nutrient

T_2

Figure 4-40. Scheme of an autoclave for hydrothermal single crystal growth.

shows up to 600–700 bar (= 60–70 MPa) a retrograde behavior (Figure 4-41); thus the temperature coefficient is negative. Only at higher pressures does the temperature coefficient become positive, which allows for dissolution at high temperatures and crystallization in the colder zone of the autoclave.

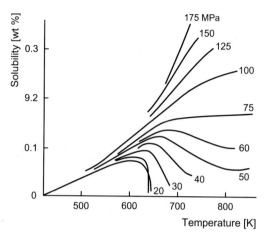

Figure 4-41. Solubility of SiO_2 in water.

Non-aqueous solvents

Although water is considered to be the most important solvothermal reaction medium, other solvents may also be used (Table 4-3).

Table 4-3. Examples for hydrothermal processes in non-aqueous environments.

Solvent	Examples
NH_3	nitrides, imides, amides, $CsOH$, Cs_2Se_2
HF	$MO_{3-x}F_x$ ($M = Mo, W$)
HCl, HBr	$AuTe_2Cl$, $AuSeCl$, $AuSeBr$, $Mo_3S_7Cl_4$
Br_2	$SbSBr$, $SbSeBr$, $BiSBr$, $BiSeBr$, $MoOBr_3$
S_2Cl_2	MoS_2Cl_3, $Mo_3S_7Cl_4$
S_2Br_2	$Mo_3S_7Br_4$
$SeBr_2$	$SbSeBr$, $BiSeBr$
$H_2S + (C_2H_5)_3NHCl$	β-Ag_2S
C_2H_5OH	SbI_3, BiI_3
CS_2	monoclinic Se
CCl_4	$SeCl_4$, $TeCl_4$
C_6H_6	selenium
CH_3NH_2	CH_3NHLi

Non-aqueous solvents (with the exception of ammonia) are not very common for solvothermal processes. However, in cases in which the solubilities are not high enough in water, in which the solvent itself is part of the reaction, or in which the product reacts with water, other solvents have to be used. Table 4-4 lists the critical data for some non-aqueous solvents.

Formic acid is different because it decomposes at high temperatures to CO and H_2O or CO_2 and H_2. Many oxides and carbonates have been synthesized in this reducing and CO_2-rich atmosphere.

Table 4-4. Critical constants of some solvents.

Solvent	Critical temperature [°C]	Critical pressure [bar]
H_2O	374.1	221.2
NH_3	132.3	111.0
Cl_2	144	77.1
HCl	51.4	83.2
CO_2	31.3	73
SO_2	157.8	78.7
H_2S	100.4	90.1
CS_2	279	79
C_2H_5OH	243	63.8
CH_3NH_2	156.9	40.7
CH_3OH	240	81
HCOOH	308	–

4.4.2 Hydrothermal Synthesis

Commercial processes which utilize hydrothermal syntheses include those for the manufacture of a wide range of zeolites (see Section 6.4.1). In addition, a number of processes have been developed for a variety of electronic, magnetic and ceramic materials. For example, in magnetic tapes chromium(IV) oxide is used due to its excellent magnetic properties. In this case, Cr(III) or Cr(VI) precursors are transformed by hydrothermal treatment at 300–400 °C and 50–800 bar to needle-like CrO_2 crystals.

For most hydrothermal syntheses only moderate temperatures (100–300 °C) at the corresponding low solution vapor pressures (subcritical region) are applied. The feed slurry or nutrient solution, usually consisting of oxides, hydroxides, and salts of the corresponding metal is fed into the autoclave. This slurry is heated, and at the desired reaction temperature the nutrient materials react and/or transform, primarily through dissolution and precipitation, to the stable compound. An example is the formation of barium titanate (see Section 2.1.1), from barium hydroxide and titanium oxide at temperatures between 150 and 250 °C (Eq. 4-14).

$$Ba(OH)_2(aq) + TiO_2 \longrightarrow BaTiO_3(s) + H_2O \qquad (4\text{-}14)$$

After cooling the autoclave, the product can be isolated by filtration, and after a series of washing steps the pure product is obtained.

Synthesis of zeolites

Zeolites are important materials for catalysis, adsorbents and molecular sieve applications (see Section 6.4.1 for more details). Zeolites are crystalline silicate materials with a well-defined and highly ordered structure and porosity, and have the general composition $M_{x/n}(Al_xSi_yO_{2(x+y)}) \cdot z\ H_2O$ (M = metal cation with the charge n). Most (but not all) zeolites are prepared via hydrothermal synthesis. The synthesis conditions depend on the desired composition of the material, the desired particle size, morphology, etc.
For the synthesis of the commercial zeolite ZSM-5, for example, a mixture of polysilicic acid, sodium hydroxide, aluminum sulfate, water, *n*-propylamine, and tetrapropylammonium bromide is heated in an autoclave to 160 °C for several days. The synthesis process is sensitive to numerous variables such as temperature, pH, origin of silica and alumina, the type of alkaline cation, reaction time, the templating agent, and the use of seeding to control rate of nucleation (▶ glossary) and particle size.
The hydrothermal approach to the synthesis of zeolites is considered to be "chimie douce". Compare the typical technical preparation conditions with the biological synthesis of structured SiO_2 matrices (Table 4-5) discussed in Section 4.3.1.

Table 4-5. Technical and biological synthesis of SiO_2-materials.

	Zeolites	Unicellular algae
Reaction time	days	shell formation within hours
Concentration of the inorganic precursor	>1 M	<0.001 M
pH	6 – 14	6 – 8
Temperature	125 – 200 °C	4 – 25 °C
Pressure	1 – 100 bar	1 bar
Structural description	compact translatorial repetitive microporous with uniform pores (typical: 0.3 – 2 nm)	shell hierarchical micro- and macroporous (pore diameters 5 nm to μm)

4.4.3 Hydrothermal Leaching

The Bayer process is an important and large-scale industrial process that uses hydrothermal leaching to extract and precipitate high-grade aluminum hydroxide from bauxite ore. Bauxite is a mixture of $Al(OH)_3$ and $AlOOH$, and can contain large quantities of silica and Fe_2O_3. Extraction is carried out under hydrothermal conditions to remove these components by reacting bauxite with a concentrated sodium hydroxide solution. The soluble sodium aluminate complex $Na[Al(OH)_4]$ is formed (Eq. 4-15), while Fe_2O_3 is insoluble under these conditions.

$$Al(OH)_3 + NaOH \longrightarrow NaAl(OH)_4(aq)$$

$$AlOOH + NaOH + H_2O \longrightarrow NaAl(OH)_4(aq)$$

(4-15)

After filtration, $Al(OH)_3$ is precipitated from the $Na[Al(OH)_4]$ solution by cooling, diluting, and seeding with $Al(OH)_3$. The pure $Al(OH)_3$ obtained is calcined to yield corundum, Al_2O_3, which is mainly used for the production of aluminum metal.

4.5 Sol–Gel Processes

The goal of this section is to present the general principles of sol–gel processing as well as its potential for material synthesis. This process allows the obtainment of solid products by gelation rather than by crystallization or precipitation, which have been discussed in Sections 4.1 and 4.2. The sol–gel process can be described as the creation of an oxide network by progressive polycondensation reactions (▶ glossary) of molecular precursors in a liquid medium, or as a process to form materials via a sol, gelation of the sol and finally removal of the solvent. This method is considered as a "chimie douce" or soft chemical approach to the synthesis of metastable (amorphous) oxide materials (▶ glossary).

Before turning to the mechanisms by which solids are eventually formed, the processing steps and applications of the final materials, the terms "sol" and "gel" must be defined:

- A *sol* is a stable suspension of colloidal (▶ glossary) solid particles or polymers in a liquid. The particles can be amorphous or crystalline. Note the difference to aero*sols* (▶ glossary) discussed in Section 3.3, which are suspensions of particles or droplets in a *gas* phase.
- A *gel* consists of a porous, three-dimensionally continuous solid network surrounding and supporting a continuous liquid phase ("wet gel"). In "colloidal" ("particulate") gels, the network is made by agglomeration of dense colloidal particles (▶ glossary), whereas in "polymeric" gels the particles have a polymeric sub-structure made by aggregation of sub-colloidal chemical units. In general, the sol particles can be connected by covalent bonds, van der Waals forces, or hydrogen bonds. Gels can also be formed by entanglement of polymer chains. In most sol–gel systems used for materials syntheses, gelation (= formation of the gels) is due to the formation of covalent bonds and irreversible. Gel formation can be reversible when other bonds are involved in gelation.

Figure 4-42 presents a scheme of the different processing routes leading from the sol to a variety of materials. Powders can be obtained by spray-drying of a sol. Gel fibers can be drawn directly from the sol, or thin films can be prepared by standard coating technologies such as dip- or spin-coating, spraying, etc. Here, gelation occurs during the preparation of the film or fiber due to rapid evaporation of the solvent.

Gelation can also occur after a sol is cast into a mold, in which case it is possible to make monolithic objects of a desired shape. Drying by evaporation of the pore liquid gives rise to capillary forces that causes shrinkage of the gel network. The resulting dried gel is called a xerogel ("xero" means dry). Compared to the original wet gel its volume is often reduced by a factor of 5 to 10. Due to the drying stress, the monolithic gel body is often destroyed and powders are obtained. When the wet gel is dried in a way that the pore and network structure of the gel is maintained even after drying, the resulting dried gel is called an "aerogel" (see also Section 6.3). Aerogels are usually obtained after supercritical drying processes, in which the interface between liquid and vapor is avoided during drying, and therefore no capillary pressure exists. Dense ceramics or glasses can be obtained after heat treatment of xerogels or aerogels to a temperature high enough to cause sintering.

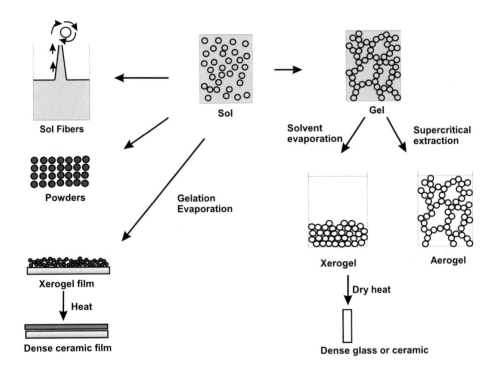

Figure 4-42. Sol–gel processing options.

Typical applications of sol–gel materials include all kinds of coatings (see below), catalysts and catalyst supports, ceramic fibers, electroceramic powders, insulating materials, high-purity glasses, etc. The many processing options allow a unique access to multicomponent oxide systems, as demonstrated in the following example.

The perovskite (▶ glossary) lead zirconate titanate, $PbZr_{1-x}Ti_xO_3$, (PZT) is an important ferroelectric (▶ glossary) and piezoelectric (▶ glossary) ceramic material. The properties of PZT depend very critically on the exact stoichiometry; optimal values are obtained for $x = 0.47$.

PZT powders can be obtained by the ceramic method (Section 2.1.1), for example, by heating an intimate mixture of PbO, TiO_2 and ZrO_2 to 750–1100 °C for several hours. Owing to the volatility of lead oxide at the required high temperatures, the precise stoichiometry is difficult to obtain. PZT thin films are difficult to make by the ceramic method, because the film thickness cannot be controlled and the required temperatures for the formation of the PZT phase are too high for many substrates. PZT films can be prepared by the CVD method. However, the problem to control exactly the stoichiometry is similar to that of the ceramic method. A typical flow chart for the preparation of PZT films by sol–gel processing is shown in Figure 4-43.

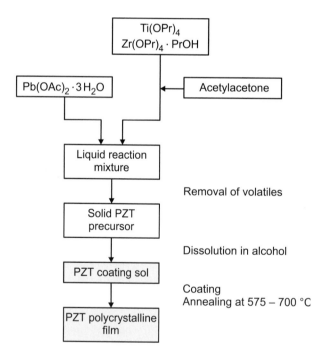

Figure 4-43. Processing steps for the preparation of polycrystalline PZT films.

The precursors for the preparation of PZT by the sol–gel method according to Figure 4-43 are $M(OPr)_4$ (M = Ti, Zr) and lead acetate. The role of acetylacetone to modify the alkoxide precursors will be discussed later (see Eq. 4-29). The water necessary for the hydrolysis reactions is brought in by the hydrated lead acetate. In the resulting coating sol, with particle sizes between 6 and 10 nm, the metals are homogeneously distributed. The sol is applied to a substrate by conventional coating techniques (such as dip- or spin-coating, doctor-blading (see Section 2.1.4), etc.). Subsequent heat treatment at relatively mild conditions (low temperatures and short heating periods) results both in the removal of the organic groups by pyrolysis and thermolysis reactions and the formation of a crystalline PZT film up to 1.5 µm thick. Thicker films are obtained by repeating the coating and annealing cycle. Figure 4-44 shows a metallic printing drum coated with PZT as an example of a technical application of such films.

Figure 4-44. Metallic printing drum coated with a 8-µm thin film of ferroelectric $PbZr_{0.53}Ti_{0.47}O_3$ (PZT). Patterning is achieved by application of a directed electric field. This creates a remnant charge in the PZT layer that is used to adsorb the printing ink. After use, the printing pattern can be electrically erased, and the drum can be used again.

PZT powders or aerogels can be made by essentially the same method (without the coating step). The gels can also be used for making ceramic fibers, similar to the preceramic polymers that will be discussed in Chapter 5. Due to the high melting points of ceramic materials, ceramic fibers cannot be made by processing the ceramic material itself. The thin PZT fiber shown in Figure 4-45 was prepared by first preparing a PZT gel from alkoxide and carboxylate precursors similar to the process shown in Figure 4-43. The properties of gels allow their extrusion through a spinneret. The green fiber (▶ glossary) is then sintered at 925 °C. The process of fiber spinning is described in more detail in Section 5.5.

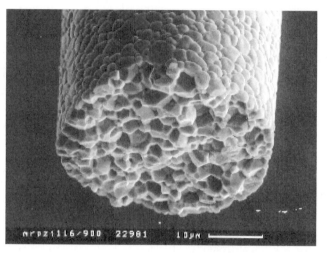

Figure 4-45. Cross-section of a dense, polycrystalline PZT fiber with the composition $PbZr_{0.53}Ti_{0.47}O_3$. The grain size (▶ glossary) is 2–4 µm.

The example of PZT shows that sol–gel processing even allows the preparation of ceramic materials with complex compositions. Another example of such a compound is the high-temperature superconductor yttrium barium copper oxide (▶ glossary; YBCO, see also Section 2.1.1).

In the following, we will first discuss some of the fundamental physical principles that govern sol and gel formation (Section 4.5.1), followed by a detailed treatment of the individual steps in the sol–gel processing of silicate materials (Section 4.5.2). In silicate systems, these steps were particularly well investigated; the fundamental issues are common to sol–gel processing of other materials that will be treated in Section 4.5.3. In the final section (Section 4.5.4), different types of inorganic–organic hybrid materials will be introduced which are composed of inorganic and organic building blocks and are easily prepared by sol–gel processing.

4.5.1 The Physics of Sols

The stability and coagulation of sols (colloids; ▶ glossary) is of great importance to sol–gel chemistry since the structure of the final oxide network (gel) depends mainly on the size and shape of the aggregating species: the sol particles. Therefore, some elementary aspects will be outlined in the following paragraphs.

The easy agglomeration of fine particles is caused by attractive van der Waals forces and/or forces (V_A) that tend to minimize the total surface or interfacial energy of the system. Van der Waals forces result from dipole–dipole interactions, and are rather weak; thus the attractive potential extends over distances of only a few nanometers.

In order to prevent aggregation, repulsive forces of comparable dimensions are required. This can be done by:

- creating *electrostatic repulsion* between particles. This repulsion results from charges adsorbed on the particles. Electrostatic stabilization occurs when the electrostatic repulsive forces overcome the attractive van der Waals forces between the particles.
- adsorbing a thick organic layer ("*steric barrier*"). Steric stabilization can occur in the absence of an electric barrier and is particularly efficient in dispersing high concentrations of particles. Surfactant molecules can adsorb at the surface of the particles (see also Section 4.3 and Chapter 7). Their hydrophobic chains will then extend into the solvent and interact with each other (Figure 4-46).

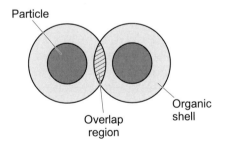

Particle

Overlap region

Organic shell

Figure 4-46. Representation of the steric barrier by surfactant molecules adsorbed to the particle surface.

The solvent–chain interaction increases the free energy of the system and creates an energy barrier to the closer approach of the particles. The thickness of the layer must therefore be in a range (usually <3 nm) that the closest possible distance between the particles is outside the range for a sufficient van der Waals attraction. Furthermore, the motion of the chains extending into the solvent becomes restricted and produces an entropic effect when the particles come into closer contact with each other.

Steric barriers are very often used for the stabilization of nano-sized particles (see Chapter 7). The electrostatic approach is mostly used to stabilize colloidal (▶ glossary) dispersions in aqueous systems.

The repulsive electrostatic barrier results from an electric *double layer*: the surface of the particle is covered with ionic groups that control the potential of the surface. Counterions in solution in the vicinity of the particle screen the surface charges.

For hydrous oxides, the charge-determining ions are H^+ and OH^-, which create the charge at the particle surface by protonating or deprotonating M–OH groups (Eqs. 4-16 and 4-17).

$$M-OH + H^+ \longrightarrow M-HO_2^+ \qquad\qquad (4\text{-}16)$$

$$M-OH + OH^- \longrightarrow M-O^- + H_2O \qquad\qquad (4\text{-}17)$$

The pH at which the particle is electrically neutral is called the *point of zero charge* (PZC). At pH >PZC, Eq. 4-17 dominates, and the particle is negatively charged (surrounded by positively charged counterions in solution), whereas at pH <PZC the particle is positively charged (Eq. 4-16). Typical values for the PZC are: MgO 12, Al_2O_3 9.0, TiO_2 6.0, SnO_2 4.5, SiO_2 2.5. Note that the acidity of the surface M–OH groups depends somewhat on the particle size and the degree of condensation (see below). The magnitude of the surface potential ϕ_o depends on the difference between pH and PZC. That potential attracts the counterions present in solution.

The double layer is schematically shown in Figure 4-47 for a positively charged surface. The repulsive potential V_R drops linearly through the tightly bound layer of water and counterions (called the Stern layer). Beyond the so-called Helmholtz plane, the linear dependence changes into a region in which the electrostatic potential varies exponentially with the distance from the particle (h).

When an electric field is applied to a sol (colloid; ▶ glossary), the charged particles move towards the electrode with the opposite charge ("electrophoresis"). When the particle moves, it carries along the adsorbed layer and part of the cloud of counterions, while the more distant portion of the double layer is drawn to the opposite electrode. The "slip plane" or "shear plane" separates the region of fluid that moves with the particle from the region that flows freely. The potential at the slip plane is the zeta (ζ) potential (ϕ_ζ). The pH at which ϕ_ζ is zero is called the isoelectric point (IEP), which in general is not equal to the PZC. The stability of the colloid (▶ glossary) correlates with the magnitude of the ζ-potential. The higher the ζ-potential the more stable are

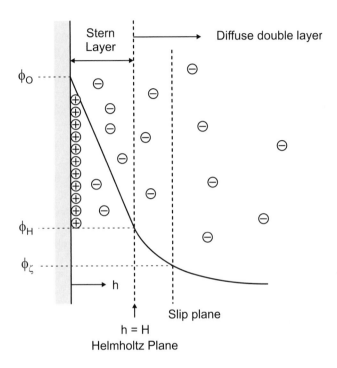

Figure 4-47. Electric double layer over a positively charged surface. h is the distance to the particle surface and V_R the repulsive potential.

the sols; stability requires a repulsive potential $\geq$ 30–50 mV.

For the same surface potential, the repulsive barrier is greater for larger particles. This effect results from the different dependencies of the attractive potential V_A and the repulsive potential V_R on the particle size. It explains why stable sols can be formed. Once the initial nuclei have grown to sufficient size, the repulsive barrier becomes large enough to prevent coagulation.

Coagulation of sol particles results from

- decreasing the surface potential ϕ_o (by changing the pH), or

- increasing the electrolyte concentration in the solution. As the concentration of counterions increases, the double layer is compressed, because the number of charges required to balance the surface charge is now available in a smaller volume surrounding the particle. The colloid will eventually coagulate because the attractive force is unchanged, while the repulsive barrier is reduced.

In some cases, a coagulated colloid can be redispersed. This process is called peptization. This can be done by removing the counterions that caused coagulation (washing) or by adsorbing charge-determining ions that re-establish the double layer.

4.5.2 Sol–Gel Processing of Silicate Materials

The coordination number of silicon is generally four, although coordination expansion can occur in transition states (see below). Compared to the transition metals, silicon is less electropositive and therefore it is not very susceptible for a nucleophilic attack. This makes silicon compounds quite stable and easy to handle.

Sol–gel processing proceeds in several steps which will subsequently be discussed in sequence:

1. Hydrolysis and condensation of the molecular precursors and formation of sols.
2. Gelation (sol–gel transition).
3. Aging.
4. Drying.

Hydrolysis and condensation reactions

Silica gels certainly belong to the best-investigated materials in inorganic chemistry. Various types of precursors can be used to prepare silica gels by sol–gel processing. The basic chemical principle behind sol–gel processing is the transformation of Si–OR and Si–OH-containing species (Figure 4-48) to siloxane compounds. To obtain a particle or even a gel, the number of siloxane bonds has to be maximized and consequently the number of silanol and alkoxide groups has to be minimized.

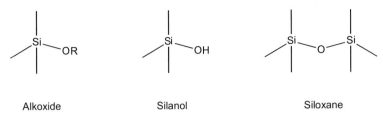

Figure 4-48. Silicon–oxygen groups relevant to sol–gel processing.

The chemical reactions during sol–gel processing can be formally described by three equations (Eq. 4-18). Condensation can happen either via an alcohol producing mechanism, or via a water producing mechanism.

$$\equiv\text{Si–OR} + \text{H}_2\text{O} \longrightarrow \equiv\text{Si–OH} + \text{ROH} \qquad \text{Hydrolysis}$$

$$\equiv\text{Si–OH} + \equiv\text{Si–OR} \longrightarrow \equiv\text{Si–O–Si}\equiv \;+\; \text{ROH}$$

$$\equiv\text{Si–OH} + \equiv\text{Si–OH} \longrightarrow \equiv\text{Si–O–Si}\equiv \;+\; \text{H}_2\text{O}$$

Condensation

(4-18)

The most common precursors are aqueous solutions of sodium silicates ("water glass") and silicon alkoxides, e.g., Si(OMe)_4. The formula of water glass is often given as "Na_2SiO_3". In reality, however, water glass solutions contain mixtures of different silicate species. The most important differences between these two types of precursors are:

- The solvent in water glass-based reactions is always water, while the alkoxides are employed either neat or dissolved in an organic solvent (mostly an alcohol).
- The reactive groups in aqueous silicate systems are silanol groups. In alkoxide-based systems, hydrolysis reactions are required to convert the Si–OR groups into Si–OH groups. As a consequence, gelation is initiated in aqueous silicate systems by pH changes, and by addition of water to the precursors in alkoxide-based systems.
- Alkoxide-based systems are more complex because more parameters are involved which influence the sol–gel reactions (see below). This gives more possibilities to control the morphology and properties of the obtained materials.

The basic chemical reactions are the same for both types of silicon-containing precursors and for monomeric, oligomeric or polymeric species. However, two chemical different situations have to be considered with regard to the reaction mechanisms: reactions under acidic or basic conditions, respectively. To understand the difference it is necessary to remember that the point of zero charge (PZC) of Si–OH-containing species is between pH 1.5 and 4.5 (the higher the degree of condensation of the silica species the lower the PZC). Acidifying the solution to a pH below the PZC means that the siliceous species are positively charged, and increasing the pH above the PZC (more basic) means that the species are negatively charged.

Under acidic conditions, the oxygen atom of a Si–OH or a Si–OR group is protonated in a rapid first step (Eq. 4-19). A good leaving group (water or alcohol) is created by the protonation. Additionally, electron density is withdrawn from the central silicon atom, making it more electrophilic and thus more susceptible to attack by water (hydrolysis reaction) or silanol groups (condensation reaction).

$$\equiv Si-OX \; + \; H^+ \; \rightleftharpoons \; \equiv Si-O^{+}\!\!\begin{smallmatrix} ,H \\ \diagdown X \end{smallmatrix}$$

X = R, H

$$Y-OH \; + \; \equiv Si-O^{+}\!\!\begin{smallmatrix} ,H \\ \diagdown X \end{smallmatrix} \; \longrightarrow \; Y-O-Si\equiv \; + \; HOX$$

Y = H,   ≡Si

hydrolysis reaction: **X = R, Y = H**

condensation reaction: **X = R or H, Y = Si≡**

$$(4\text{-}19)$$

Under basic conditions, to which we generally refer when the reactions occur at a pH >2, the reaction proceeds by nucleophilic attack of either an OH^- or an $Si–O^-$ ion to the silicon atom (Eq. 4-20). The entering OH^- or SiO^- are formed by dissociation of H^+ from a water molecule or a Si–OH group.

In hydrolysis reactions, the hydroxide anion attacks the silicon atom by an S_N2-type mechanism in which OH^- displaces OR^-. In the condensation reaction, a nucleophilic silanolate ion attacks the neutral silicate species and displaces OH^- or OR^-.

$$\equiv Si\text{-}OX \;+\; YO^- \;\rightleftharpoons\; \left[\begin{array}{c} OY \\ | \\ \text{}_{\text{\tiny{///}}}Si\text{—} \\ | \\ OX \end{array} \right]^- \;\rightleftharpoons\; Y\text{-}O\text{-}Si\equiv \;+\; XO^-$$

$$Y = H, \;\equiv Si$$

hydrolysis reaction: $X = R, \; Y = H$

condensation reaction: $X = R \text{ or } H, \; Y = Si\equiv$

$$(4\text{-}20)$$

One has to keep in mind that all mechanisms discussed are, in principle, reversible. This means that Si–O–Si bonds can be cleaved by alcohols and OH^-, and silanol groups can react with alcohol to alkoxy groups. The degree to which the back reaction may occur mainly depends on the pH and the amount of water.

With these general mechanisms in mind we will now continue to discuss how various parameters influence the reaction rates of the hydrolysis and condensation reactions.

Silicon alkoxides. The overall reaction for sol–gel processing of silicon tetraalkoxides is given in Eq. 4-21.

$$Si(OR)_4 \;+\; 2\,H_2O \;\longrightarrow\; SiO_2 \;+\; 4\,ROH \qquad\qquad (4\text{-}21)$$

The most common tetraalkoxysilanes used for sol–gel processing are tetramethoxysilane (TMOS) and tetraethoxysilane (TEOS). Since many alkoxysilanes are immiscible with water, alcohols are used as solvents to homogenize the reaction mixture. However, alcohols not only serve as solvent but are also reactants due to the reversibility of the reactions. A mixture of a tetraalkoxysilane in water and alcohol would react very slowly. Therefore, acid or base catalysis is necessary to start the hydrolysis and condensation reactions.

The sol–gel chemistry of silicon alkoxides as described by the overall reaction (Eq. 4-21) or the hydrolysis and condensation reactions (Eq. 4-18) seems to be straightforward and easy to understand. However, the whole picture is much more complex since hydrolysis and condensation reactions compete with each other during all steps of the sol–gel process and are additionally influenced to a different degree by the parameters discussed in the following paragraphs. Figure 4-49 shows some of the species that can be obtained in the initial stages of the reaction of $Si(OR)_4$. Each species can further react by either hydrolysis or condensation.

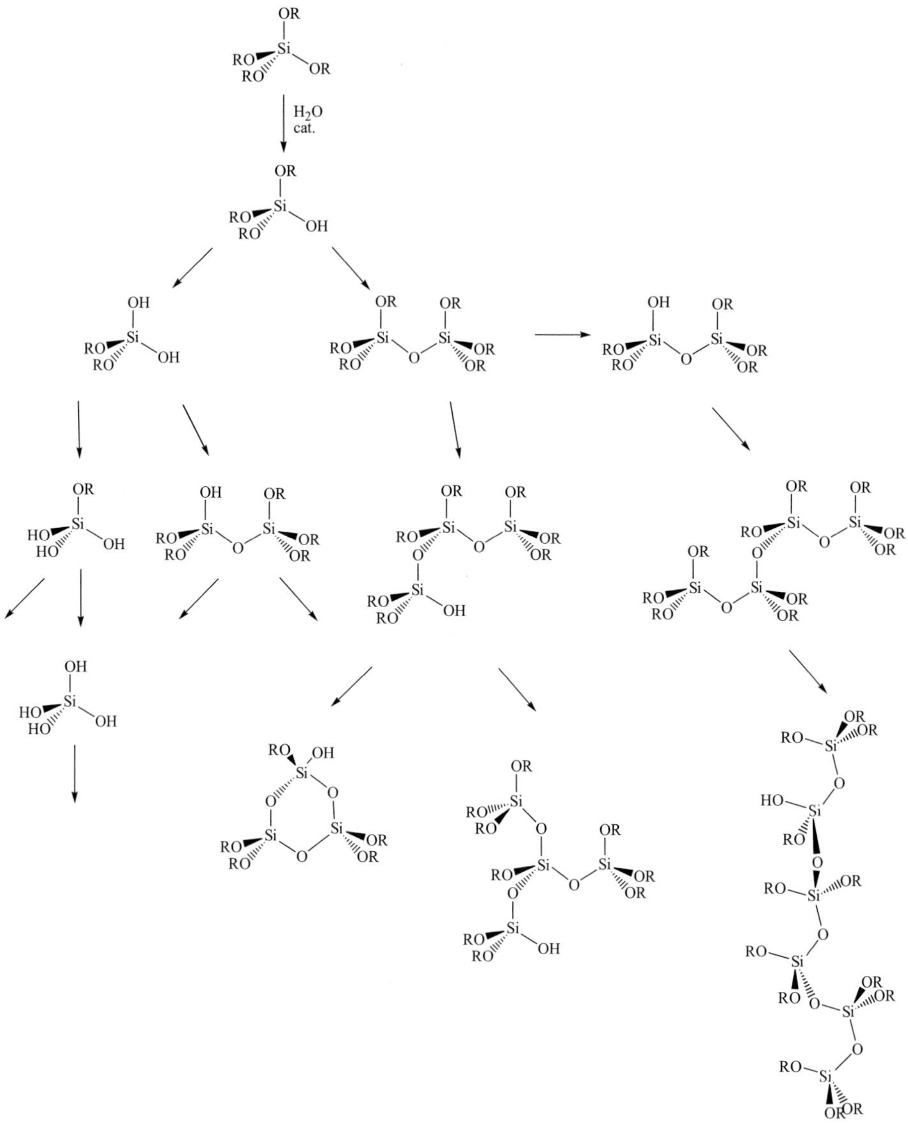

Figure 4-49. Some initial intermediates during sol–gel processing of $Si(OR)_4$. Note that each intermediate can, in principle, undergo either hydrolysis or condensation reactions.

Furthermore, these reactions can occur on chemically different silicon atoms once the trimer stage has been reached (the silicon atom in the center is chemically different to the terminal silicon atoms). From Figure 4-49 it is obvious that the systems are very complex and many different routes from the $Si(OR)_4$ precursor to the final silica gel are possible. The chemical parameters discussed below determine which route is taken. The final properties of the obtained gel very much depend on the structural evolution.

The most important factor for the later material properties is the relative rate of the hydrolysis and condensation reactions. The influence of the different parameters on the network formation is very complex, since many parameters change progressively as polycondensation (▶ glossary) proceeds. Therefore, it must be well understood that we are discussing not a static, but a continuously changing system.

Because of this complexity, a detailed understanding of the parameters influencing the reaction rate and the course of the reaction is necessary to tailor the properties of the final material. Parameters which influence hydrolysis and condensation and whose deliberate variation are used for materials design are:

- the kind of precursor(s);
- the alkoxy group to water ratio (R_w);
- the kind of catalyst;
- the kind of solvent;
- the temperature;
- the pH; and
- the relative and absolute concentration of the components in the precursor mixtures.

a) Steric and inductive effects of the precursor(s)

The hydrolytic stability of alkoxysilanes is influenced by steric factors. Any branching of the alkoxy group or increasing of the chain length lowers the hydrolysis rate of the alkoxysilanes. That means, that the reaction rate decreases in the order

$$Si(OMe)_4 > Si(OEt)_4 > Si(O^nPr)_4 > Si(O^iPr)_4.$$

Inductive effects of the substituents attached to a silicon atom are very important, because they stabilize or destabilize the transition states during hydrolysis and condensation. The electron density at the silicon atom decreases in the following order:

$$\equiv Si-R > \equiv Si-OR > \equiv Si-OH > \equiv Si-O-Si$$

For acid catalysis, the electron density at the silicon atom should be high because the positive charge of the transition state is then stabilized best. Therefore, the reaction rates for hydrolysis and condensation under acidic conditions increase in the same order as the electron density. For base catalysis a negatively charged transition state has to be stabilized. Therefore, the reaction rates for hydrolysis and condensation increase in the reverse order of the electron density.

This has several consequences, for example:

- As hydrolysis and condensation proceed (increasing number of OH and OSi substituents), the silicon atom becomes more electrophilic. This means, for example, that in acidic media, monomeric $Si(OR)_4$ hydrolyzes faster than partially hydrolyzed $Si(OR)_{4-x}(OH)_x$ or oligomeric species (which have more Si–O–Si bonds), and vice versa in basic media.
- More branched (i.e., more highly condensed) networks are obtained under basic conditions and chain-like networks under acidic conditions, because reactions at central silicon atoms (i.e., atoms with two or three Si–O–Si bonds) are favored at high pH, and reactions at terminal silicon atoms (i.e., atoms with only one Si-O-Si bond) are favored at low pH.
- Organically substituted alkoxysilanes $RSi(OR)_3$ are more reactive than the corresponding $Si(OR)_4$ under acidic conditions and less reactive under basic conditions.
- The acidity of a silanol group increases with the number of Si–O–Si bonds at the silicon atom. Note that this is one of the reasons why the PZC changes with the degree of condensation.

b) The alkoxy group/H_2O ratio (R_w)

The overall reaction for sol–gel processing of silicon tetraalkoxides (Eq. 4-21) implies that two equivalents of water ($R_w = 2$) are needed to convert $Si(OR)_4$ to SiO_2. Four equivalents of water ($R_w = 1$) are needed for the complete hydrolysis of $Si(OR)_4$ if no condensation would take place. Increasing the water content (i.e., lowering R_w) generally favors the formation of silanol groups over Si–O–Si groups, especially since the condensation reaction is, in principle, reversible. Even with an $R_w > 2$, only mixtures of hydroxylated species $[SiO_x(OH)_y(OR)_z]_n$ ($2x+y+z = 4$) are created. As a general rule, an $R_w \gg 2$ favors the condensation reaction, and an $R_w \leq 2$ favors the hydrolysis reaction. The R_w, together with the kind of catalyst, influences the properties of the silicate material very strongly.

c) The catalyst (pH value)

Sol–gel reactions of silicon alkoxides are typically catalyzed by an acid or a base. As discussed above, the reaction mechanisms for acid or base catalysis are very different. Additionally, the reaction rates for hydrolysis and condensation have a different dependence on the pH (Figure 4-50).

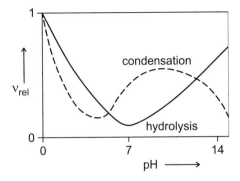

Figure 4-50. Dependence of the relative rates of hydrolysis and condensation reactions of Si(OR)$_4$ on the pH.

The minimal reaction rate for hydrolysis is at pH = 7, and for condensation around pH 4.5. For the condensation reaction, the minimum pH corresponds to the IEP of silica, where the sol particles can approach each other without a large electrostatic barrier. The pH value is the decisive parameter for the relative rates of hydrolysis and condensation of tetraalkoxysilanes (Si(OR)$_4$). Under acidic conditions (pH <5) hydrolysis is favored, and condensation is the rate-determining step. A great number of monomers or small oligomers with reactive Si–OH groups is simultaneously formed. In contrast, hydrolysis is the rate-determining step at pH >5 and hydrolyzed species are immediately consumed because of the faster condensation.

d) The solvent
The solvent is important to homogenize the reaction mixture, especially at the beginning of the reaction. The polarity, dipole moment, viscosity, protic or non-protic behavior of the solvent influence the reaction rate and thus the structure of the final sol–gel material.

Polar and particularly protic solvents (H$_2$O, alcohols, formamide) stabilize polar siliceous species such as [Si(OR)$_x$(OH)$_y$]$_n$ by hydrogen bridges; non-polar solvents (tetrahydrofuran, dioxane) are usually used for organotrialkoxysilanes [R'Si(OR)$_3$] or incompletely hydrolyzed systems.

Aqueous silicic acid solutions. A generalized overall reaction for sol–gel processing of silicic acid is given in Eq. 4-22.

$$Si(OH)_4 \longrightarrow SiO_2 + 2\,H_2O \tag{4-22}$$

Silicic acid, $Si(OH)_4$, which is formed when water glass is reacted with an acid or by ion exchange of the sodium ions for H^+, is only stable in solution at pH <7. At pH >7 deprotonated species, mainly $H_3SiO_4^-$, are found in diluted solutions ($<10^{-4}$ mol·l^{-1}), and mainly condensed anionic species are present in concentrated solutions (>0.1 mol·l^{-1}).

Contrary to sol–gel processing of alkoxide precursors, there are mainly three possibilities to control the morphology and properties of the products: the concentration of the precursor, the presence of salts and the pH of the solution.

Three important pH regimes have to be distinguished: pH <2, pH 2–7 and pH >7. At pHs lower than about 2, the silicic acid species are positively charged, and according to the mechanism given in Eq. 4-19, the reaction rate of the condensation is proportional to the concentration of the H^+ ions. Between pH 2 and 7 the reaction rate of condensation is proportional to the concentration of the OH^- ions (see Eq. 4-20). This means, that the condensation rates are increasing when the pH is increased. At pHs higher than 7, the rates of solubility and for re-dissolution of particles are maximal. The solution now contains mainly anionic species which reject each other. Therefore, the presence of electrolytes has a strong influence on the gelation behavior (see also Section 4.5.1).

The sol–gel transition (gelation)

We will now discuss how the sol particles aggregate to give gel networks. In the initial stage of the sol–gel reactions, small three-dimensional oligomeric particles are formed, with Si–OH groups on their outer surface. The oligomers serve as nuclei. They may either grow, or agglomerate at a certain size, depending on the experimental conditions (Figure 4-51). Agglomeration of the oligomeric particles does not necessarily result in gelation; instead, larger particles with a polymeric sub-structure can be formed. Whether the larger particles (with a polymeric or dense substructure) remain suspended in solution (i.e., whether a stable sol is obtained) or agglomerate themselves to form a three-dimensional network (i.e., whether gelation occurs) again depends on the system and the experimental conditions which influence the stability of the sol.

As the sol particles aggregate and condense, the viscosity of the sol gradually increases and a gel is eventually formed (provided that the sol is sufficiently concentrated). The sol–gel transition (gel point) is reached when a continuous network is formed.

From a practical point of view, the gel time (t_{gel} = time at which the gel point is reached after starting hydrolysis and condensation reactions) is determined by turning the reaction vessel upside down. Before the gel point has been

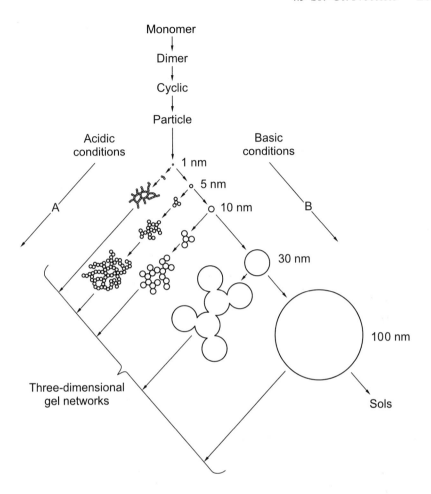

Figure 4-51. Structural development of silica gels.

reached, the colloidal (▶ glossary) solution behaves like a more or less viscous liquid, i.e., the liquid will flow out of the vessel. At the gel point, the viscosity increases sharply, and a form-stable, elastic gel body is obtained. Since all liquid is retained in the gel body, no liquid can flow out of the vessel if it is turned upside down. For the same reason, the volume of the gel in this stage is the same as that of the original precursor solution.

The t_{gel} is generally lowered by all parameters that increase the rate of condensation reactions, as discussed above. These parameters thus allow to deliberately influence the gel times. Typical t_{gel} values for $Si(OMe)_4$ in the presence of 0.05 mol of a catalyst are 92 h with HCl as the catalyst, H_2SO_4 106 hours, NH_4OH 107 hours, HI 400 hours. Without an catalyst, t_{gel} would be about 1000 hours.

The simplest picture of gelation is that the particles grow by aggregation or condensation until they collide to give clusters of particles. The clusters become bigger and bigger by repeated collisions. In this picture, which is described mathematically by the percolation theory, the gel is formed when the last link between two giant clusters of particles is formed. Percolation theory offers a simple description of gelation. Figure 4-52 illustrates percolation on a two-dimensional lattice.

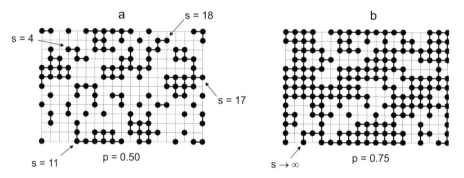

Figure 4-52. Site percolation on a square lattice: (a) for p = 0.50, and (b) for p = 0.75 (p is the fraction of filled spheres).

Starting with an empty grid, spheres (representing the sol particles) are placed randomly on the intersections of the grid lines. The fraction of filled sites is designated p. If two neighboring sites are filled, they are joined by a bond. This process is called *site* percolation. It produces clusters of certain sizes (s). As more and more sites are filled, the average cluster size (s_{av}) increases. In Figure 4-52a, there is a broad range of cluster sizes, but none of them is a "spanning cluster", i.e., a cluster that reaches across the vessel that contains it. Formation of the spanning cluster is equivalent to gelation. The percolation threshold p_c is defined as the value of p at which the spanning cluster first appears. The value of p_c can be calculated mathematically; for example p_c = 0.198 for a face-centered cubic lattice filled with spheres of the same size. In Figure 4-52b, p is clearly larger than p_c, the percolation threshold has been crossed. Note, however, that:

- at the gel point a certain number of unbounded clusters is still present. This is important for the aging of gels (see below); and
- the last bond resulting in the formation of the spanning cluster is not different to the previously formed bonds, i.e., gelation is not a special thermodynamic event.

A different version of percolation, which is more appropriate as a model for gelation, is *bond* percolation. All sites are initially filled with spheres, and the bonds are filled in randomly.

An alternative description of gelation is given by kinetic growth models. These also explain the different microstructures upon changing the reaction conditions. Depending on the conditions, growth in silicate systems may occur predominantly by condensation of clusters with monomers or with other clusters. The rate of the condensation reactions may be diffusion- or reaction-limited. Note that the term "cluster" is used with different meanings. In the kinetic growth models, "cluster" is used equivalent to "particles" or "oligomeric species" in the preceding discussion.

As has been discussed before, hydrolysis of silicon alkoxides is much faster than condensation under acidic conditions. Since all species are hydrolyzed at an early stage of the reaction, they can condense to form small oligomeric species (clusters) with reactive Si–OH groups. Under these conditions, reactions at terminal silicon atoms are favored for electronic reasons. This results in polymer-like gels, i.e., small clusters undergo condensation reactions with each other to give a polymer-like network with small pores. This process is called reaction-limited cluster aggregation (RLCA).

Monomer-cluster growth requires a continuous source of monomers. Hydrolysis is the rate-determining step under basic conditions. The hydrolyzed species are immediately consumed by reaction with existing clusters because of the faster condensation reactions. Furthermore, the rate of hydrolytic cleavage of (terminal) Si–O–Si bonds, is much higher than under acidic conditions. This additionally insures that a source of monomers is available. This model is called reaction-limited monomer cluster growth (RLMC), or Eden growth. Condensation of clusters among each other is relatively unfavorable because this process requires inversion of one of the silicon atoms involved in the reaction. Due to a different mechanism, reaction at central silicon atoms of an oligomer unit is favored (see above). The resulting network has a particulate character with big particles and large pores (colloidal gels).

The formation of larger particles, mainly in aqueous systems, is also favored by Ostwald ripening (▶ glossary) by which small particles dissolve and larger particles grow by condensation of the dissolved species. Solubility of a particle is inversely proportional to its radius. The solubility of small particles (<5 nm) therefore is rather high. Growth stops when the difference in solubility between the smallest and the largest particles in the system becomes only a few ppm. Solubility depends on the given conditions (temperature, pH of the solution, etc.). For example, silica particles grow to a size of about 5–10 nm at pH >7, but growth stops at a size of 2–4 nm at lower pH. At higher temperatures, larger particles are obtained because the solubility of silica is higher.

An example of how the growth mechanisms discussed above can be used to obtain special morphologies of silica materials, is the production of smooth monodispersed silica spheres by the so-called Stöber method. TEOS is reacted at high pH (with ammonia as catalyst) and an R_w between 0.5 to 0.05 (large excess of water). The particle diameter can be accurately controlled through the process parameters (temperature, concentrations, etc.). Silica spheres with diameters between tenths of a micron and a micron can be produced (Figure 4-53). Growth of the silica spheres represents a good example for RLMC growth. Small primary particles form by nucleation (▶ glossary) and growth; repeated dissolution and re-precipitation insures that a source of monomers is available, and that reaction-limited conditions exist. Because of the low R_w it is likely, that the growing primary particles are fully hydrolyzed. Larger monosized spherical aggregates are formed with the primary particles as "monomers" (also by RLMC).

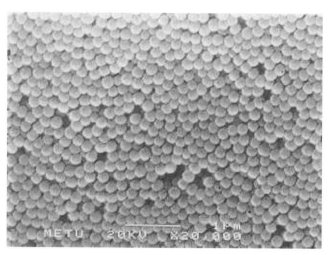

Figure 4-53. Monodisperse silica spheres with a diameter of about 0.2 μm made by the Stöber process.

Aging

The sharp increase in viscosity at the gel point freezes in a particular network structure. Thus, gelation is related to glass-forming processes (see Section 4.1). However, this structure may change appreciably with time, depending on the temperature, solvent or pH conditions, or upon removal of the pore liquid. It is very important to realize that the chemical reactions are not finished with gelation, and structural rearrangements take place in the wet gels (i.e., in the gels still containing their pore liquid). This phenomenon increases the stiffness of the gels and is called "aging". Aging is due to several processes:

- The gel network still contains a continuous liquid phase. The pore liquid initially is a sol, i.e., it contains condensable particles or even monomers (see Figure 4-52), which eventually condense to the existing network. The chemical reactions that cause gelation continue long beyond the gel point. This causes gradual changes in the structure and properties of the gels.
- The gel network originally is still very flexible. This allows neighboring M–OH or M–OR groups to approach each other and to undergo condensation reactions. This causes contraction of the network and expulsion of pore liquid. This spontaneous shrinkage of some gels is called "syneresis". Syneresis continues as long as the gel network exhibits sufficient flexibility. The driving force is the reduction of the large solid–liquid interface in the gels.
- Hydrolysis and condensation reactions are, in principle, reversible. Therefore, mass is dissolved from thermodynamically unfavorable regions, mostly regions with a high positive curvature or small particles. The solutes condense to thermodynamically more favorable regions, particular in pores, crevices, particle necks, etc. This process ("ripening" or "coarsening") results in the reduction of the net curvature, disappearance of small particles, and filling up of small pores (the processes are similar to those discussed in Section 2.1.4 for sintering processes).

Drying

The resulting wet gels obtained by the hydrolysis and condensation processes are sometimes called *aquagels, hydrogels*, or *alcogels*, depending on the solvent in the pores. When the pore liquid is replaced by air without decisively altering the network structure or the volume of the gel body, *aerogels* are obtained. These materials will be discussed in Section 6.3. A *xerogel* is formed by conventional drying of the wet gels, i.e., by temperature increase or pressure decrease, with concomitant large shrinkage (and mostly destruction) of the initially uniform gel body.

The evaporation of the liquid from a wet gel proceeds in a very complex way in which three different stages can be distinguished:

1.   The gel shrinks by the volume that was previously occupied by the liquid. The liquid flows from the interior of the gel body to its surface. If the network is compliant, as it is in alkoxide-derived gels, the gel deforms. Upon shrinkage, OH-groups at the inner surface approach at each other and can react with each other. For example, new siloxane bridges are formed in SiO_2 gels. As drying proceeds, the network becomes increasingly stiffer and the surface tension in the liquid rises correspondingly because the pore radii become smaller.

2.   This stage of the drying process begins when the surface tension is no longer capable of deforming the network and the gel body becomes too stiff for further shrinkage. The tension in the gel becomes so large that the probability of cracking is highest. In this stage of drying, the liquid/gas interface retreats into the gel body. Nevertheless, a contiguous funicular liquid film remains at the pore walls, i.e., most of the liquid still evaporates from the exterior surface of the gel body.

3.   Here, the liquid film is raptured. Eventually, there is only liquid in isolated pockets which can leave the network only by diffusion via the gas phase.

Two processes are important for the collapse of the network. First, the slower shrinkage of the network in the interior of the gel body results in a pressure gradient which causes cracks. Second, larger pores will empty faster than smaller during drying, i.e., if pores with different radii are present, the meniscus of the liquid drops faster in larger pores. The walls between pores of different size are therefore subjected to uneven stress (▶ glossary), and crack (Figure 4-54).

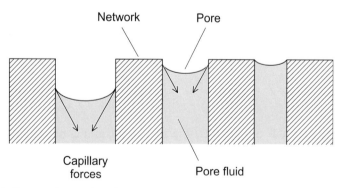

Figure 4-54. Contracting surface forces in pores of different size during drying.

For these reasons, xerogel powders are usually obtained when wet gel bodies are conventionally dried. Although strategies were developed to obtain crack-free xerogel bodies, the large shrinkage cannot be avoided. Due to the shrinkage problem, one of the most important applications of sol–gel materials is for films and coatings, where shrinkage is easier to control.

4.5.3 Sol–Gel Chemistry of Metal Oxides

The fundamental issues of sol–gel processing were mainly investigated for silica-based systems. One of the reasons is that network structures with no long-range order form easily (see also Section 4.1). Nevertheless, any other metal oxide can, in principle, be prepared by sol–gel processing. Although the principles of sol and gel formation, aging, etc. are the same, there are some differences in the chemistry of the precursors.

As in silicate sol–gel processes, inorganic or metal–organic (alkoxide) precursors can be used; an organic modification of the precursors is also possible to some degree. The chemistry of these precursors and ways to influence their reactivity is discussed in the following paragraphs.

Inorganic precursors (metal salts)

Many metal salts are hydrolytically unstable, i.e., they form oxide/hydroxide precipitates from aqueous solutions when the pH is increased. Hydrolysis reactions occur because water molecules coordinated to metal ions are more acidic than in the non-coordinated state due to charge transfer from the oxygen to the metal atom. When the pH is increased (i.e., if a base is added), the series of equilibria already shown in Eq. 4-10 is shifted to the right. Water (aqua) ligands ($M–OH_2$) are thus converted to hydroxo ($M–OH$) and oxo ($M=O$ or $M–O–M$) ligands. The general dependence of the deprotonation equilibria on the charge of the metal and the pH of the solution has already been discussed in Section 4.2.

For example, when aluminum salts are dissolved in water, the cationic metal complex $[Al(H_2O)_6]^{3+}$ exists only below pH 3. As the pH is increased, the water ligands are deprotonated, and the ions $[Al(OH)_x(H_2O)_{6-x}]^{(3-x)+}$ are formed. Mononuclear species with x = 0–4 are only stable in very dilute solutions; at higher concentrations, polynuclear species are formed by condensation reactions, i.e., by formation of Al–O–Al links.

Depending on the kind of bridging group formed between the metal centers, two condensation processes are distinguished:

1. *Olation* is a condensation process in which a *hydroxy* bridge is formed. For coordinatively saturated precursor species, olation proceeds by nucleophilic substitution in which the hydroxy group is the nucleophile and a water molecule is the leaving group. Since water is the leaving group, the kinetics of olation is related to the lability of the aqua ligand. The ability to dissociate from the metal center depends on the size, electronic configuration and Lewis acidity of the metal ion. In general, the smaller the charge and the larger the size of the metal ion, the greater the rate of olation. Several types of OH bridges can be formed by olation as shown in Eq. 4-23.

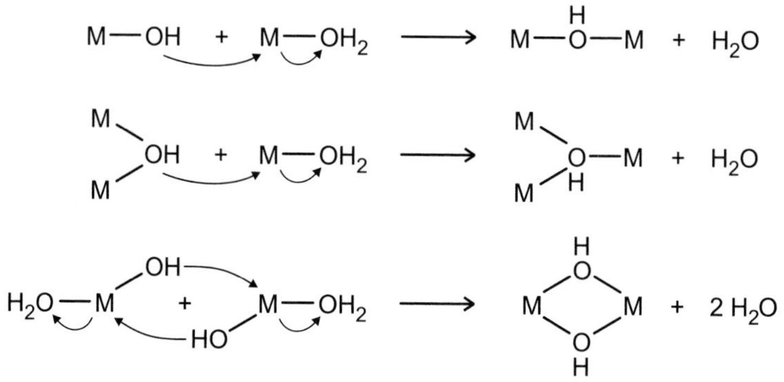

$$(4\text{-}23)$$

2. *Oxolation* is a condensation process in which an *oxo* bridge is formed. If the metal is coordinatively unsaturated, oxolation occurs by rapid nucleophilic addition (Eq. 4-24).

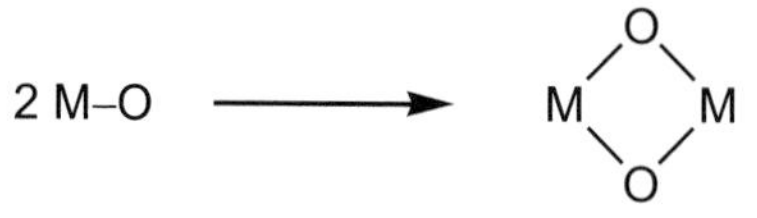

$$(4\text{-}24)$$

Otherwise, oxolation is a two-step addition/elimination process (Eq. 4-25). Under basic conditions, the first step is catalyzed, because a M-OH group is deprotonated and the resulting M–O⁻ is a stronger nucleophile. In the presence of an acid, the second step is catalyzed, because protonation of a terminal M–OH group creates $M–OH_2^+$, and water is a good leaving group. Thus, the rate of oxolation reactions is smallest at the isoelectric point. Owing to catalysis by both H^+ and OH^-, oxolation occurs over a wider pH range than olation. The mechanisms of base and acid catalysis are similar to what has been discussed for silicate systems (Eq. 4-19 and 4-20).

$$M\text{-OH} + M\text{-OH} \longrightarrow$$

(4-25)

Whether gelation or precipitation (Section 4.2) eventually occurs, depends not only on the reaction conditions but also on the reaction kinetics (which is, of course, influenced by the reaction conditions). Gelation is favored by slower reactions.

There is an important stereochemical difference between silicate systems and most transition metal systems. As has been discussed in Section 4.1, all silicate materials are composed of corner-sharing tetrahedra. When a silica network grows, the question that decides on the morphology of the gels obtained is whether condensation occurs preferentially at the end of a chain or at a central atom. In the former case, polymeric gels are obtained, and in the latter case particulate gels. For transition metals this issue is more complicated and hardly understood in detail in most cases. First, transition metals have higher coordination numbers, and the larger polyhedra (octahedra, square antiprisms, etc.) can share not only corners but also (more typically) edges and faces, i.e., there are more options to connect the polyhedra with each other. Second, there are stereoisomeric sites for hydrolysis and condensation. Consider condensation of the octahedral ion $[Al(OH)(H_2O)_5]^{2+}$ by an olation mechanism (Figure 4-55). Once the dimer $[(H_2O)_4Al(OH)_2Al(H_2O)_4]^{4+}$ is formed, there are two stereochemical possibilities for further hydrolysis and condensation: either one of the aqua ligands in the plane of the two bridging OH groups is deprotonated (labeled *) or the aqua ligand *cis* to both bridging OH groups (labeled ⁺). The latter is more acidic. Condensation of another monomer $[Al(OH)(H_2O)_5]^{2+}$ by olation at the former site would eventually result in a ribbon-like structure of edge-sharing octahedra, while condensation at the latter site results in condensed structures.

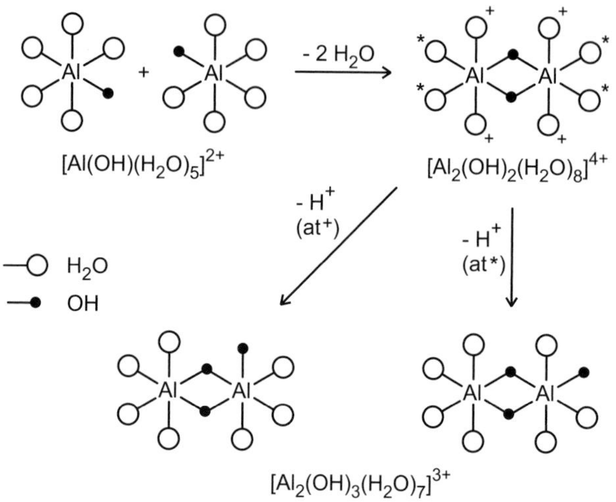

Figure 4-55. Stereochemical possibilities for the condensation of $[Al(OH)(H_2O)_5]^{2+}$ by an olation mechanism.

It should be noted at this point that the previous discussion was somewhat simplified, because the counter-anion (X) of the metal salt precursor (MX_n) was considered not to interact with the metal center during hydrolysis and condensation reactions. In reality, the counter-ion can compete with the aqua ligands for coordination to the metal center. As will be discussed later for the modification of metal alkoxides by carboxylate ligands, an anionic ligand (X) may remain coordinated to the metal through all stages of the overall process and even turn up in the final product. The stability of a M–X bond in a metal complex depends on several electronic and steric factors; a more detailed discussion is beyond the scope of this book (see textbooks on coordination chemistry for details). For example, in basic zirconium salts of the overall composition $Zr(OH)_2X_2(H_2O)_n$ (all of which have oligomeric or polymeric structures) NO_3^-, SO_4^{2-} or HPO_4^{2-} are coordinated to zirconium, while ClO_4^- or Cl^- are not able to displace the coordinated water molecules at the zirconium atom and are not involved in the formation of condensed species. A strong metal-anion interaction (coordination) can influence sol–gel processing in several ways:

- The hydrolysis and condensation reactions may proceed differently when different salts of the same metal are employed.
- Strong coordination of a counter-ion blocks coordination sites and leads to a smaller degree of condensation (fewer M–O–M links per metal atom).
- Coordinated counter-ions can direct the site of nucleophilic attack (*cis* or *trans* to the coordinated X, for example) during hydrolysis and condensation reactions and thus influence the microstructure and morphology of the gels or precipitates.
- The counterions may influence the electric double layer of the sol particles and the ionic strength of the solution, and hence the aggregation of the sol particles.

Alkoxide precursors

There are some important differences between metal alkoxides and $Si(OR)_4$:

- Owing to their lower electronegativity, metals alkoxides are stronger Lewis acids than silicon alkoxides. Nucleophilic attack at the metal is thus facilitated, and the hydrolysis rates are strongly increased. For example, the hydrolysis rate of $Ti(OR)_4$ is about 10^5 times faster than that of $Si(OR)_4$ with the same alkoxide substituents.
- Most metals have several stable coordination numbers, or the expansion of the coordination sphere in transition states is easier.

Owing to both effects, the reactivity of some tetravalent iso-propoxides in hydrolysis reactions increases in the order:

$$Si(O^iPr)_4 <<< Sn(O^iPr)_4, Ti(O^iPr)_4 < Zr(O^iPr)_4 < Ce(O^iPr)_4.$$

The reactivity of many metal alkoxides towards water is so high that precipitates are formed spontaneously. While the reactivity of alkoxysilanes has to be promoted by catalysts, the reaction rates of metal alkoxides must be moderated to obtain gels instead of precipitates.

Hydrolysis of metal alkoxides occurs by an addition/elimination mechanism (Eq. 4-26). The mechanisms of the condensation reactions are very similar to what was discussed before for inorganic precursors. Both oxolation (Eq. 4-25) and olation (Eq. 4-23) are possible; a ROH molecule may be the leaving group instead of H_2O. Oxolation with elimination of an alcohol is also called alcoxolation (Eq. 4-27).

$$\begin{array}{c}
\text{H}\diagdown \\
\quad\text{O} \\
\text{H}\diagup
\end{array}\quad +\quad \text{M–OR}\quad\longrightarrow\quad
\begin{array}{c}
\text{H}\diagdown \\
\quad\text{O} \longrightarrow \text{M–OR} \\
\text{H}\diagup
\end{array}$$

(4-26)

$$\longrightarrow\quad \text{HO–M}\longleftarrow
\begin{array}{c}
\text{H} \\
\text{O}\diagup \\
\quad\diagdown\text{R}
\end{array}\quad\longrightarrow\quad \text{M–OH} + \text{ROH}$$

$$\text{M–OH} + \text{M–OR}\quad\longrightarrow\quad
\begin{array}{c}
\text{H} \\
\text{O} \\
\text{M}\diagup\quad\diagdown\text{M}\diagup\text{OR}
\end{array}
\quad\longrightarrow\quad
\begin{array}{c}
\text{O} \\
\text{M}\diagup\quad\diagdown\text{M}
\end{array}
+ \text{ROH}$$

(4-27)

In principle, the same parameters influence the rates of the hydrolysis and condensation reactions of metal alkoxides for a given metal as those already discussed for silicon alkoxides. An additional factor is the degree of oligomerization of the alkoxide precursors. Silicon alkoxides are always monomeric, while metal alkoxides may be associated via μ_2- or μ_3-OR bridges (the subscript denotes the number of metal atoms coordinated to the bridging ligand). The reason for the association is that the metals do not reach their full coordination number by the terminal alkoxide groups. For example, the usual coordination number of titanium is six, but there are only four alkoxide ligands in $Ti(OR)_4$. When neat or dissolved in non-polar solvents, coordination expansion occurs by association via OR bridges.

The degree of association depends on:

- the size of the metal: the tendency to oligomerize increases with the size of the metal. For example, the average degree of oligomerization in $M(OEt)_4$ (the oligomerization processes are equilibria) is 2.9 for M = Ti, 3.6 for Zr, 3.6 for Hf, and 6.0 for Th.
- the size of the groups R: larger groups R favor smaller units because of steric hindrance. For example, $Ti(OEt)_4$ in ethanol has a trimeric structure, while $Ti(OPr^i)_4$ is monomeric in propanol solution.

The degree of oligomerization influences not only the solubility of the metal alkoxides (highly associated oligomers may even be insoluble) but also the reaction kinetics. Coordinatively unsaturated species have a higher reactivity in hydrolysis and condensation reactions. For example, the monomeric species $Ti(OPr^i)_4$ is hydrolyzed more rapidly than trimeric $Ti(OEt)_4$ despite the larger alkoxide ligands. When the degree of oligomerization of two metal alkoxides

$M(OR)_n$ with different groups R is the same, then the size of R has the same influence on the reaction kinetics, as was discussed for silicon alkoxides.

Association via alkoxide bridges is not the only way for coordination expansion. When the alkoxides are dissolved in polar solvents such as alcohols, addition of solvent molecules may occur and compete with association. For example, hydrolysis of $Zr(OPr^n)_4$ in PrOH is very fast and results in a precipitate, while hydrolysis of the same alkoxide in cyclohexane gives a homogeneous gel. The explanation of this difference is that OR-bridged oligomeric species are formed in cyclohexane solution, while an alcohol molecule is coordinated to $Zr(OPr^n)_4$ in propanol solution. Upon hydrolysis, the coordinated ROH is more easily cleaved than the OR bridge.

Neither association to oligomeric species nor solvate formation is observed for silicon alkoxides.

The high reactivity of metal alkoxides can be moderated by their chemical modification. The general approach is to replace one or more alkoxide ligands by groups which are less easily hydrolyzed (i.e., which are more strongly bonded to the metal) and additionally block coordination sites at the metal. The most common ligands are carboxylate (Eq. 4-28) or β-diketonate groups (Eq. 4-29), but any other anionic bidentate ligand (▶ glossary) can, in principle, be used.

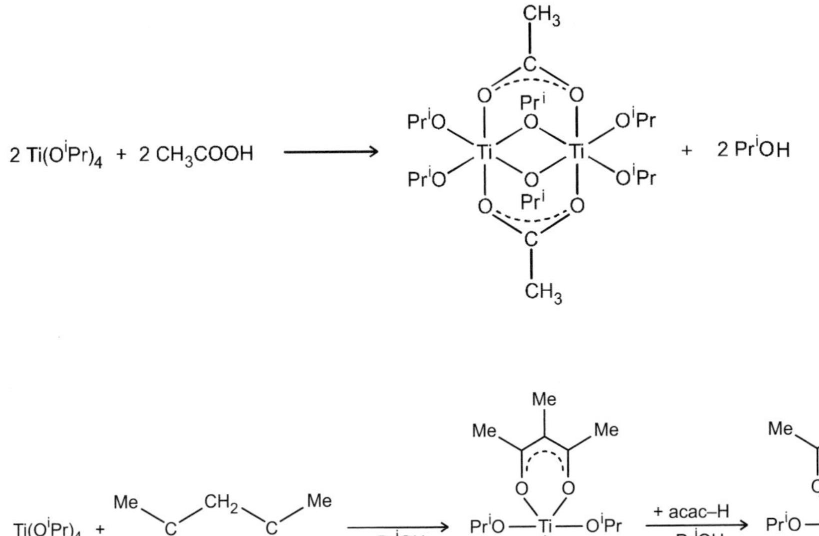

$$2\ Ti(O^iPr)_4\ +\ 2\ CH_3COOH \longrightarrow \qquad +\quad 2\ Pr^iOH$$

(4-28)

$$Ti(O^iPr)_4\ + \qquad \xrightarrow[-\ Pr^iOH]{} \qquad \xrightarrow[-\ Pr^iOH]{+\ acac–H}$$

(acac–H)

(4-29)

The bidentate ligands can be chelating (▶ glossary; i.e., bonded to the same metal, Eq. 4-29) or bridging (Eq. 4-28). Bidentate ligands are inherently more stronger bonded than comparable monodentate ligands and therefore are less readily hydrolyzed than the remaining OR groups. The bidentate ligands are introduced by substitution reactions, as shown in Eqs. 4-28 and 4-29.

The substitution of OR ligands by anionic bidentate ligands has several consequences:

- A new precursor is created in which the remaining alkoxide groups have a different reactivity in hydrolysis and condensation reactions than the parent alkoxide. The situation is comparable to alkoxysilanes: $Si(OMe)_4$ has a different reactivity than $MeSi(OMe)_3$, for example.
- The substitution of a mono-anionic monodentate ligand (OR^-) by a mono-anionic bidentate ligand (▶ glossary) maintains the charge balance within the precursor. However, an extra coordination site is blocked. This additionally reduces the reactivity in hydrolysis and condensation reactions but also introduces building blocks with a different connectivity. The bidentate ligand may stereochemically direct the hydrolysis or condensation site as discussed before for coordinated counter-ions (Figure 4-55). The formation of gels instead of precipitates is facilitated by the lower degree of crosslinking of the network. Furthermore, the microstructure of the materials is influenced.
- Functional or non-functional organic groups are introduced by the bidentate ligands (see Section 4.5.4).

4.5.4 Inorganic–Organic Hybrid Materials

One of the major advances of sol–gel processing is undoubtedly the possibility of synthesizing hybrid inorganic–organic materials. These are called *hybrid* materials because they combine, to some extent, the properties of inorganic and organic compound in *one* material. The idea for developing such materials is similar to that of composite materials (▶ glossary), where two or more materials are combined that differ in form and (mostly) chemical composition. While macroscopic constituents with defined phase boundaries are used for composite materials (▶ glossary), molecular *building blocks* of different composition (inorganic and organic) are instead combined in inorganic–organic hybrid materials.

Sol–gel processing is the only way to make such materials because of the mild processing conditions. The high-temperature synthesis route to ceramic materials, for example, does not allow the incorporation of thermally labile organic moieties. Inorganic–organic hybrid materials of another type and structure have been referred to previously in Section 2.4 (intercalation of inorganic layered structures by organic molecules or polymers).

The inorganic–organic hybrid materials made by sol–gel processing have been called ORMOSILs (*or*ganically *mo*dified *sil*icates), ORMOCERs (*orga*nically *mo*dified *cer*amics), CERAMERs (*cera*mic poly*mers*), or POLYCER-AMs (*poly*meric *ceram*ics). As these names already imply, these hybrid materials are composed of building blocks found in inorganic oxidic glasses or ceramics, silicones, or organic polymers (Figure 4-56). The concept is to combine properties of organic groups and polymers (functionalization, ease of processing at low temperatures, toughness) with properties of glass- or ceramic-like materials (hardness (▶ glossary), chemical and thermal stability) in order to generate new and synergetic properties not accessible otherwise.

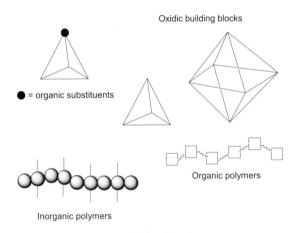

Figure 4-56. A selection of often used building blocks.

There is a wide range of possibilities to vary the composition and structure, and thus the properties of the materials:

- chemical composition of the organic and inorganic precursors;
- ratio of the inorganic to organic components;
- structure of the building blocks; and
- distribution of the building blocks (statistical, block-like, etc.).

Two different approaches can be used for the incorporation of organic groups into an inorganic network, namely embedding of organic molecules into gels without chemical bonding, and incorporation of organic molecules via covalent bonding.

Embedding of organic molecules

This is achieved by dissolving the molecules or dispersing the particles in the precursor solution. The gel matrix is formed around them and traps them. A variety of organic or organometallic molecules can be employed, such as dyes, catalytically active metal complexes, sensor compounds, or biomolecules (Figure 4-57, left). If sol–gel processing of alkoxides is performed in the solution of an organic polymer, the inorganic network (formed by sol–gel processing) and the organic network interpenetrate but are not bonded to each other (Figure 4-57, right).

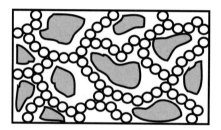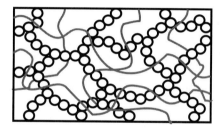

Figure 4-57. Embedding of molecules or particles (left) into an inorganic gel matrix; interpenetrating inorganic–organic polymer networks not covalently connected with each other (right). The pearls-on-a-string features represent the inorganic (sol–gel-de-rived) network.

An interesting application for this type of sol–gel materials are biosensors. A biosensor is a sensor in which a biological species, e.g., an enzyme or antibody, is used to detect certain other species. Biomolecules are very sensitive molecules which often react by denaturation to changes of the pH, higher temperatures, and organic solvents. The requirements necessary for the processing of biomolecules seem to be in disagreement to the typical synthesis conditions of inorganic materials.

However, the mild reaction conditions during sol–gel processing, particularly the low temperatures and the possibility of using buffered aqueous solutions, allow the entrapment of enzymes and other biological species such as antibodies (or even whole cells) in sol–gel materials. The biomolecule is stabilized against denaturation due to the adsorbed water on the hydrophilic surface inside the gel. Additionally, the fine porosity of the gel protects the biomolecule against influences from outside, but permits nutrients and reactants to penetrate.

Stabilization of entrapped molecules is a general phenomenon which is also observed for sensing molecules or catalytically active species.

Incorporation of organic groups via covalent bonding

Very important modifications of sol–gel materials are based on a covalent linkage of the organic groups (Figure 4-58). In silicate systems it is possible to use organotrialkoxysilanes [R'Si(OR)$_3$], diorganodialkoxysilanes [R'$_2$Si(OR)$_2$] or even (RO)$_3$Si–R'–Si(OR)$_3$ as precursors for sol–gel processing in the same way as tetraalkoxysilanes. In the organically substituted derivatives, the group R' is bonded through a Si–C link to the network-forming inorganic part of the molecule (Eq. 4-30). Since Si–C bonds are hydrolytically stable, the organic groups are retained in the final material.

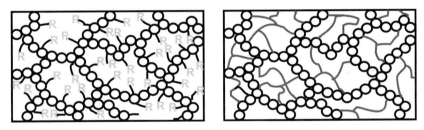

Figure 4-58. Incorporation of organic groups by covalent linkage into an inorganic gel matrix (left); dual inorganic–organic hybrid polymer networks connected with each other by covalent bonds (right). The pearls-on-a-string features represent the inorganic (sol–gel-derived) network.

$$R'Si(OR)_3 + {}^3/_2\,H_2O \longrightarrow R'SiO_{3/2} + 3\,ROH \qquad (4\text{-}30)$$

The (partial) replacement of Si(OR)$_4$ by R'Si(OR)$_3$ in the precursor mixture has several consequences (which are similar as already discussed for metal alkoxides substituted by bidentate ligands; ▶ glossary):

- the crosslinking density of the inorganic network is decreased;
- the organically substituted trialkoxysilane exhibits a different hydrolysis and condensation behavior;
- the inorganic network is modified and/or functionalized; and
- organic substituents change the polarity of the system.

Connectivity. The connection of [SiO$_4$] tetrahedral by four Si–O–Si bonds per silicon allows the formation of a three-dimensional network (see quartz or silica). Each substitution of an oxygen atom by a carbon atom lowers the cross-linking density of the gel network. For organo*di*alkoxysilanes, only linear chain-like products can be obtained, as in silicones (see Section 5.2). For orga-no*tri*alkoxysilanes it is still difficult to obtain three-dimensional gel networks. Sometimes only oligomeric cage compounds are obtained, such as the so-called POSS (polyhedral oligomeric silsesquioxanes). Some examples of POSS are shown in Figure 4-59. Similar units can also be formed during sol–gel processing of tetralkoxysilanes, with X = OH or OR.

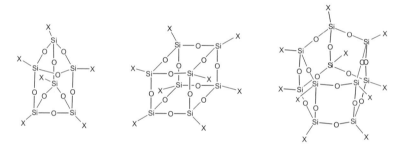

Figure 4-59. Molecular structures of (XSiO$_{1.5}$)$_n$ cages for n = 6, 8, 10; X = R, H, OH, etc.

Because of the difficulties in obtaining three-dimensional gel networks from solutions of organically substituted alkoxysilanes, they are typically used as mixtures with the corresponding tetraalkoxysilane. The mechanical properties of the resulting materials are strongly influenced by the degree of crosslinking (▶ glossary) between the building blocks (i.e., by the portion of [SiO$_4$] and [RSiO$_3$] units) as discussed for silicones in Section 5.2.

Hydrolysis and condensation behavior. When (organically) modified alkoxides are reacted instead of the parent tetraalkoxides, the hydrolytically stable groups R change the reactivity of the silicon atom to which they are directly bonded. A general rule is that bulkier organic substituents result in slower sol–

gel reactions. However, the electronic effects of the organic group are more important. As discussed in Section 4.5.2, the electron density at the central silicon atom is typically increased due to the presence of the organic substituent (+I-effect).

If only the organotri- and dialkoxysilanes are employed as precursors, these factors just influence the overall rate of the hydrolysis and condensation reactions. However, if mixtures of the organically substituted alkoxysilanes and tetraalkoxysilanes are used, the reaction conditions must be chosen very carefully to obtain homogeneous materials and avoid phase separation. Due to the different mechanisms and transition states, the organically substituted compounds react faster than the tetraalkoxysilane in acid-catalyzed systems and slower in base-catalyzed systems (Figure 4-60).

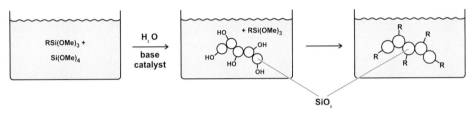

Figure 4-60. When a $Si(OMe)_4/R'Si(OMe)_3$ mixture is reacted under base-catalyzed conditions, the gel network is predominantly formed from $Si(OR)_4$ because it reacts faster. The $R'SiO_{3/2}$ units then condense onto the network.

Under basic conditions, the inorganic network is formed from the faster-reacting tetramethoxysilane. Only in a later stage of the reaction does the organically substituted trimethoxysilane condense onto the surface of the pre-formed silica network, which results in an *in-situ* modification of the gel network. Performing the same reaction in an acidic medium would reverse everything, i.e., the organotrimethoxysilane will react first, followed by the tetramethoxysilane. Therefore, different materials will be obtained by working in solutions of different pH.

Modification and functionalization. The choice of the organic group R for modification of the inorganic network is nearly unlimited. Table 4-6 lists some typical organotrialkoxysilanes, their properties, and some potential applications. Their synthesis will be discussed in Section 5.2. Diorganodialkoxysilanes $[R'_2Si(OR)_2]$ are used typically in the silicone industry (see Section 5.2).

Table 4-6. Some organofunctional trialkoxysilanes used for the preparation of inorganic–organic hybrid materials by sol–gel processing, and their function.

$R'Si(OR)_3$	Function
	polymerizable group for the preparation of hybrid polymers
	crosslinking (▶ glossary) sites
	group for organic polyaddition reactions; generation of hydrophilic diols by opening of the epoxide ring
	hydrophilicity, coupling sites, coordination sites for metals
	hydrophobicity
	crosslinking site in thiol-ene UV cure systems, coordination site for metals
	chromophoric substituent (non-linear optic NLO dye (▶ glossary)
	fluorescent substituent (pyrene derivative)
	coordination of transition metal complexes, catalysis

While Si–C bonds are hydrolytically stable and thus allow to introduce organic substituents into the gel materials, metal–carbon bonds in transition metals are usually cleaved by water. However, the organic groups can be introduced via the bidentate ligands (▶ glossary; Eqs. 4-28 and 4-29) already discussed in Section 4.5.3. Usually, these complexing ligands are used to moderate the reactivity of the alkoxide precursors, but they can also be used for an organic functionalization. Figure 4-61 shows some examples for organically modified transition metal-based precursors. In most cases these compounds are not monomeric, but oligomeric.

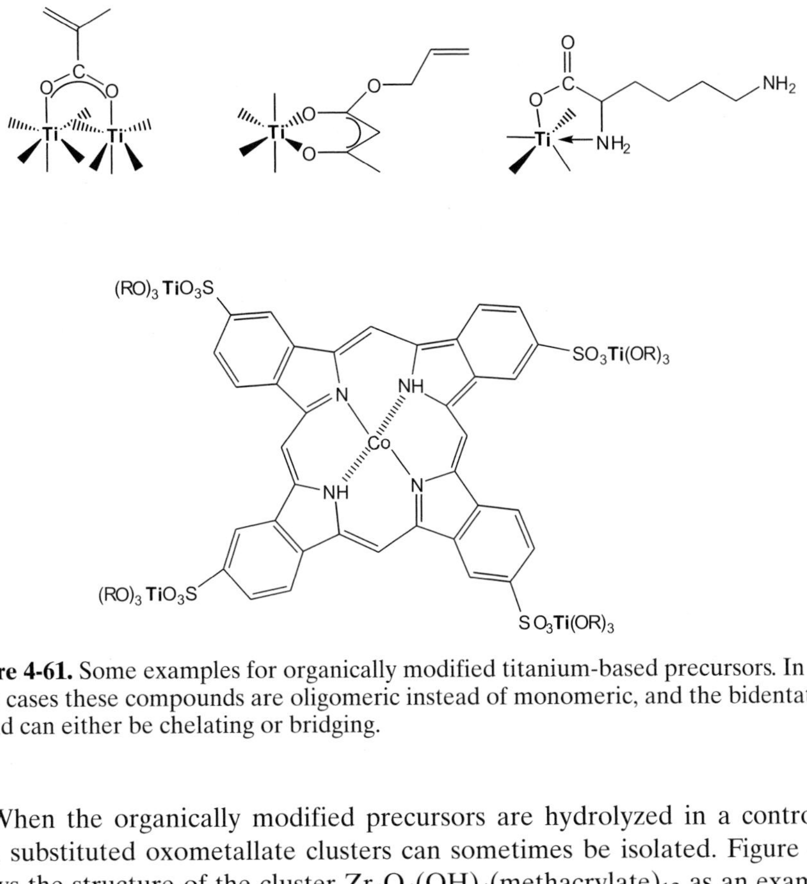

Figure 4-61. Some examples for organically modified titanium-based precursors. In most cases these compounds are oligomeric instead of monomeric, and the bidentate ligand can either be chelating or bridging.

When the organically modified precursors are hydrolyzed in a controlled way, substituted oxometallate clusters can sometimes be isolated. Figure 4-62 shows the structure of the cluster $Zr_4O_4(OH)_4(methacrylate)_{12}$ as an example, which was obtained by controlled hydrolysis of methacrylate-substituted $Zr(OPr)_4$. The cluster consists of a highly condensed $Zr_6O_4(OH)_4$ cluster core, the surface of which is covered by the organic ligands. Clusters of this type are the transition-metal analogues of the $(XSiO_{1.5})_n$ cages (Figure 4-59) in silicate chemistry. Both show in which way the primary units are connected to form three-dimensional structures. In silicate chemistry, the cages are formed by corner-sharing of tetrahedral units, in the cluster shown in Figure 4-62, square anti-prismatic units are condensed via shared faces.

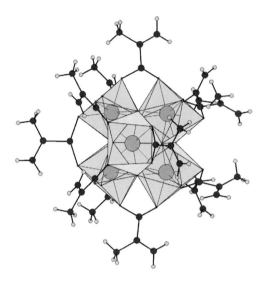

Figure 4-62. The molecular structure of $Zr_6O_4(OH)_4(methacrylate)_{12}$. The methacrylate ligands wrap the highly condensed cluster core consisting of six face-sharing $[ZrO_8]$ square antiprisms. The sixth polyhedron cannot be seen in this perspective.

These examples show that the modification of inorganic gel networks by covalent linkage of all kinds of organic groups (left side of Figure 4-58) offers a huge variety of possibilities to modify or functionalize sol–gel materials.

Inorganic–organic hybrid polymers

A related possibility is the formation of dual inorganic–organic hybrid polymer networks (Figure 4-58, right), in which cluster or polymer-type inorganic structures are linked by organic groups or polymer fragments. There are three principal different approaches to prepare such materials (Figure 4-63):

1.   Formation of hybrid polymers from compounds of the type $[(RO)_nM]_xY$, in which Y is an organic group linking two ($x = 2$) or more ($x > 2$) metal alkoxide units. The structure of the preformed organic building block is retained in the final material. For example, organic groups of variable length (e.g., saturated or unsaturated hydrocarbon chains, or polyaryls)

substituted with Si(OR)₃ groups at both ends or polymers with grafted
Si(OR)₃ groups can be used (Figure 4-64).

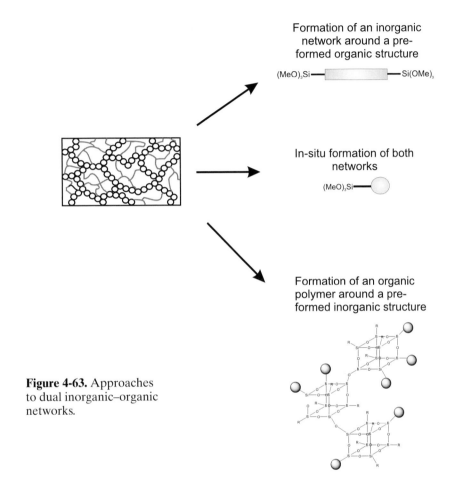

Formation of an inorganic
network around a pre-
formed organic structure

$(MeO)_3Si$ ——▭—— $Si(OMe)_3$

In-situ formation of both
networks

$(MeO)_3Si$ ——◯

Formation of an organic
polymer around a pre-
formed inorganic structure

Figure 4-63. Approaches
to dual inorganic–organic
networks.

2. Formation of hybrid polymers from functionalized inorganic building
    blocks. Here, the preformed inorganic structures are crosslinked by po-
    lymerization (▶ glossary) reactions of the organic functions. For exam-
    ple, the methacrylate substituted cluster shown in Figure 4-62 or vinyl-
    modified POSS (Figure 4-59, X = vinyl) can be used.
3. Formation of hybrid polymers from bifunctional molecular precursors
    (RO)ₙM–X–A bearing an inorganic (RO)ₙM, and an organic (A) func-
    tionality. The latter must be capable of undergoing polymerization or
    crosslinking (▶ glossary) reactions. Usually, these precursors are

reacted first with water. Formation of hybrid polymers from bifunctional molecular precursors $(RO)_nM–X–A$ bearing an inorganic
$(RO)_nM$, and an organic (A) functionality. The latter must be capable
of undergoing polymerization or crosslinking (▶ glossary) reactions.
Usually, these precursors are reacted first with water.

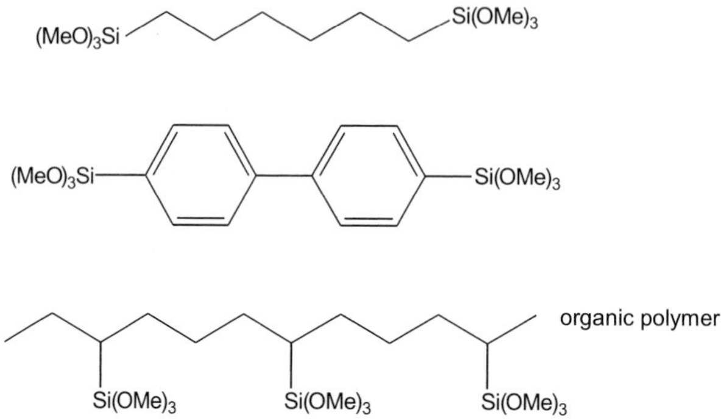

Figure 4-64. Precursors with preformed organic linkages.

Formation of hybrid polymers from bifunctional molecular precursors
$(RO)_nM–X–A$ bearing an inorganic $(RO)_nM$, and an organic (A) functionality.
The latter must be capable of undergoing polymerization or crosslinking (▶
glossary) reactions. Usually, these precursors are reacted first with water.

An example for a material prepared by the latter approach is shown in Figure 4-65. A scratch-resistant coating with good adhesion on polymeric organic
substrates is obtained by sol–gel processing of a mixture of vinyl and mercapto
substituted trialkoxysilanes (see Table 4-6). A sol is first formed by reaction of
the silane mixture with water. After coating, the film is photochemically cured.
In this step an organic link is formed by addition of the SH group to the double
bond (Eq. 4-31).

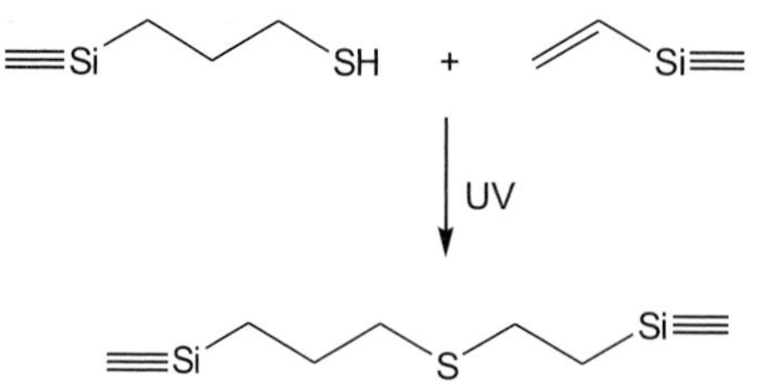

(4-31)

Figure 4-65. Scratch-resistant coating on polycarbonate (left half of slide coated, right half uncoated).

The methacrylate-substituted silane $(MeO)_3Si(CH_2)_3OC(O)CMe=CH_2$ is often used for this purpose. After sol–gel processing, the organic network is formed by polymerization (▶ glossary) of the methacrylate substituents.

Films and coatings represent the earliest commercial applications of sol–gel materials, since most of the disadvantages of the sol–gel technology such as large shrinkage, high costs of the starting materials, etc. can be overcome or are less important for thin films. Typical applications for inorganic–organic hybrid sol–gel materials are also coatings or films. For example, they are used for the protection of surfaces (e.g., against corrosion, abrasion, scratching, etc.), for optical applications (e.g., absorption, emission, or reflection of radiation), as chemically active layers (e.g., for sensing or catalysis), as membranes, as diffusion barriers, or to modify the surface properties.

The special properties of a surface originate from the fact, that atoms on a surface have a higher energy than those within the bulk. For a liquid on the surface of a substrate typically three different interfaces and thus three different interfacial energies have to be distinguished: γ_{SV}, the energy of the surface–vapor interface, γ_{SL}, the energy of the surface–liquid interface; and γ_{LV}, the energy of the liquid–vapor interface. If a layer of liquid spreads on a smooth and planar solid surface, the change in energy is given by (Eq. 4-32):

$$\Delta E = \gamma_{SL} + \gamma_{LV} - \gamma_{SV} \tag{4-32}$$

If $\Delta E < 0$, the energy of the system is reduced so the liquid will spread spontaneously (Figure 4-66a); otherwise, the solid–liquid–vapor interface will be characterized by a contact angle, Θ (Figure 4-66b and c) The balance of tensions at the point of intersection leads to a relationship between the surface tensions that is known as Youngs equation (Eq. 4-33):

$$\gamma_{SV} = \gamma_{SL} + \gamma_{LV} \cos\Theta \tag{4-33}$$

If $\Theta \cong 0$ (as in Figure 4-66 a), the liquid spreads easily and the solid will be covered with a liquid film. If $\Theta > 0$, the liquid does not really wet the surface and the contact angle is determined by the balance of forces at the intersection of solid–vapor and liquid–vapor interfaces.

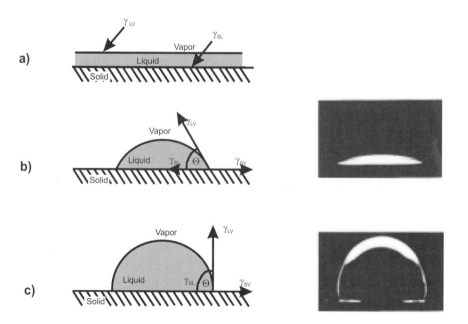

Figure 4-66. Different wetting behavior on a surface: a) the liquid spreads easily over the surface; b) the liquid wets the surface partly; and c) the liquid does not wet the surface. Corresponding to the drawings, a water droplet on a hydrophilic (b) and a hydrophobic surface (c) is shown.

Surfaces can be modified deliberately to control their polarity. This allows not only an achievement of hydrophobicity/hydrophilicity, but even oleophobicity/oleophilicity can be obtained for coatings for antisoiling and antifogging applications.

Imagine you are asked to modify a surface or substrate to render it scratch-resistant, and at the same time insure that it is not wetted by organic solvents. According to the building block approach, it should be possible to add an oleophobizing component to the precursor solution used for making scratch resistant coatings (see above). This is indeed possible. The antisoiling coating shown in Figure 4-67 was prepared by adding fluoro-substituted alkoxysilanes (see Table 4-6) to the mixture of vinyl and mercapto-substituted trialkoxysilanes used for the scratch-resistant coating in Figure 4-65.

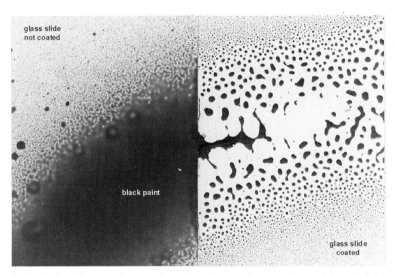

Figure 4-67. Antisoiling coating on glass; the picture shows a glass slide, of which only the right half is coated. The coated part is not wetted by the sprayed paint.

4.6 Further Reading

L. Addadi, S. Weiner, "Control and design principles in biomineralization" *Angew. Chem. Int. Ed. Engl.* **31** (1992) 153–169.

C. J. Brinker, G. W. Scherer, *Sol–Gel Science: The Physics and Chemistry of Sol–Gel Processing*, Academic Press, Boston, 1990.

D. Evered, S. Harnett (Eds.), *Cell and Molecular Biology of Vertebrate Hard Tissue* (Ciba Foundation Symposium 136), Wiley, Chichester, 1988.

J. H. Fendler, F. C. Meldrum, "The colloid chemical approach to nanostructured materials", *Adv. Mater.* **7** (1995) 607–632.

L. L. Hench, J. K. West, "The sol–gel process", *Chem. Rev.* **90** (1990) 33–72.

R. K. Iler, *The Chemistry of Silica*, Wiley, New York, 1979.

W. L. Johnson, "Bulk glass-forming metallic alloys: science and technology", *Mater. Res. Soc. Bull.* **24**(10) (1999) 42–56.

L. C. Klein (Ed.), *Sol–Gel Technology for Thin Films, Fibers, Preforms, Electronics and Specialty Shapes*, Noyes Publishers, Park Ridge, 1988.

G. Krampitz, G. Graser, "Molecular mechanisms of biomineralization in the formation of calcareous shells", *Angew. Chem. Int. Ed. Engl.* **27** (1988) 1145–1160.

R. A. Laudise, E. D. Kolb, "Hydrothermal synthesis of single crystals", *Endeavour* **28** (1969) 114–117.

J. Livage, M. Henry, C. Sanchez, "Sol–gel chemistry of transition metal oxides", *Progr. Solid State Chem.* **18** (1988) 259–341.

J. Livage, C. Sanchez, F. Babonneau, "Molecular precursor routes to inorganic solids" in *Chemistry of Advanced Materials – an Overview*, Wiley-VCH, New York, 1998, Eds. L. V. Interrante, M. Hampden-Smith, pp. 389–448.

S. Mann, *Biomimetic Materials Chemistry*, Wiley-VCH, Weinheim, 1996.

S. Mann "Biogenic inorganic materials" in *Inorganic Materials*, Wiley, 1992, Eds. D. W. Bruce, D. O'Hare, pp. 238–294.

S. Mann, J. Webb, R. J. P. Williams (Eds.), *Biomineralization, Chemical and Biochemical Perspectives*", VCH, Weinheim, 1989.

E. Matijevic, "Monodispersed metal (hydrous) oxides – a fascinating field of colloid science", *Acc. Chem. Res.* **14** (1981) 22–29.

C. C. Perry "Biomaterials" in *Chemistry of Advanced Materials: An Overview*, Wiley-VCH, New York, 1998, Eds. L. V. Interrante, M. J. Hampden-Smith, pp. 499–562.

A. C. Pierre, *Introduction to Sol–Gel Processing*, Kluwer, Boston, 1998.

A. Rabenau, "The role of hydrothermal synthesis in preparative chemistry" *Angew. Chem. Int. Ed. Engl.* **97** (1985) 1026–1040.

R. Reisfeld, C. K. Jorgenson (Eds.), *Chemistry, Spectroscopy, and Applications of Sol–Gel Glasses*, Springer, Berlin, 1992 (*Struct. Bonding*, Vol. 77).

C. Sanchez, J. Livage, "Sol–gel chemistry from metal alkoxide precursors", *New J. Chem.* **14** (1990) 513–521.

U. Schubert, N. Hüsing, A. Lorenz, "Hybrid inorganic–organic materials by sol–gel processing of organofunctional metal alkoxides", *Chem. Mater.* **7** (1995) 2010–2027.

J. E. Shelby, *Introduction to Glass Science and Technology*, The Royal Society of Chemistry, Cambridge, 1997.

K. Simkiss, K. M. Wilbur, *Biomineralization*, Academic Press, San Diego, 1989.

J. Stávek, M. Šípek, I. Hirasawa, K. Toyokura, "Controlled double-jet precipitation of sparingly soluble salts. A method for the preparation of high added value materials", *Chem. Mater.* **4** (1992) 545–555.

T. Sugimoto, "Preparation of monodispersed colloidal particles", *Adv. Colloid Interf. Sci.* **28** (1987) 65–108.

C. Suryanarayana (Ed.), *Non-Equilibrium Processing of Materials*, Pergamon, Amsterdam, 1999.

W. Vogel, *Glaschemie*, Springer, Berlin, 1992.

J. Wen, G. L. Wilkes, "Organic/inorganic hybrid network materials by the sol–gel approach", *Chem. Mater.* **8** (1996) 1667–1681.

K. T. Wilke, *Kristallzüchtung*, VEB Deutscher Verlag, Berlin, 1973.

J. Zarzycki (Ed.), *Materials Science and Technology, Vol. 9. Glasses and Amorphous Materials*, VCH, Weinheim, 1991.

5 Preparation and Modification of Inorganic Polymers

5.1 General Aspects

Silicones (polysiloxanes, Section 5.2) are the commercially most successful types of "inorganic polymers", i.e., polymers that have a backbone containing other elements than carbon, nitrogen, and oxygen, to which organic or organometallic substituents are attached. Other inorganic polymers (Figure 5-1) with a high commercial potential as specialty polymers are polyphosphazenes (Section 5.3) and polysilanes (Section 5.4). The technical interest in polycarbosilanes (Section 5.5) and polysilazanes (Section 5.6) is their use as *preceramic polymers*, i.e., polymers that give ceramic materials upon pyrolysis. This option is of great interest, because the use of polymer technology offers certain processing advantages over conventional ceramic processing routes (see Section 2.1.4). New materials are needed in all fields of modern technology. Because of the almost limitless combination of elements, there should be other inorganic polymers with useful properties. Section 5.7 will provide an outlook on polymers with other inorganic backbones, which may turn out to have some useful properties.

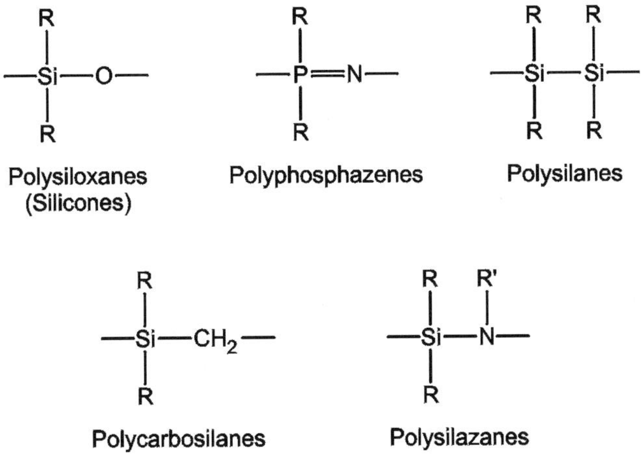

Figure 5-1. The repeat units in silicones, polyphosphazenes, polysilanes, polycarbosilanes, and polysilazanes.

Polymers are chemically characterized by the sequence of atoms forming the main chain ("backbone"), such as $\cdots$Si–O–Si–O$\cdots$ in silicones, and by the nature of the groups ("side groups") attached to (all or some) atoms of the backbone (the groups R and R' in Figure 5-1). The main chain may be branched or crosslinked.

Before turning to the chemistry of the particular classes of inorganic polymers (Sections 5.2–5.7), some common aspects will be discussed.

There is a variety of methods for the preparation of *organic* polymers, such as polymerization ($\blacktriangleright$ glossary) of multiple bonds, polyaddition, or polycondensation reactions ($\blacktriangleright$ glossary). Since there are no thermodynamically stable double bonds for the elements of the third and higher periods, preparation methods involving double bonds are not applicable. There are only two general methods to make the backbone of inorganic polymers:

- *Polycondensation reactions* (Eq. 5-1)

$$\text{Y–A–B–X} \longrightarrow \text{[A–B–]}_n + \text{X–Y} \tag{5-1}$$

where A and B are the elements forming the inorganic backbone, and X and Y are the eliminated groups. The spectrum of X–Y ranges from H_2O or HCl to Me_3SiOR, H_2 and NaCl. A major drawback of polycondensation reactions is that small cyclic oligomers (mainly six- and eight-membered rings) are thermodynamically more favorable than polymers. One major reason for this is the steric hindrance generated by the side groups. Irrespective of the chain conformation of the polymer, the substituents are always closer to their neighbors along a chain than they are in a small cyclic oligomer. The polycondensation reactions must therefore be performed under kinetically controlled conditions to obtain polymers instead of cyclic oligomers.

- *Ring-opening polymerization* (ROP) (Eq. 5-2)

$$\tag{5-2}$$

Although rings are thermodynamically more favorable, there are methods to open the rings and attach the segments to each other. The obtained polymer may be stable at temperatures where the rate of depolymerization to the cyclic oligomers is very slow (although it would depolymerize when heated to higher temperatures). The cyclic oligomers will not ring-open and polymerize when the repulsion of the substituents is too large.

Examples for both types of reactions will be given in the following sections. Substitution, or chemical modification of the substituents are other options to modify the properties of inorganic polymers.

5.1.1 Polymeric Materials

The main interest in inorganic polymers is that they may combine the advantages and minimize the disadvantages of biological or carbon-based polymers and inorganic materials. Organic polymers are light, tough, and easy to fabricate, but in general they cannot be exposed to high temperatures, strongly oxidizing conditions, or high-energy radiation.

In inorganic polymers, combinations of properties can be achieved and tuned by the combination of the properties of the inorganic backbone and the organic substituents. The backbone of inorganic polymers can provide heat-, fire-, or radiation-resistance, or electrical conductivity, as well as flexibility. The substituents control properties such as solubility, liquid crystallinity (▶ glossary), optical properties, hydrophobicity or hydrophilicity, adhesion or biological compatibility. They also provide possibilities for generating crosslinks between the polymer chains (see below).

The question whether a polymer is an oil, an elastomer or a grease, etc., depends not only on the nature of its repeating unit(s) (backbone plus side groups), but also on the three-dimensional structure of the macromolecule, the degree of polymerization (its molecular mass), the degree of branching, and its crystallinity. Many polymers are amorphous, i.e., their three-dimensional structure exhibits no long-range order as in a crystal. However, a polymer may partially crystallize by packing of chains into a stereoregular (ordered) arrangement.

The flexibility of the main chain determines the degree of either random coiling or "rigid rod" character in solution. Highly flexible, long-chain molecules become entangled in solution and markedly increase the viscosity. In the solid state, non-crosslinked macromolecules interact strongly with their neighbors, either through entanglements or through the formation of ordered (micro)crystalline domains owing to interchain packing. The presence of branched structures and any molecular feature that eliminates symmetry or linear order disfavors microcrystallite formation that have a profound influence on materials properties.

An important characteristic of any polymeric material is the temperature dependence of its physical properties. An amorphous polymer will be a glass at temperatures below its glass transition temperature (T_g) (see Section 4.1), and an elastomer above it.

Factors that affect the T_g and other physical properties of polymers are:

- Elements and bond types in the main chain determine the torsional barrier and hence the intrinsic mobility of the chain. For example, some of the lowest T_g values known (–100 to –130 °C) are obtained with polymers that possess Si–O or P–N backbone bonds. These bonds have very low barriers to torsion. Furthermore, only every second atom along the polymer chain in these polymers bears substituents.
- Shape and rigidity of the side groups affect chain flexibility by steric interactions between them. Bulky substituents which sterically interfere when the backbone bonds are twisted, will have high T_gs.
- The size and shape of the substituents may create free volume between themselves and their counterparts on neighboring chains. The greater the free volume, the easier will it be for macromolecular motion to occur. Hence, systems with large free volumes are expected to have low T_gs.
- Side-group regularity and symmetry along the chain favor efficient chain-packing that lead to reduced free volume and the formation of microcrystallites.

5.1.2 Preceramic Polymers

The fabrication of ceramics from preceramic polymers involves the steps shown in Figure 5-2.

The main advantages of this route, compared to conventional syntheses of ceramics, are the lower reaction temperatures and the option to prepare amorphous or crystalline powders depending on the pyrolysis conditions. A technological advantage is that polymer processing technologies can be used for shaping (molding, fiber drawing, coating, infiltration, etc.). The shaped preceramic polymer is pyrolyzed to the final ceramics. The green body (▶ glossary) shrinks during pyrolysis, but must retain its shape. The shrinkage is mainly due to the change in density from polymer (~1 g·cm^{-3}) to ceramic (~2–3 g·cm^{-3}) and to the mass loss that occurs during polymer pyrolysis. Pyrolysis usually results in amorphous ceramics, which optionally can be converted into thermodynamically stable ceramic phases by heat treatment at high temperatures.

molecular compounds

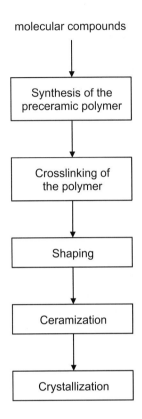

Figure 5-2. Processing steps in the preparation of ceramics from preceramic polymers.

Applications of preceramic polymers

Typical applications of preceramic polymers include the following:

- For the fabrication of *ceramic fibers* a polymer is needed, which can be processed by melt-spinning or dry-spinning techniques (see Section 5.5).
- Preceramic polymers can be used for the preparation of *ceramic films.* After coating, the polymer layer is decomposed to a ceramic material. Typical applications for ceramic films are the improvement of the oxidation resistance or the tribological properties (▶ glossary) of the substrate.

- The pyrolysis of large-volume polymeric green bodies (▶ glossary) is very difficult, because of the shrinkage problem and the release of gases during pyrolysis. However, preceramic polymers are used to form the *ceramic matrices in composite materials* (▶ glossary). The polymer is infiltrated in porous structures such as fiber filaments, and the infiltrated part is subsequently pyrolyzed.
- As *low-loss binders* the preceramic polymers could replace the widely used organic binders in ceramics processing (see Section 2.1.4). The result would be a stronger part that has far fewer defects, because the binder is for the most part converted to ceramic.
- The gases formed during pyrolysis allow to produce *porous materials*, which can be used as ceramic membranes, for example.

Not every inorganic polymer is a useful preceramic polymer. The requirements depend somewhat on the intended application and are partially contradicting. The chemical challenge is to design a preceramic polymer with a good compromise of properties.

The properties of preceramic polymers

The properties of an ideal preceramic polymer are:

- The starting monomers should be commercially available and reasonably cheap.
- The polymer synthesis should be simple and proceed in high yields.
- The preceramic polymer must be processable by conventional polymer-processing techniques. It therefore should be liquid, or a fusible or soluble solid. It must have the correct rheological properties (▶ glossary) for the intended shaping process.
- The polymer should be stable at ambient temperature.
- Pyrolysis of the preceramic polymer should give a high yield of ceramic residue: Ceramic yields (▶ glossary) of 60–75 % are acceptable, but higher yields are desirable. For this reason, hydrogen and methyl groups are preferred substituents owing to their low mass.
- The polymer should have a high molecular mass to avoid evaporation during pyrolysis. The structure should be branched, because many linear inorganic polymers decompose thermally by evolution of volatile products of low molecular mass.
- To retain its shape during pyrolysis, the polymer must be infusible or rendered infusible by extensive crosslinking (▶ glossary) during the

initial stages of the pyrolysis. This requires appropriate reactive substituents.

- There should be minimal gas evolution during pyrolysis. Excessive formation of gaseous compounds lowers the ceramic yield (▶ glossary), increases shrinkage of the ceramic body, creates porosity (which may not be wanted), and even may cause cracking or rapture of the ceramic part.
- The volatile products of pyrolysis should be non-hazardous and nontoxic.

The correct ratio of the elements in the preceramic polymer does not guarantee that a stoichiometric ceramic phase will be obtained on pyrolysis, because one of the elements may be partially lost as volatile products. One approach to achieve the desired elemental composition is to use chemical or physical combinations of two different polymers in the appropriate ratio, for example a polycarbosilane giving silicon-rich SiC in combination with another giving carbon-rich SiC. The stoichiometry can also be controlled by an appropriate atmosphere during pyrolysis. The gas used may be either inert or reactive. Important reactive gases are, for example, ammonia or nitrogen for the formation of nitrides, hydrogen for SiC, or water for oxides.

An approach to overcome the shrinkage problem in the fabrication of large-volume ceramic bodies is the active filler–controlled polymer pyrolysis process. The polymer is loaded with an "active filler (▶ glossary)" which undergoes a volume expansion upon reaction with the decomposition products of the polymer or a reactive gas atmosphere. This allows the near net shape manufacturing of components with complex geometries. Suitable active fillers (▶ glossary) are elements or compounds forming carbides, nitrides or oxides, such as Al, B, Si, Ti, $CrSi_2$, or $MoSi_2$, which exhibit a high specific volume increase upon reaction.

5.1.3 Crosslinking

The macromolecules initially obtained by some preparation methods are very often linear or weakly branched polymeric chains. For many uses, this structure needs to be changed after some initial processing steps by "crosslinking" reactions, by which bonds are formed between different chains. The process of crosslinking (▶ glossary) is also called "vulcanization" (▶ glossary) or "curing". Crosslinking has several important consequences for the materials properties, and is thus one of the most important aspects of polymeric science.

- The macromolecules in an amorphous un-crosslinked polymer at a temperature above T_g have considerable freedom of motion and are thus in a quasi-liquid state. An average of only 1.5 crosslinks per polymer chain severely restricts these motions. Crosslinks prevent chains from sliding past each other when the material is stretched. This raises the tensile strength and generates rubber–elastic properties from materials that would otherwise be gums. Heavy crosslinking provides *rigidity* and *strength* to the system. When a crosslink exists for every two or three chain atoms, the material becomes inflexible and infusible.

- An average of only 1.5 crosslinks per chain is sufficient totally to prevent *dissolution* of soluble polymers, although the polymer may swell. Increasing the number of crosslinks per chain will progressively reduce the degree of swelling. Swelled polymers may contain up to 95 % of their mass as solvent, yet have definite size and shape.

- Polymers in their non-crosslinked state are easily fabricated but are prone to decompose at moderately high temperatures by evolution of volatile products of low molecular mass. Crosslinking *improves thermal stability.*

- The proper degree of crosslinking is essential for the *conversion of polymers to ceramics.* A compromise has to be found between adjustment of the right rheological properties (▶ glossary) for the intended application and the requirements for obtaining high ceramic yields (▶ glossary). This problem will be discussed in more detail in Section 5.5 for ceramic fibers.

Thus, one of the main challenges in making useful polymeric materials involves the development of crosslinking methods. Some general concepts for the crosslinking of inorganic polymer chains are exemplarily shown in Figure 5-3; we will repeatedly come back to this issue in the following sections.

Substitution reactions. These are only rarely used, because polymers with displaceable substituents are not readily prepared and handled. This method is sometimes used in polyphosphazene chemistry. A variation of this method is the metathetical (▶ glossary) exchange of groups. An example is the crosslinking of OCH_2CF_3-substituted polyphosphazene chains by treatment with $NaOCH_2(CF_2)_xCH_2ONa$ (by which $P-OCH_2(CF_2)_xCH_2O-P$ links are formed and $NaOCH_2CF_3$ is released).

Addition reactions. These are more widely used for all classes of inorganic polymers. The most general option is the addition of reactive element–hydrogen bonds (SH, SiH, etc.) to unsaturated groups. An industrially used example (see

Section 5.2) is the crosslinking of polysiloxanes by reacting Si–H and Si–CH=CH$_2$ entities in different chains. This results in the formation of Si–CH$_2$CH$_2$–Si links. A related example is given in Eq. 4-31.

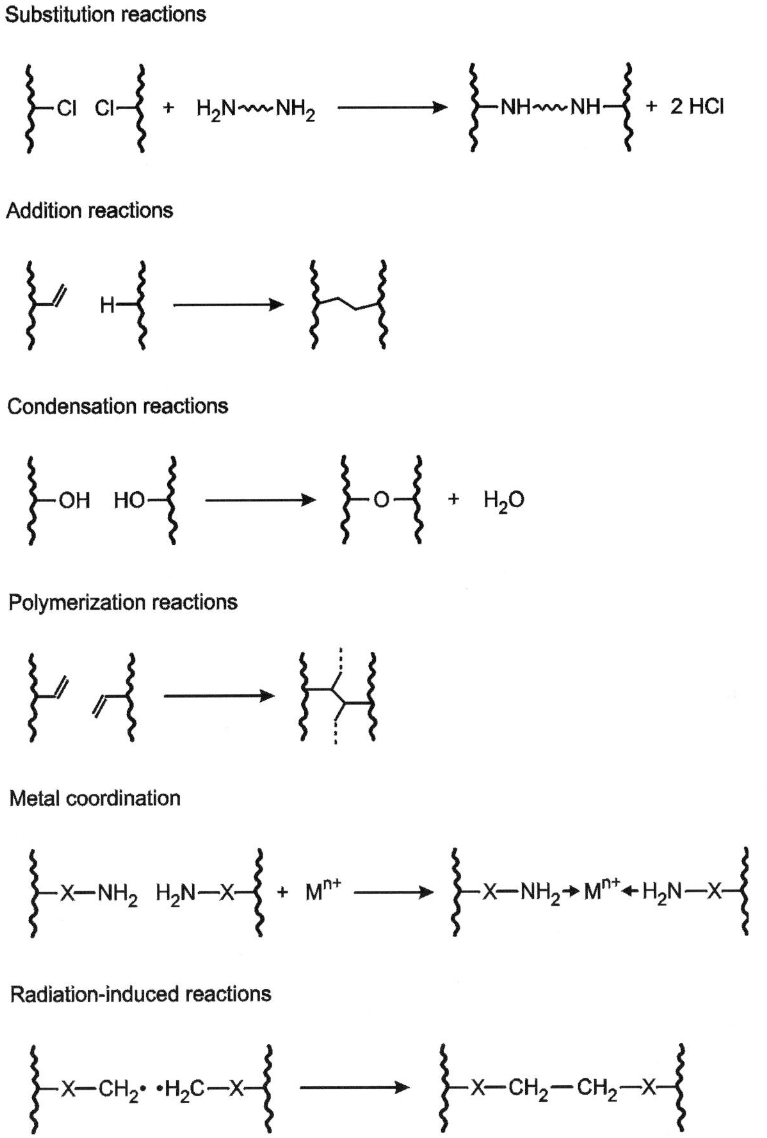

Substitution reactions

Addition reactions

Condensation reactions

Polymerization reactions

Metal coordination

Radiation-induced reactions

Figure 5-3. General methods for the linking of inorganic polymer chains.

Condensation reactions. These are very important for silicon-based materials by which the very stable Si–O–Si bonds are formed by elimination of water starting from Si–OH groups. Another option is the formation of Si–NR–Si or P–NR–P bonds by elimination of amines starting from Si–NHR or P–NHR substituted polymer chains.

Polymerization (▶ glossary) *or polyaddition reactions.* These have proven especially effective. Almost any unsaturated organic group in the side chain can be used for polymerization, but vinyl crosslinking is most widely employed. *Crosslinking via metal coordination.* This is restricted to rather special applications and is nearly exclusively used in polyphosphazene chemistry.

Radiation-induced crosslinking. This includes irradiation with UV, X-ray, γ or electron radiation, and has many advantages, mainly because most inorganic backbones are stable to irradiation and transparent to visible and near-UV light (polysilanes are different, see Section 5.4). Furthermore, no functional side groups are required. The underlying chemistry is radiation-induced homolytic cleavage of C–H or C–C bonds. The thus generated carbon radicals combine to give covalent bonds. However, irradiation techniques are difficult to incorporate into traditional ceramic processing and therefore only make sense for the fabrication of fibers or coatings.

5.2 Polysiloxanes (Silicones)

Silicones have been used technologically for almost 50 years, and may be produced as oils, elastomers (rubbers), or resins. About 60 % of the produced silicones worldwide are oils (including emulsions and greases), 30 % elastomers and 10 % are resins.

5.2.1 Applications and Properties of Silicones

The many applications are based on a number of interesting properties:

- Silicones exhibit excellent thermal stability up to 200–300 °C, and have low T_gs, for instance –123 °C of pure poly(dimethylsiloxane). Their physical properties change only slightly on changing the temperature.

- Silicones have a very good oxidation stability. They are only oxidized by air at $T > 200–300\,°C$, the final inorganic product being silica. They also show prolonged resistance to ultraviolet irradiation and weathering.
- Most silicones act as very good electrical insulators. The volume resistivity (▶ glossary) (10^{14} to 10^{16} Ω·cm), dielectric strength (▶ glossary) (>15 kV·mm^{-1}) and dielectric loss (▶ glossary) tan δ (<10^{-5} at 100 Hz) are nearly unchanged between 20 °C and 200 °C.
- The excellent water repellence are mainly due to methyl substituents. The contact angle (see Figure 4-66) for water droplets can reach 110° (comparable to paraffins). Closely connected with the hydrophobic properties are antistick properties against oils, fats, or organic polymers.
- Silicone oils have a very high spreadability on nearly all organic and inorganic materials, except fluoropolymers. Their surface tension (21 mN·m^{-1} for poly(dimethylsiloxane)) is in the same range as that of tensides ($\approx$30 mN·m^{-1}).
- Silicones are chemically and physiologically inert.
- Liquids and gases permeate more rapid through silicone rubbers than through other polymers. For example, silicones are 10 times more permeable for oxygen than natural rubber or low-density polyethylene. A good oxygen permeability (▶ glossary) is important for medical applications (for example, contact lenses).
- Silicones retain their flexibility even at low temperatures, and they do not become brittle.

Silicone oils

Silicone oils are used as capacitor and transformer fluids, hydraulic oils, compressible fluids for liquid springs, and lubricants. Other uses are as heat transfer media in heating baths and as components in car polish, sun-tan lotion, lipstick, shampoo, and other cosmetic formulations. Their low surface tension leads to their use as release agents, antifoaming agents and hydrophobizing agents, such as water-repelling impregnations for textiles or buildings. Their complete nontoxicity allows them to be used in cooking oils and the processing of fruit juices, for example.

Silicone elastomers

Silicone elastomers obtain their final properties only after curing ("vulcaniz-ing", ▶ glossary). There are four main classes: *room-temperature vulcanizing* elastomers (RTV), which can consist of one (RTV-1) or two components (RTV-2), *liquid rubber* (LR), and *high-temperature vulcanizing* formulations (HTV; also called *heat curing* formulations, HC). Silicone rubbers in general are unmatched by any other synthetic or natural rubbers in retaining their inertness, flexibility, elasticity and strength up to 250 °C (heat-stabilized silicone rubbers) and down to –120 °C. General aspects of crosslinking (▶ glossary) have already been discussed in Section 5.1; the specific reactions employed in the production of silicones will be treated at the end of this section after the chemistry of silicones has been developed.

RTV-1 silicones. These have good adhesive properties, because they wet most substrates. They are therefore mainly used as adhesives or coatings in cars, buildings, industrial plants and household items, and in mechanical and electri-cal engineering. Adhesion can be improved, if necessary, by adding an adhesion promoter or by pretreating the substrate surface with a "primer". Both are mostly silanes with reactive organic groups. Each substrate/silicone combina-tion requires a special primer. Vulcanization (▶ glossary) of RTV-1 silicones occurs on contact with humidity.

RTV-2 silicones. These find use for sealing and encapsulating electrical and electronic components, for elastic damping (vibration absorption), plugs, rub-ber stamps, rams for screen printing, etc. They are also often used for making accurate molds and to provide rapid, accurate, and flexible impressions for den-tures and inlays. Silicone foams are, for example, used for fire protection or for the isolation of cable passages. In contrast to RTV-1 formulations, RTV-2 sili-cones adhere to almost no material. However, reactive organic groups can be incorporated into the silicones to provide them with adhesion properties.

The two components of RTV-2 silicones have to be intimately mixed in a cer-tain ratio before use. The mixture then must be used within a certain period of time, typically between half an hour and several hours. The viscosity of the mixture steadily increases after mixing until the formulation is no longer processable.

A representative application is the making of accurate molds, not only for the replication of art objects (Figure 5-4) and decorations, but also for the production of prototypes and models in various industries. For all these appli-cations it is important that a negative of the object is made which reproduces every detail, particularly the texture of its surface. RTV-2 silicones are easy to apply, have a high elasticity, do not stick, and provide a precise copy of the object.

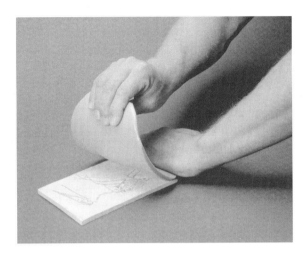

Figure 5-4. RTV-2 silicones allow accurate copies of objects to be made because they do not adhere to inorganic and organic materials.

LR formulations. These were developed particularly for the quick, automated and cheap production of small elastic objects in a high number of pieces by injection molding. LR formulations also consist of two components which are mixed before injection into the mold. Vulcanization (▶ glossary) occurs in the heated mold (180–200 °C) within seconds. Typical objects prepared from LR silicones are O-rings, plugs, contact mats, nipples, membranes, diaphragms, tube couplings, corrugated bellows, or gaskets.

A typical example for the application of such silicones is shown in Figure 5-5. Keyboard switches made of silicone rubber have become an indispensable part of phones, pocket calculators, remote controls, etc. The mat under the push button has to establish electrical contact at a certain pressure, and in many cases the rubber contact itself is touched. This application requires the material to endure a continuous mechanical stress (▶ glossary; $>10^7$ switching cycles). It must have a high specific volume resistivity (▶ glossary), pigmentability, and printability (for lettering the contact).

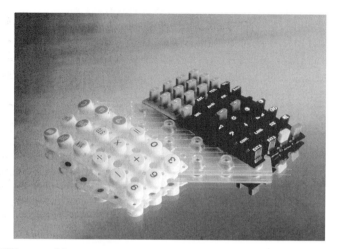

Figure 5-5. Silicone rubber contact.

Keyboard switches can be made as integrated composition of silicone rubber and conducting silicone rubber (Figure 5-6), the latter being produced by filling the silicone with carbon black.

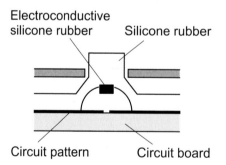

Electroconductive
silicone rubber   Silicone rubber

Circuit pattern   Circuit board

Figure 5-6. The working principle of rubber contacts.

HTV silicone rubbers. These have found wide spread applications for many uses. These include transparent tubings in food industry and medicine, electrical insulations, gaskets, belting, rollers, plug-and-socket connectors, keyboards, oxygen masks, soft tissue prostheses, etc. The uncured silicones are rubbery materials. Their forming processes are the same as in the rubber industry, such as compression or injection molding, extrusion, calandering, etc.

Silicone resins

Silicone resins are used in the insulation of electrical equipment and machinery, as laminates for printed circuit boards, and for the encapsulation of components such as resistors and integrated circuits. Non-electrical uses include high-temperature binders in paints and coatings for industrial plants, as well as household items, such as cooking ware, ovens, etc.

5.2.2 Structure of Silicones

Silicones are composed of four structural units (Figure 5-7):

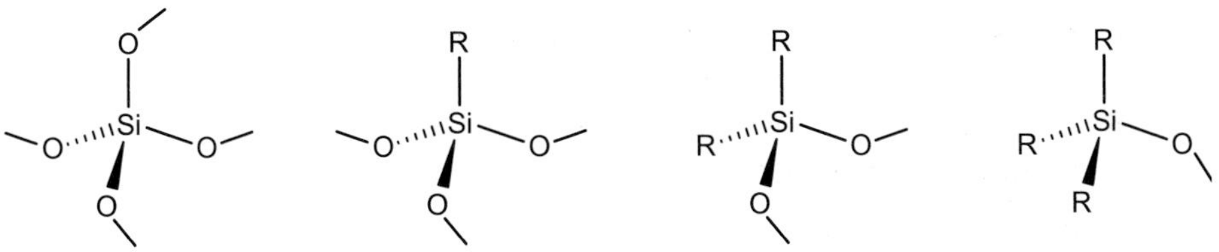

Figure 5-7. Structural units of silicones (R = organic group).

There are 15 possible combinations: QT, QD, QM, TD, TM, DM, TT, DD, MM, QQ, QTD, QTM, QDM, TDM, and QTDM. MM represents disiloxanes, and QQ SiO_2 or silicates. Mainly DM, QM, TD, TM, TDM, and TT are realized in technical products. Materials of the QT and TT type are the domain of sol–gel chemistry (see Section 4.5).

The properties of silicones are determined by:

- the type and proportion of the structural units;
- the type of the organic groups R;
- the distribution of the organic groups if there is more than one kind of R; and
- the chain length and chain length distribution.

The properties of phosphazene polymers (Section 5.3) are to a very large extent determined by the properties of the organic substituents. In contrast, the properties of silicones are mostly influenced by the distribution of the structural units. Each of them has different functions. M units terminate chains or three-dimensional entities. A higher proportion of M units therefore results in a lower molecular mass of the silicone. The combination of D units results in chains, while each Q and T unit is a branching point. Q units are only employed in silicone resins and for some crosslinking reactions (▶ glossary) of silicone elastomers (see below). A small amount of T units in a TD silicone results in light crosslinking of the polysiloxane chains, but no three-dimensional network is formed. Viscosity increases with an increasing proportion of T units.

The high thermal stability of silicones is mainly due to the strength of the Si–O bonds (420–500 kJ·mol^{-1}). The Si–O–Si backbone of silicones has a unique dynamic flexibility. Compared to organic polymers (with a C–C–C backbone) polysiloxanes have longer bond distances (Si–O 164 pm, C–C 154 pm), and every second atom along the chain (the oxygen atom) has no substituent. There are only low intermolecular forces between the polysiloxane chains. This is, inter alia, the reason for the very high spreadability and the low viscosity of silicone oils, and also explains the small temperature dependency of the physical properties of silicone oils and elastomers.

Methyl groups are predominant in silicones, and can be partially replaced by other groups to achieve special materials properties:

- Phenyl groups disrupt crystallinity and therefore improve flexibility at low temperatures. For example, cured dimethylsiloxane elastomers lose their elasticity at about –50 °C. Formulations containing about 8 mol% PhMe-SiO groups instead of Me$_2$SiO, or 5 mol% Ph$_2$SiO, retain their elasticity down to –110 °C. Phenyl groups increase the already high thermal stability and weatherability of silicones, raise their refractive index, are less prone to oxidation, and additionally strengthen the Si–O bonds. In silicone resins, they improve the processability and the compatibility with organic polymers.
- Vinyl groups and hydrogen substituents are important for organic crosslinking reactions (see below).
- CH$_2$CH$_2$CF$_3$ (and other fluorinated alkyl) groups reduce swelling of silicone rubbers in many solvents ("fluorosilicones").

A variety of other organofunctional groups is used for the preparation of specialty silicones, the most important groups are of the type (CH$_2$)$_n$X with

$$X = -SH, -NR_2, -O-H_2C-HC\overset{O}{\underset{}{\diagup\!\!\diagdown}}CH_2, -OH, -O-\overset{\overset{O}{\|}}{C}-R$$

(see also Section 4.5 on sol–gel materials). In silicone chemistry, these groups mainly act as coupling sites, for example for providing links to organic polymer structures. However, organofunctional substituents can also determine physical properties of the silicone. This aspect will be discussed in more detail for polyphosphazenes (Section 5.3), where it plays a major role. For example, many liquid crystalline (▶ glossary) silicone polymers have been prepared by substituting the silicon atoms in polysiloxanes with mesogenic (▶ glossary) side groups.

5.2.3 Preparation of Silicones

Müller–Rochow process

Silicones are prepared from the corresponding silanes $R_{4-n}SiX_n$ as precursors, where X almost always is Cl. The methyl derivatives are produced in the Müller–Rochow process ("direct process") in which methyl chloride is reacted with elemental silicon at 280–300 °C and 200–400 kPa in the presence of copper as the catalyst and promoters (Eq. 5-3) in a fluidized-bed reactor. The actual catalyst (or at least the precursor to active surface species) is η-Cu_3Si highly dispersed at the silicon surface. Zinc is mainly used as promoter, in addition to small amounts of other elements.

$$Si + CH_3Cl \xrightarrow{\text{Cu}} (CH_3)_x SiCl_{4-x} \qquad (5\text{-}3)$$

A mixture of methyl(chloro)silanes is formed. The process has been optimized that 80–90 % Me_2SiCl_2 is obtained, the most important precursor for the preparation of silicones, particularly for silicone elastomers. The second largest product is $MeSiCl_3$ (3–15 %), followed by Me_3SiCl (2–5 %), $Me(H)$-$SiCl_2$ (0.5–4 %) and a variety of other mono- and disilanes. The spectrum of products and the reproducibility is influenced by the processing conditions such as temperature, pressure, catalyst and promoters, inhibitors, silicon purity, particle size distribution of the solid reactants, homogeneity of the fluidized-bed, dust removal from the reactor, and purity of the methylchloride. A total of 1.25 million tons of silanes were produced by the Müller–Rochow process in 1995, and an annual growth of about 5 % is expected.

The Müller–Rochow process can also be used for the preparation of phenyl-substituted silanes. Silanes with functional organic groups of the general type Cl_3Si–X–A with a great variety of groups A connected to the silicon atom by an inert spacer X are easily prepared by indirect routes. The spacer X usually is a $(CH_2)_n$ (n = 2, 3) chain. Chlorosilanes Cl_3Si–X–A or the derived alkoxysilanes $(RO)_3Si–X–A$ (prepared from the chlorosilanes by reaction with alcohols) have found wide spread industrial applications, not only as components in silicones, but also as adhesion promoters, for derivatizing surfaces, or for the immobilization of substrates.

The more general preparation routes for these compounds are summarized in Eqs. 5-4 to 5-6 (Y = Cl or OR).

$$Cl_3SiH + CH_2=CH–X–A \longrightarrow Cl_3Si–CH_2–CH_2–X–A \tag{5-4}$$

$$Y_3Si–(CH_2)_n–A + A' \longrightarrow Y_3Si–(CH_2)_n–A' + A \tag{5-5}$$

$$Y_3Si–(CH_2)_n–A + Y–X–A' \longrightarrow Y_3Si–(CH_2)_n–A(Y)–X–A' \tag{5-6}$$

The hydrosilylation of alkenes or alkynes, i.e., the addition of Si–H bonds to double or triple bonds, is a rather general method for the formation of Si–C bonds. In practice, $HSiCl_3$ (obtained by reaction of HCl gas with elemental silicon) is added to unsaturated organic compounds (Eq. 5-4). For example, the technically very important silanes $Cl_3Si(CH_2)_3Cl$, $Cl_3Si(CH_2)_2CN$ or Cl_3Si-$(CH_2)_3OC(O)CMe=CH_2$ are prepared that way, starting from allylchloride, acrylonitrile or allylmethacrylate, respectively.

Substitution of a group A by a group A′ (Eq. 5-5) or chemical modification of a functional group (Eq. 5-6) is often the easier way to introduce more complex functional groups. For example, the technically important thio-substituted compound $[(EtO)_3Si(CH_2)_3]_2S_4$ (used as adhesion promoter between rubber and silica fillers) is prepared by reaction of $(EtO)_3Si(CH_2)_3Cl$ (A = Cl) with Na_2S_4 (A′ = S_4). An example for the modification by addition reactions to functional groups A (Eq. 5-6), such as (meth)acrylate, isocyanate or epoxide groups, is given in Eq. 5-7.

$$\tag{5-7}$$

Hydrolysis, methanolysis, and polycondensation reactions

The conversion of the chlorosilanes to polysiloxanes is exemplarily discussed for Me_2SiCl_2. Processing of other chlorosilanes proceeds analogously.

Reaction of Me_2SiCl_2 with water results in the formation of cyclic and OH-terminated linear dimethylsiloxane oligomers (Eq. 5-8).

$$(n+m)\ Me_2SiCl_2 \xrightarrow[-HCl]{H_2O} [\ Me_2Si(OH)_2]$$

$$\xrightarrow[-H_2O]{} HO(-Me_2SiO)_n-H + (Me_2SiO)_m$$

$$(5\text{-}8)$$

The prime reaction is the hydrolysis of the Si–Cl groups to Si–OH groups (Eq. 5-9).

$$\equiv Si\text{–}Cl + H_2O \rightleftharpoons \equiv Si\text{–}OH + HCl \qquad (5\text{-}9)$$

However, the reaction cannot be stopped at this point. As has already been discussed in Section 4.5, hydrolysis (Eq. 5-9) and condensation reactions (Eq. 5-10) usually cannot be decoupled from each other, particularly in the presence of HCl (formed as a by-product) which acts as a catalyst.

$$2 \equiv Si\text{–}OH \rightleftharpoons \equiv Si\text{–}O\text{–}Si \equiv + H_2O \qquad (5\text{-}10)$$

Both the hydrolysis and the condensation reactions (Eqs. 5-9 and 5-10) are equilibria. Thus, cyclic and linear products, and linear products with different chain lengths are in equilibrium, and their ratio is influenced by the reaction conditions. A very important parameter is the HCl concentration and its contact time with the siloxane. Rapid removal or neutralization of the HCl results in siloxanediols $HO[SiMe_2O]_nH$ with short chain lengths (n). The cyclic products are dominated by $(Me_2SiO)_4$ ("D_4").

Me_2SiCl_2 is completely converted into OH-terminated linear siloxanes in the technically used continuous hydrolysis process. The cyclic products are continuously distilled off and fed back to the starting silane. Acid-catalyzed ring-opening polymerization (ROP) of the cyclosiloxanes $(SiMe_2O)_n$ (see below) and reaction with Me_2SiCl_2 gives α,ω-dichlorsiloxanes, $Cl[SiMe_2O]_mSiMe_2Cl$, which are then hydrolyzed. A flow chart of the continuous hydrolysis process is shown in Figure 5-8.

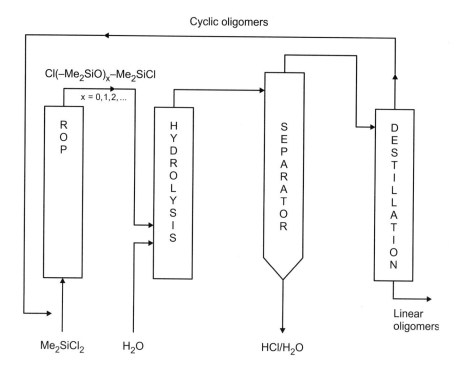

Figure 5-8. Continuous hydrolysis process.

The HCl obtained at the hydrolysis of Me_2SiCl_2 is reacted with methanol to give methyl chloride, which is fed back in the Müller–Rochow process. Chlorine is thus recycled.

An option in which chlorine from the Si–Cl groups is directly converted in methyl chloride is the reaction of Me_2SiCl_2 with methanol ("methanolysis", Eq. 5-11). The cyclic products can be recycled in a continuous process similar to that in Figure 5-8.

$$(n+m)\ Me_2SiCl_2 \xrightarrow[-MeCl]{MeOH} HO(-Me_2SiO)_n-H + (Me_2SiO)_m \qquad (5\text{-}11)$$

To obtain highly polymeric silicones, the OH-terminated dimethylsiloxane oligomers obtained by the hydrolysis or methanolysis reaction have to undergo polycondensation reactions (▶ glossary), in which the oligomer units are coupled (Eq. 5-12). The polycondensation can be catalyzed by strong acids or bases, or by solid catalysts. The presence of precursors of chain-terminating

[Me$_3$SiO–] or crosslinking groups [MeSi(O–)$_3$] must be carefully avoided for the production of linear polysiloxanes.

$$\text{(5-12)}$$

Equilibration and ring-opening polymerization of cyclosiloxanes

If cyclic oligosiloxanes or linear polysiloxanes are exposed to a base or acid catalyst, an equilibrium mixture between cyclic and linear species is obtained, as mentioned above. The process of simultaneous cleavage and re-formation of siloxane bonds is called equilibration.

Acid- or base-catalyzed ROP reactions are performed under non-equilibrium conditions. In the acid-catalyzed reaction, the Si–O–Si bond of a cyclic or linear siloxane is *broken* by the electrophilic attack of the Lewis acid (proton) to the oxygen atom. This generates species terminated by an OH group at one end and by an active electrophilic center at the other. The group A in Eq. 5-13 may be the counter-ion of the acid (X) or an oxonium ion ($\equiv$Si–O$^+$H$_2$ generated by substitution of X by water, which is present in the system owing to the subsequent condensation reactions). Two Si–O–Si bond *forming* mechanisms can occur simultaneously. The active center can add to another Si–O–Si bond (addition polymerization mechanism) or hydrolysis/condensation reactions can take place. All these reactions are basically equilibria.

KOH is the most widely used catalyst for the anionic polymerization ($\blacktriangleright$ glossary). The OH$^-$ ion attacks a silicon atom and thus generates a $\equiv$Si–OH and a $\equiv$SiO$^-$ (silanolate) terminated end. The silanolate group than attacks another Si–O–Si group, etc. (living anionic polymerization). Before the residual cyclosiloxanes are removed from the reaction mixture, the catalyst must be destroyed. Otherwise the equilibria would be shifted towards the cyclic products (see above).

Both equilibration and acid- or base-catalyzed ROP reactions play an important role for the regulation of chain lengths, and the production of copolymers from two differently substituted polysiloxanes. They allow the very reproducible production of silicone products. Technically important examples are:

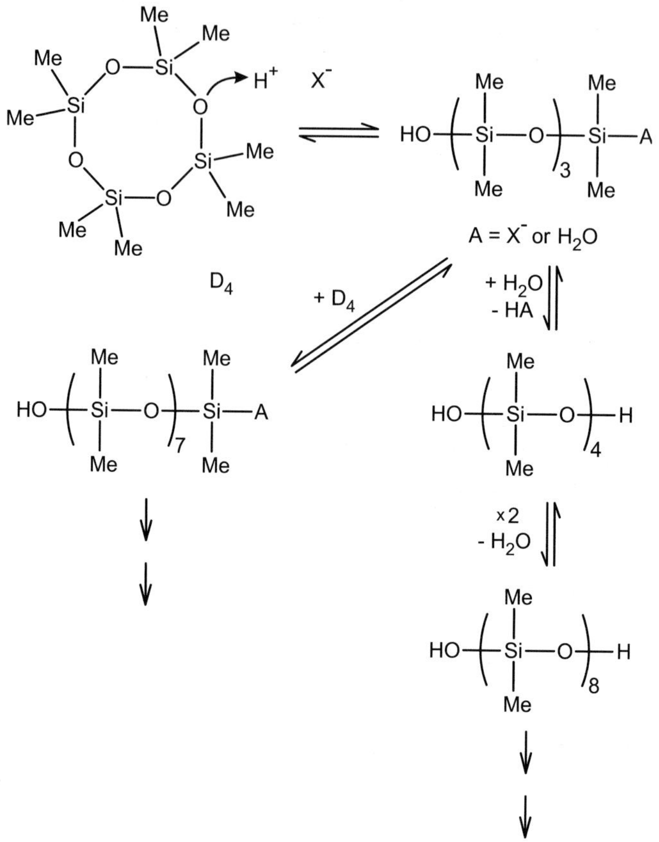

(5-13)

- Silicone oils with very defined chain lengths and viscosities are made by shaking suitable proportions of $Me_3Si-O-SiMe_3$ (as the precursor for end-stopping $-OSiMe_3$ units) and either a cyclo(dimethylsiloxane) or a linear poly(dimethylsiloxane) with a small quantity of 100 % H_2SO_4 (Eq. 5-14). The reason why this reaction proceeds towards polymers (although all reactions are equilibria) is that the reaction rate of Si–O–Si bond cleavage in acidic media decreases in the following order: D_4 > MM > MDM > MDDM. Thus, cleavage of a Si–O–Si bond in D_4 is preferred to that of an oligomeric or polymeric chain.

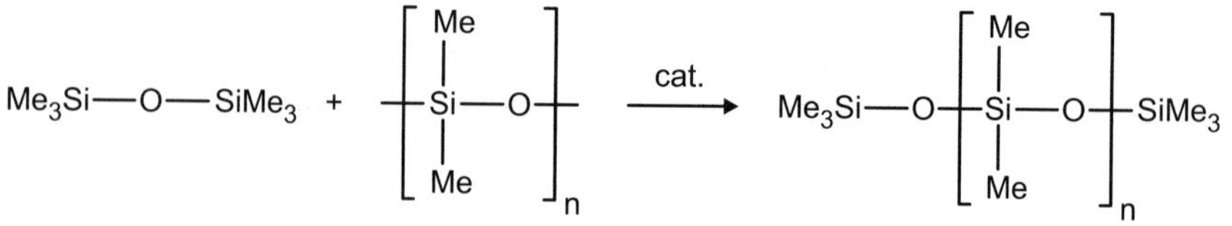

(5-14)

- Endblocks with reactive groups (X) such as Cl, OR, OAc, –ON=CR$_2$ or –NR′C(O)R can be introduced by a related reaction (Eq. 5-15). Polymers with such groups are needed for the production of silicone elastomers (see below). The acid-catalyzed ring-opening of cyclosiloxanes in the presence of Me$_2$SiCl$_2$ in the continuous hydrolysis process (Figure 5-8) is another example.

$$Me_2SiX_2 + m \ (Me_2SiO)_n \longrightarrow X–(Me_2SiO)_{nm}–SiMe_2X \quad (5\text{-}15)$$

- Polysiloxanes containing a certain number of Si–H groups are needed for the production of silicone elastomers (see below). Their synthesis by equilibration of SiMe$_3$-terminated poly- or oligo(methylsiloxanes) and cyclic or linear *dimethylsiloxanes* is shown in Eq. 5-16. The proportion of Si–H groups in the equilibrated polymer determines the later crosslinking density (▶ glossary), and thus the physical properties of the elastomers. The clearly defined incorporation of these groups is therefore a very important issue. Other functional groups can be incorporated by the same method.

$$(5\text{-}16)$$

- When the cyclosiloxanes are distilled off from an equilibrium mixture between cyclic and linear species, the equilibrium is continuously shifted, and the polymer is eventually completely degraded. This depolymerization reaction can be used for the recycling of silicone products. The thus obtained cyclosiloxanes can again be converted to linear polysiloxanes using the processes discussed above.

Curing ("vulcanizing")

Silicone oils are SiMe$_3$-terminated poly(dimethylsiloxanes), Me$_3$SiO[SiMe$_2$O]$_n$-SiMe$_3$. Their properties depend only on the chain length (n). There is a linear dependence between the logarithm of the viscosity and the logarithm of the average chain length. Viscosity at room temperature for n = 50 is about 70 mPa·s, and about 10^5 mPa·s for n = 1000. This changes only slightly with temperature, and even silicone oils with a molecular mass of 10^5 are still fluids.

Organic polymers with a much smaller molecular mass are already solid. The reason for the low viscosity are the weak intermolecular interactions between poly(dimethylsiloxane) chains.

Crosslinking (▶ glossary) of the polysiloxane chains is necessary to obtain silicone materials with a higher viscosity, such as elastomers, rubbers, or resins (see above). This requires functional groups bonded directly to silicon, or as part of the organic substituents. General methods for crosslinking (▶ glossary) were already discussed in Section 5.1; the following paragraphs will describe specifically which of these processes are used industrially to obtain the various silicone varieties.

Silicone *elastomers* (silicone rubbers) are crosslinked poly(dimethylsiloxanes) of high molecular weight (10^4–10^7). Most silicones are reinforced by a filler (▶ glossary) (mostly silica [see Section 3.3], but also silicates, $CaCO_3$, $BaSO_4$, etc.) to increase their modulus, tensile strength, tear strength and abrasion resistance. The reinforcing effect depends on the size, structure and surface properties (type and amount of surface OH groups, adsorbed water, surface modification) of the particles and the filler loading. The surface interaction determines how effective the elastomer properties are controlled.

Silicones of the RTV-1 type (see Section 5.2.1) are prepared by reaction of OH-terminated polysiloxanes with silanes of the type $RSiX_3$ (Eq. 5-17) or by the ring-opening reaction shown in Eq. 5-15. Curing occurs upon exposure to humidity (Eq. 5-18). The curing rate increases with increasing humidity. Depending on the cleaved HX molecule, basic (cleavage of an amine, HX = NH_2R), neutral (cleavage of an oxime or an amide, HX = $HON{=}CR_2$ or $R'NHC(O)R$) or acidic (cleavage of a carboxylic acid, HX = HOOCR) systems are distinguished. Curing proceeds by diffusion of atmospheric moisture. Therefore, fast surface curing is obtained, but hardening of unexposed material can be quite long, especially for thick samples. This is favorable to adhesion, because there is a long period of time for spreading and wetting of the elastomer on the substrate. Cure accelerators have been developed to applications which require shorter curing periods.

$$HO{-}\left[\begin{array}{c} Me \\ | \\ Si{-}O \\ | \\ Me \end{array}\right]_n{-}H \; + \; 2\,RSiX_3 \; \longrightarrow \; R{-}\begin{array}{c} X \\ | \\ Si{-}O \\ | \\ X \end{array}{-}\left[\begin{array}{c} Me \\ | \\ Si{-}O \\ | \\ Me \end{array}\right]_n{-}\begin{array}{c} X \\ | \\ Si{-}R \\ | \\ X \end{array} \; + \; 2\,HX$$

(5-17)

$$(5\text{-}18)$$

The two main crosslinking reactions in RTV-2 formulations are shown in Eqs. 5-19 and 5-20.

$$(5\text{-}19)$$

$$(5\text{-}20)$$

In the condensation–crosslinking process (Eq. 5-19), the first component is an OH-terminated polysiloxane, and the second a tetra(alkoxy)silane. The latter is mixed with a catalytically active tin compound in one package, and the former with water in another. Curing begins upon intimate mixing. Since no atmospheric water is required, the surface and the bulk cure at the same rate. The eliminated alcohol must be removed to avoid the back reaction. The main advantage of this crosslinking (▶ glossary) method is the variable ratio of the two components.

In the addition-crosslinking process (Eq. 5-20), Si–H groups in one polysilox-ane chain are catalytically added to Si–CH=CH$_2$ groups in another chain. In contrast to Eq. 5-19, this crosslinking is irreversible. There is typically one crosslink for every 200–2000 Si atoms. A statistical distribution of the vinyl groups and the Si–H groups along the siloxane chain is achieved by equilibra-tion reactions (e.g., Eq. 5-16). The mixing ratio between both components is determined by the reactive groups. The crosslinking (▶ glossary) is strongly influenced by the temperature. Vulcanization (▶ glossary) of the *liquid rubber (LR) formulations* proceeds by the same process at higher temperature (in the heated mold). There is almost no shrinkage upon curing, because the reaction proceeds without the loss of mass. The main disadvantage of the addition–crosslinking method is that the catalyst (usually 5 to 50 ppm of a Pt compound) can be poisoned by materials that are in contact with the silicone.

In *HTV formulations*, mainly vinyl groups are crosslinked. The solid rubbers are vulcanized by radical reactions using 1–3 % of an aroyl or alkyl peroxide (Eq. 5-21). The mixture is typically heated to 150 °C for 10 min at the time of pressing or molding and then cured for 1–10 h at 250 °C. Crosslinking by cou-pling of Si–CH$_2$$^{•}$ radicals is also possible.

(5-21)

While the chain-forming Me$_2$SiCl$_2$ is the main component for the prepara-tion of silicone oils and elastomers, *silicone resins* are prepared by using phenyl- or methyltrichlorosilane as the base component. The rigid skeleton is made more flexible by addition of Me$_2$SiCl$_2$, PhMeSiCl$_2$, and Ph$_2$SiCl$_2$. Me(H)SiCl$_2$, CH$_2$=CHSiCl$_3$ and CH$_2$=CH(Me)SiCl$_2$ enable hardening by crosslinking reactions (as discussed above), and Me$_3$SiCl serves to provide end groups. The hydrolyzed mixture of chlorosilanes is washed with water to remove HCl and partly polycondensed to a stage at which the resin is still soluble. The resins are normally applied in this form. After evaporation of the solvent, the resin is crosslinked to a three-dimensional siloxane network by heating above 150 °C in the presence of a catalyst to condense the silanol groups. Additional crosslinking (▶ glossary) via vinyl groups, either by radical polymerization or by addition of Si–H groups, as for the elastomers, is also possible.

Polysiloxane copolymers

Totally different materials properties are obtained when polysiloxane structures are combined with organic polymer structures. There are basically two types of copolymers, *block* and *graft* copolymers (Figure 5-9).

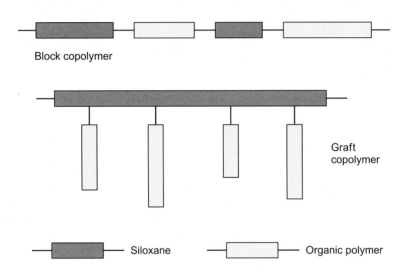

Figure 5-9. Arrangement of the polymer blocks in block and graft copolymers.

In block copolymers, shorter or longer siloxane units are connected by blocks of the organic polymer. In graft copolymers, there are continuous polysiloxane chains, which are either connected or substituted by organic polymer blocks.

Copolymers between polysiloxanes and nearly all types of organic polymers have been made, and some are applied industrially. A detailed discussion of the preparation and properties of these copolymers is beyond the scope of this book. Therefore, only a few examples are outlined.

Siloxane–alkylene oxide copolymers are widely used as emulsifiers and stabilizers for foams of organic polymers. Silicone–polyurethane copolymers have better mechanical, fatigue and physiological properties than silicone homopolymers, and are therefore used for medical applications, such as blood pumps or heart replacements. Silicone–polycarbonate copolymers are thermoplastic and can be injection molded and extruded (see Figure 2-16). They have high mechanical and tear strengths, while maintaining a high degree of oxygen and water permeability (▶ glossary).

The most general method to make such copolymers is to react silicones with appropriate organofunctional substituents with organic monomers (or oligomers containing the reactive end groups). The substituted silicones can be prepared by equilibration reactions (Eq. 5-16) or by cocondensation of the corresponding monomeric chlorosilanes in the appropriate ratio. The preparation of a silicone–polymethacrylate copolymer (Eq. 5-22) may serve to illustrate the general approach.

$$(5\text{-}22)$$

5.3 Polyphosphazenes

Several hundred different polyphosphazenes have been synthesized, with a wide range of physical and chemical properties and with high molecular weights. Polyphosphazenes exhibit a very broad spectrum of useful chemical, mechanical, optical, and biological properties. Their properties vary from fluids to vulcanizable elastomers and glasses, from crystalline to amorphous, from water-soluble to hydrophobic, from bioinert to bioactive, and from electrical insulators to conductors.

5.3.1 Properties and Applications of Polyphosphazenes

High-performance elastomers, which often contain fillers (▶ glossary), are the technically most advanced applications, and their utility parallels that of silicones. For example, a commercially available polyphosphazene with OCH_2CF_3 and $OCH_2(CF_2)_nCF_3$ substituents (and a small amount pendent allyl groups for curing) is a soft gum which can be processed by conventional rubber technology. The polymer is cured with peroxides or by radiation. The cured polymer has a tensile strength of 7–14 MPa (depending on the filler; ▶ glossary), can be elongated by 75–200 %, adheres very well to metals and fabrics, has excellent vibration damping properties and a hydrophobic surface, is resistant to fungi, ozone and a broad range of fluids, is non-flammable, shows an excellent weatherability, and is resistant to liquid oxygen. It maintains its properties between –65 °C and +175 °C. Because of these properties the elastomer is used for seals, gaskets, and shock mounts in demanding automotive, aerospace, and petroleum applications.

The backbone of polyphosphazenes consists of (NPR_2) or $(NPRR')$ repeat units. The substituents R (or R′) can be organic or inorganic moieties. Although the bonding in the backbone is formally represented as a series of alternating single and double bonds (Eq. 5-23), all the bonds along the chain are equal in length. There is no conjugation comparable to organic unsaturated polymers because of the participation of d-orbitals at phosphorus in the π-bond.

$$
\left[\begin{array}{c} R \\ | \\ -P=\overline{N}- \\ | \\ R \end{array}\right]_n \longleftrightarrow \left[\begin{array}{c} R \\ | \\ =P-\overline{N}= \\ | \\ R \end{array}\right]_n \longleftrightarrow \left[\begin{array}{c} R \\ \overset{\oplus}{|} \\ -P-\overline{N}- \\ | \\ R \end{array}\right]_n
$$

$$
\left[\begin{array}{c} R \\ | \\ -Si-\overline{O}- \\ | \\ R \end{array}\right]_n \longleftrightarrow \left[\begin{array}{c} R \\ \overset{\ominus}{|} \\ -Si=\overset{\oplus}{O}- \\ | \\ R \end{array}\right]_n \tag{5-23}
$$

Polyphosphazenes are isoelectronic to polysiloxanes (see Eq. 5-23 and Section 5.2). This explains the similarity of some of their physical properties, for example, their elasticity down to low temperatures. In both polyphosphazenes and polysiloxanes the substituents are only attached to every second backbone atom, which results in the high flexibility of the polymer chain. There is no torsional barrier of the skeleton, because there are no p_π–p_π-bonds. The conformation of a polyphosphazene chain is therefore mainly determined by interactions between the substituents at neighboring phosphorus atoms. It can be seen from Figure 5-10 that the repulsion is minimized in the *cis–trans* planar conformation compared to the *trans–trans* planar conformation or any other conformation between. The repeating distance found in nearly all polyphosphazenes studied so far is about 50 pm, which is the value expected for a configuration near *cis–trans* planar.

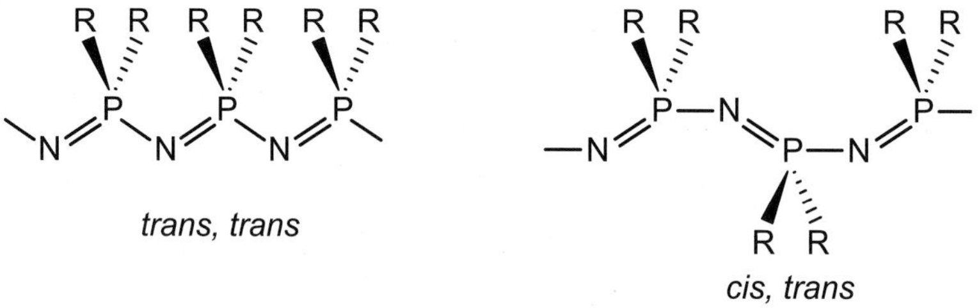

trans, trans

cis, trans

Figure 5-10. *Trans, trans* and *cis, trans* conformation of a polyphosphazene chain.

While the properties of silicones are largely determined by the type and proportion of the structural units, the properties of polyphosphazenes are mainly varied by the types of substituents, and their sequencing along the chain if there is more than one type. The influence of a particular substituent on the properties of a polymer is probably quite general. However, the substituents are more easily varied in polyphosphazenes than in any other polymer.

Enough is now known about the effect of the substituents to allow to some extent the prediction of the polymer properties. For example, polyphosphazenes have low glass transition temperatures (T_g) when the substituents are small (R = F: –96 °C, Cl: –66 °C, OMe: –74 °C) or when the substituents themselves are very flexible ($OCH_2CH_2OCH_2CH_2OMe$: –84 °C, OPr: –100 °C). Large and inflexible substituents sterically interfere with each other as the P–N skeleton undergoes twisting motions and therefore impose restrictions on the flexibility of the macromolecule. For example, if the substituents are OPh or OC_6H_4–p–Ph the T_g is –8 °C or +93 °C, respectively.

A selection of substituents and their influence on other materials properties is given in Table 5-1.

Table 5-1. The influence of selected substituents R on the properties of polyphosphazenes $[NPR_2]_n$.

R	Property/Application
NHMe, glucosyl, glyceryl	hydrophilicity
$OCH_2CH_2OCH_2CH_2OCH_3$	Li^+ conductivity in the solid state, water-solubility, elasticity, low T_g
OR (R = alkyl)	elasticity
$NHCH_2CH=CH_2$	crosslinking (▶ glossary) sites
$OC_6H_4–p–OC(O)CH=CHPh$	photosensitivity
$OCH_2(CF_2)_xCF_3$	low T_g, increased solvent resistance, hydrophobicity
OCH_2CF_3	high transmission coefficient for O_2, solubility in organic solvents, bioinert materials
OR (R = aryl)	thermoplasticity, solubility in organic solvents
$[OC_6H_4–p–COO^-]_2Ca^{2+}$	microencapsulation of bioactive species
$OCH_2CH_2N(Et)C_6H_4–N=N–C_6H_4$-p-$NO_3$	χ^2 NLO (▶ glossary) activity
$O(CH_2CH_2O)_3C_6H_4–p–C_6H_4–p–$OMe	iquid crystallinity (▶ glossary)
OC_6H_4Ph	high refractive index
$OC_6H_4–PPh_2$	coordination of metal complexes
$OC_6H_4–p–N=CH(CH_2)_3CHO$	immobilization of enzymes
NHCH(R)COOEt	biodegradable materials

The refractive index is increased by increasing the number of electrons in the substituents, for instance by biaryl substituents. Rigid, mesogenic (▶ glossary) substituents can give rise to liquid crystallinity. Other properties, such as solubility, chemical stability, hydrophobicity or hydrophilicity, ionic conductivity, non-linear (▶ glossary) optical activity, photochromic properties and biological behavior can also be introduced by appropriate substituents. Membranes have been developed by amphiphilic polymers with both NHMe (hydrophilic) and OCH_2CF_3 (hydrophobic) substituents.

Most polyphosphazenes are stable to water. However, if a P–R bond is cleaved by hydrolysis, a P–OH group is generated. The proton of this group can migrate to a nitrogen atom to give a saturated phosphazane (–P(O)R–NH– repeating units) that is susceptible to hydrolytic cleavage of its skeleton. If hydrolysis is very slow, as for R = NHCH(R)COOEt, hydrolytic sensitivity is a useful property, because it allows the use of such polyphosphazenes as biodegradable materials in medicine.

5.3.2 Preparation and Modification

The most general and most widely used method for the preparation of polyphosphazenes is the ROP/substitution sequence shown in Eq. 5-24.

$$(5\text{-}24)$$

Hexachlorocyclotriphosphazene, $(NPCl_2)_3$, is prepared on an industrial scale by the reaction of PCl_5 with NH_4Cl in an organic solvent such as chlorobenzene or tetrachlorethane. The purified, molten compound is heated in a sealed vessel in the absence of moisture (which would convert the P–Cl bonds to P–O–P bonds) for several days at 200–250 °C to induce ROP. The polymerization can also be conducted in an organic solvent at lower temperatures in the presence of a Lewis acid catalyst. It is assumed that ROP is induced by ionic species generated, for example, by heterolytic cleavage of a P–Cl bond or by abstraction of Cl⁻ by the Lewis acid. ROP of $(NPCl_2)_3$ results in linear poly(dichlorophosphazene), $[NPCl_2]_n$, soluble in benzene, toluene, or tetrahydrofuran. The control of the molecular weight of this polymer is difficult, and the polydispersity (▶ glossary) is large.

The chloro-substituted polymer $[NPCl_2]_n$, known since the end of the 19th century as "inorganic rubber", is still very moisture-sensitive. It also reacts rapidly with a variety of nucleophiles such as metal alkoxides, amines, or organometallic reagents by substitution of the P–Cl groups. A wide range of groups can thus be introduced to give polymers which are stable towards water. With some reactive reagents, all the chloride atoms of the polymer are replaced within minutes at room temperature! In any case, the substitution reaction has

to be driven to completion, because residual P–Cl bonds would result in moisture-sensitive polymers.

Certain bulky or insufficiently nucleophilic organic groups do not replace all
chlorine atoms at mild reaction conditions. However, after a bulky substituent
has partially substituted the P–Cl groups, the controlled introduction of a second (smaller) type of substituents is possible. For example, when $[NPCl_2]_n$ is
reacted with diethylamine (even in excess), only one chlorine atom per phosphorus is replaced. The remaining chlorine can then be substituted by treatment with methylamine or short-chain alkoxides, for example.

$[NPCl_2]_n$ can be deliberately reacted with a deficiency of the reagent. The
remaining chlorine substituents can then be reacted with another nucleophile
resulting in mixed-substituted polymers $[NPRR']_n$. An example is shown in Eq.
5-25. Alternatively, both reagents may be allowed to compete simultaneously
for the available P–Cl groups. The disposition of substituents along the polyphosphazene chain depends on the directing characteristics of the substituents
already present. The order in which two or more different substituents are
introduced will also affect the substitution pattern. However, there is no strict
control of the substitution pattern in most cases. Polymers with two or more
different substituents generally have a better elasticity because they crystallize
less easily.

$$\left[\begin{array}{c} Cl \\ | \\ -P=N- \\ | \\ Cl \end{array}\right]_n \xrightarrow[-\ NaCl]{+\ NaOR} \left[\begin{array}{c} OR \\ | \\ -P=N- \\ | \\ Cl \end{array}\right]_n \xrightarrow[-\ HCl]{+\ R'NH_2} \left[\begin{array}{c} OR \\ | \\ -P=N- \\ | \\ NHR' \end{array}\right]_n$$

(5-25)

The length of the polymer chain is usually unaffected by the replacement of
Cl by OR or NRR' groups. Organometallic reagents result in more complicated
reactions. For example, in the reaction of $[NPCl_2]_n$ with RMgX or LiR, replacement of chlorine by the group R is accompanied by cleavage of P–N bonds in
the skeleton. Coordination of the organometallic molecules to the nitrogen
atoms of the polyphosphazene backbone is necessary for the cleavage reaction.
Thus, any feature that increases electron density at the nitrogen atoms and thus
favors coordination will also favor cleavage of the backbone. Chlorine atoms at
phosphorus lower the electron density due to their electron-withdrawing properties and thus protect the skeleton against cleavage. As an organometallic substitution reaction proceeds, this protective effect is successively lost, and there
is an increased probability of P–N bond cleavage.

Due to the restrictions in the reactions with organometallic reagents, a modification of the ROP process was investigated. In this approach, the organic
substituents are introduced at the cyclic trimer level, followed by ROP. How-

ever, the tendency for ROP declines as more and more halogen atoms in the trimer are replaced by organic groups. Cyclic trimers that bear only one or two organic substituents usually polymerize almost as easily as $(NPCl_2)_3$. The remaining chlorine atoms in the partially organo-substituted polymers can be replaced by nucleophiles as discussed before. A typical reaction sequence is shown in Eq. 5-26. $(NPR_2)_3$ with two organic substituents per phosphorus atom only undergo equilibration to $(NPR_2)_4$ when heated, but no ROP.

$$(5\text{-}26)$$

The first intermediate in the synthesis of cyclic $(NPCl_2)_3$ from PCl_5 and NH_4Cl is $Cl_3P=NH$. The reaction then proceeds stepwise by elimination of HCl. In a promising new approach, Me_3SiX (X = OR, Cl) is eliminated instead of HCl. This method makes use of the high tendency to form the strong Si–O or Si–Cl bonds. When phosphoranimines of the type $R_2(X)P=NSiMe_3$ (R = alkyl, aryl, OR′) are heated, polymeric $[NPR_2]_n$ is obtained by elimination of Me_3SiX (Eq. 5-27).

$$(5\text{-}27)$$

There are several variations of this method:

• The pyrolysis of $(RO)_3P=NSiMe_3$ gives $[NP(OR)_2]_n$. The temperature required for the thermal polycondensation (▶ glossary) depends on the nature of OR (for example, R = CH_2CF_3: 200 °C; R = aryl: 60 °C). This is an alternative route to alkoxyphosphazene polymers instead of the ROP/ substitution route.

• The polycondensation of $R_2(X)P=NSiMe_3$ (R = alkyl, aryl; X = OR′, Cl) provides a direct access to alkyl- and aryl-substituted polyphosphazenes that are difficult to obtain by the ROP/substitution approach, as discussed above.

• The room-temperature polycondensation of $Cl_3P=NSiMe_3$, initiated by small amounts of PCl_5, is a new route for the synthesis of $[NPCl_2]_n$ with controllable molecular weights and narrow polydispersities (▶ glossary).

The stability of the polyphosphazene backbone also allows the modification of the substituents bonded to phosphorus. Groups can thus be introduced that are not accessible by substitution reactions. An example is shown in Eq. 5-28.

$$(5\text{-}28)$$

Other examples are the tethering of metal complexes or even enzymes to suitable functional groups present in the organic side groups.

Methyl-substituted polyphosphazenes can be deprotonated. Substituted products are formed upon reaction of the thus generated anionic P–CH$_2^-$ groups with electrophiles. The anionic sites can, for example, be used to initiate anionic polymerization reactions (▶ glossary). A variety of graft polymers have been prepared in this manner. An example (polyphosphazene-graft-polystyrene) is given in Eq. 5-29.

$$(5\text{-}29)$$

Side-group reactivity can also be utilized for crosslinking reactions (▶ glossary). All reactions shown schematically in Figure 5-3 can be used for this purpose.

5.4 Polysilanes

Although the first polysilanes (polysilylenes) were prepared in the first half of
the 20th century, they have only attracted scientific and technical interest since
the late 1970s. Polysilanes are now sold commercially as polymeric precursors
to SiC ceramics by the Yajima process. This process will be discussed in Section
5.5. Polysilanes are promising polymeric materials mainly as UV photoresists
(▶ glossary) for microelectronics and materials for wave-guides (▶ glossary),
but also as photo- and charge-conducting materials, thermochromic materials
(▶ glossary), NLO materials (▶ glossary), or radical photoinitiators for
organic polymerization reactions (▶ glossary).

5.4.1 Properties and Applications of Polysilanes

Polysilanes range from highly crystalline, hard, brittle solids to glassy amorphous materials and rubbery elastomers, from insoluble solids to those soluble
in organic solvents or even in water. The T_gs are between $-75\,°C$ and $120\,°C$.
When purified, most polysilanes are inert to oxygen at ambient temperatures,
and reasonably stable towards hydrolysis. Some are thermally stable to almost
$300\,°C$ in an inert gas atmosphere.

Unlike silicones or polyphosphazenes, polysilanes are photochemically
degraded (see below). This enables their use as photoinitiators in a variety of
vinyl polymerization reactions (▶ glossary). The silyl radicals generated by
scission of Si–Si bonds can add to C=C double bonds and start polymerization
processes. Compared to conventional photoinitiators, the photoinitiation efficiency of polysilanes is lower by about a factor of ten. They are, nevertheless,
interesting because they have the unique advantage that they are less susceptible to inhibition by oxygen than other photoinitiators. This makes industrial
photoprocesses, such as photopolymerization of films and coatings, less expensive, because no protection from oxygen is required.

The most promising technical application of polysilanes as specialty
polymers are as photoresists (▶ glossary) in microlithography. This is a key
process for the manufacture of microchips and other electronic components. In
this process, an image is transferred to a substrate by coating it with a uniform
layer of a photosensitive polymer, the photoresist, and then irradiating it
through a mask. Photochemical reactions may crosslink (▶ glossary) the
polymer – making it insoluble (negative resist), or degrade the polymer –
making it more soluble (positive resist). Most microlithography at present is
done with positive photoresists (▶ glossary). After the developing step, the
image is transferred through the planarization layer by etching. Figure 5-11

shows a bilayer lithographic process with the photosensitive layer on top of a planarizing layer, which is not photosensitive.

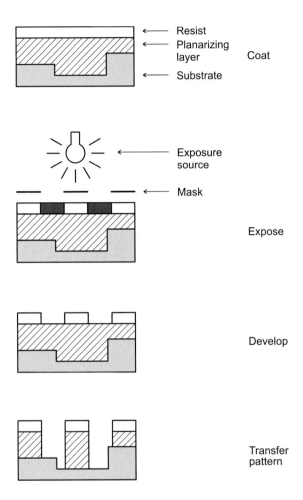

Figure 5-11. Bilayer lithographic process.

At present, visible light is used for microlithography (see Figure 3-2 for a wavelength scale). This limits the resolution to 0.5–1.5 μm due to diffraction effects. There is an increasing need for smaller resolutions on microprocessors, to increase the speed of operation, the device packing density and the number of functions that can reside on one single chip. In 1 Gbyte DRAMs, the minimum feature size is around 0.18 μm, and it will decrease in the future. This is only possible by using light of shorter wavelength in the structuring process (see Chapter 7). Polysilanes are well suited for this generation of positive photoresists (▶ glossary), since they are

- thermally and oxidatively stable, yet photochemically labile;
- strongly absorbing in the UV region, yet bleachable;
- soluble in common organic solvents and thus can be coated as films with high optical quality;
- very resistant to etching under O_2 plasma conditions; and
- resistant to X-rays, electron-beams, etc.

The exposed portion of the polysilane undergoes photochemical degradation and can be removed by washing ("developing") with solvent. The unexposed polymer is left in place and gives a positive image of the original structure. High-resolution images can also be produced directly by photoablative development. In this process, the polysilane is exposed to an excimer laser by which the polymer is degraded and volatilized. Combined with subsequent O_2 plasma etching, the overall procedure constitutes a solvent-less lithographic process.

More than in any other inorganic polymer, the physical and chemical properties of polysilanes are influenced by both the kind of the backbone and by the properties of the substituents.

Effects originating from the backbone

One of the most remarkable properties of polysilanes (and related polygermanes) is the extensive σ-*electron delocalization* along the Si–Si chain. Many of the technical uses and the remarkable properties of polysilanes result from this unusual mobility of the σ-electrons. As a consequence, their electronic and photochemical behavior is very different from that of most other inorganic or organic polymers.

Other polymers, such as polyacetylene and polythiophene, also show electron delocalization, but in these materials the delocalization involves π-electrons. The main reason for the delocalization of the σ-electrons in polysilanes is the relatively large interaction between adjacent σ-orbitals, because they are more diffuse than C–C-σ-orbitals. For the description of bonding in polysilanes the band model is therefore more appropriate than a model with localized bonds. The situation is comparable to the difference between diamond and elemental silicon.

When polysilanes absorb UV light, electrons are promoted from the σ valence band to the σ* conduction band. Because this transition is permitted, the electronic absorptions are intense, with extinction coefficients of 3×10^3 to 10^4 per Si–Si bond. As the number of silicon atoms in the chain increases, the σ/σ* energy gap becomes smaller, i.e., there is a red-shift for the σ→σ* transition. The absorption maximum of Me_3Si–$SiMe_3$ is at 220 nm, in poly(dialkylsilanes) it is shifted to 300–325 nm. Aryl substituents bonded to silicon cause a red shift of 20–30 nm per substituent due to the electronic interaction of π-orbitals with the orbitals of the Si–Si backbone (see Figure 3-2 for a wavelength scale).

Both the extinction coefficient and the wavelength of the maximum absorption depend on the chain length and reach a limit for about 40–50 silicon atoms (Figure 5-12). Therefore, polysilanes show spectral bleaching, which is the key to a number of applications. The reason is the photochemical scission of Si–Si bonds. Shorter oligosilanes are thus formed which absorb at lower wavelengths and have lower extinction coefficients.

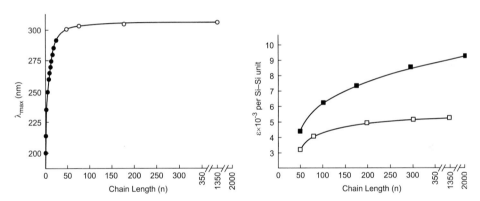

Figure 5-12. Absorption wavelength maxima (λ_{max}) and extinction coefficients (ε) as a function of the Si–Si chain length (n). ● $(SiMe_2)_n$, ○ and □ $(Si(Me)(n\text{-dodecyl}))_n$, ■ $(SiMePh)_n$.

The major reactions in photolysis reactions of polysilanes (Figure 5-13) are homolysis to give silyl radicals and silylene elimination. Bond homolysis occurs at all wavelengths absorbed by the polymer, while R_2Si is only eliminated at wavelengths <300 nm. Some chain scission with transfer of a substituent also takes place. The reactive intermediates generated by the photochemical reaction (silylenes, silyl radicals) undergo characteristic reactions, such as hydrogen abstraction, disproportionation, addition or insertion reactions.

When 254-nm UV light is used, photolysis ceases at the disilane stage because UV radiation is no longer absorbed.

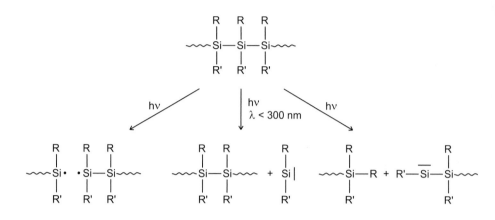

Figure 5-13. The most important photochemical reactions of polysilanes: homolysis (left), silylene extrusion (center), chain scission with transfer of a substituent (right).

The σ–σ* separation and thus λ_{max} also depends on the conformation of the polysilane chain. λ_{max} increases as the number of *trans*-Si–Si–Si–Si conformations increases. Therefore, many polysilanes show strong reversible UV thermochromism (▶ glossary) both in solution or in the solid state. For example, poly(di-*n*-hexylsilane) undergoes a striking change in UV absorption from λ_{max} = 372 nm to 317 nm when the temperature is raised above 42 °C (an absorption maximum at 317 nm is also observed in solution at room temperature). The reason for this change is a profound conformational change at the transition temperature. The low-temperature form is highly crystalline with an all-*trans*, zigzag arrangement of the polymer chain. The high-temperature form is a hexagonal columnar liquid-crystalline phase (packing of the chains) in which the polymer chain adopts an all-*gauche* helix.

A piezochromic behavior was observed for some polysilanes, which also originates from structural changes. For example, when pressure is applied to poly(di-*n*-hexylsilane) films at temperatures above 42 °C, λ_{max} changes reversibly from 317 nm to 372 nm as in the low-temperature conformation. Thus, the ordered *trans* structure is adopted, when pressure is applied.

The delocalization of σ-electrons is also the reason for the conductivity of polysilanes. Pure polysilanes are insulators, but treatment with oxidizing agents renders them electrically conducting. Conductivities as high as 0.5 S·cm^{-1} were reached by doping crosslinked "poly(silastyrene)", $[PhMeSi]_n[Me_2Si]_m$, with AsF_5. This value is in the range for semiconductors (▶ glossary). The conductivity of polysilanes arises from radical cations (a "hole") created in the polysilane chain by the oxidizing agent migrating through the polysilane.

Polysilanes also show photoconductivity when irradiated with UV light, and are excellent charge transport materials in electrophotography with high drift mobilities (~10^{-4} $cm^2 \cdot V^{-1} \cdot s^{-1}$). The activation energies associated with the charge hopping are very low for a polymeric material. Polysilanes are unique as charge transport materials because the active sites are on the polymer backbone itself. In other electrophotographic materials, charged sites are on the substituents.

Effects originating from the substituents

Some physical properties of polysilanes can be attributed to structural effects or structural changes induced by the substituents.

$[Me_2Si]_n$, $[MeEtSi]_n$, $[Et_2Si]_n$, and most $[aryl_2Si]_n$ are crystalline because of crystallization of the side chains. Such polysilanes are insoluble and infusible. Crystallinity is reduced by decreasing the symmetry or by longer alkyl chains. Thus, most poly(aryl,alkyl)silanes, the higher poly(methyl,alkyl)silanes, $[MeRSi]_n$, and the symmetrical poly(dialkylsilanes), $[R_2Si]_n$ with R = n-propyl or larger, are meltable solids or rubbery elastomers soluble in organic solvents. The minimum T_g for $[MeRSi]_n$ at about $-75\,°C$ is reached for R = n-hexyl. Crystallinity is also lower for copolymers than for the homopolymers with the same substituents. For example, the copolymer $[Me_2Si]_n[Ph_2Si]_m$ is soluble and meltable. In the technically important "poly(silastyrene)" copolymers, $[PhMeSi]_n[Me_2Si]_m$, crystallinity is at a minimum near a n:m ratio of 1.

The poor compatibility of the Si–Si backbone and the organic substituents provides morphologies with liquid crystal-like mesophases (▶ glossary). Instead of isotropic melting, polysilanes exhibit a first-order transition associated with the melting of the side group arrangements and partial disorder of the backbone conformation in addition to a low-temperature T_g.

Even water-soluble polysilanes were prepared by the proper choice of the substituents, such as R = $CH_2O(C_2H_4O)_4Me$ or aryl-OH groups.

Network polysilanes

Most work on polysilanes was done with linear polymers consisting of only R_2Si units. "Polysilynes" $[RSi]_n$ have been relatively little investigated, but it is likely that they will become increasingly important. All the evidence indicates that they have a network structure in which each silicon is linked to three other silicon atoms by σ-bonds. The polysilynes are therefore intermediate between the linear polysilanes (two Si–Si bonds per silicon atom) and elemental silicon (four Si–Si bonds). Therefore, the polysilynes should exhibit a higher electron delocalization than the linear polysilanes.

All the polysilynes are yellow or orange, and their electronic spectra differ greatly from those of linear polysilanes. For alkyl groups larger than ethyl, polysilynes are soluble in some organic solvents and can be cast into transparent films.

5.4.2 Preparation and Modification of Polysilanes

Dehalogenation of diorganodichlorosilanes (Wurtz coupling)

Most polysilanes are synthesized by dehalogenation of diorganodichlorosilanes (R_2SiCl_2 or $RR'SiCl_2$) with finely divided sodium metal above its melting point (98 °C) in an inert, high-boiling solvent such as toluene, xylene, decane, diethyleneglycol dimethylether (diglyme) or mixtures of these (Eq. 5-30). This is also the technically applied method. Other reducing agents, such as potassium metal, Na/K alloy (▶ glossary) or C_8K, have also been used, but the more reactive reagents often result in degradation of the polymer. The reaction is highly exothermic. Few functional groups can withstand such vigorous conditions; therefore, substituents are generally limited to alkyl, aryl, or silyl groups or other substituents that are intrinsically stable in a strongly reducing environment, such as fluoroalkyl or ferrocenyl.

$$RR'SiCl_2 \quad \xrightarrow[\text{T > 100 °C}]{\text{Na/solvent}} \quad \left[\begin{array}{c} R' \\ | \\ -Si- \\ | \\ R \end{array} \right]_n \quad + \quad 2\ NaCl$$

(5-30)

Cyclic oligomers (cyclosilanes) are the only products at equilibrium for the reasons discussed in Section 5.1. They cannot be converted to polymers. High molecular weight polymers are formed only when the reaction is kinetically controlled, but concomitant formation of cyclic oligomers cannot be avoided. The molecular weight distributions are usually broad and polymodal (▶ glossary). A typical distribution for poly(methylphenylsilane) consists of a moderate amount of oligomeric material ($M_w < 10^3$), mostly cyclopentasilane and cyclohexasilane, a significant amount of a polymer of intermediate molecular weight ($M_w < 5 \times 10^4$), and a small amount of a high molecular weight polymer. The oligomeric fraction is readily removed by solvent extraction, but the other two polymer fractions are inseparable. Their polymer yields depend upon the substituents on silicon and the reaction conditions, and are typically in the range of 10 to 50 %. The molecular weight and the molecular weight distribution also depend on the reaction conditions.

When mixtures of silanes are employed, either random or block-like copolymers are produced, depending on the relative reactivities of the chlorosilanes. In block copolymers, the monomeric units are arranged in blocks (for example, AAAABBBBBAAAABBBB, etc.), in random copolymers their sequence is statistical (for example, BAABABBBAAAABBABAB, etc.). Random copolymers are obtained for similar chlorosilanes, for example *n*-BuMeSiCl$_2$ and *n*-HexMeSiCl$_2$. If the chlorosilanes have rather different reaction rates, one will be consumed preferentially to generate block-like copolymers. For example, poly(silastyrene), [PhSiMe]$_n$[Me$_2$Si]$_m$, has a rather block-like structure.

The polysilynes [RSi]$_n$, are prepared by reduction of alkyl- or aryltrichlorosilanes (RSiCl$_3$) in pentane with an emulsion of Na/K alloy.

The reactions leading to the formation of polysilanes are very complex. A simplified mechanism is shown in Figure 5-14. It more closely resembles a chain growth process initiated by the sodium surface than a classical condensation reaction. The reaction of chlorosilanes with sodium probably proceeds by initial formation of a silyl anion by two single electron transfer steps via a silyl radical. The propagation step is the reaction of anion-terminated chains with dichlorosilane to add one R$_2$Si unit. This step is fast but nevertheless rate-determining, and is very probably the major reaction for chain growth. The new terminating Si–Cl bond reacts again with sodium, such that the polysilane chain grows away from the metal surface. The monomeric or polymeric radicals can also recombine.

In the presence of crown ethers or diglyme, the polymer yields are increased, and the molecular weight distributions tend toward the monomodal (▶ glossary) as more polymer is formed within the intermediate M_w range. The ethers probably clean the sodium surface by solubilizing absorbed sodium salts and also complex the sodium cations of the ≡Si⁻ Na⁺ ion pairs. Ultrasound irradiation greatly accelerates the polycondensation (▶ glossary) at temperatures lower than the melting point of the alkali metal. The reaction can then be carried out about 60 °C. Sonification offers another possibility for achieving activation and regeneration of the metal surface than thermal activation.

Other intermediates than silyl radicals may also be involved. The balance among competing mechanisms may depend on solvent, temperature, sodium

surface area, additives, and the nature of the substituents at silicon. It is therefore necessary to optimize the reaction conditions for each particular polysilane. Despite these drawbacks, this route is still the most practical.

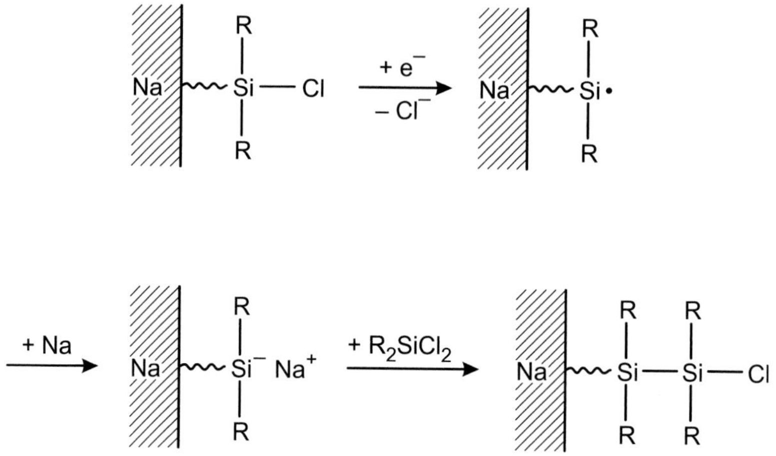

Figure 5-14. Mechanism for the dehalogenation of R_2SiCl_2 by sodium metal.

Considering that alkali metals are expensive and relatively dangerous to handle on an industrial scale, several electroreductive routes have been proposed.
The best results were obtained using a sacrificial magnesium or aluminum
anode in an undivided cell at room temperature. The mechanism is very similar
to that in Figure 5-14. The silyl anion is generated at the cathode by reduction
of a Si–Cl group ($\equiv$Si–Cl + 2 e$^-$ → $\equiv$Si$^-$ + Cl$^-$) and then reacts in solution with
another chlorosilane. At the anode, the metal is oxidized.

Anionic polymerization of masked disilenes

In contrast to olefins, disilenes $R_2Si=SiR_2$ are only stable at room temperature if they are kinetically stabilized by bulky substituents. This excludes the
possibility of preparing polysilanes from stable disilenes by polymerization (▶
glossary).

High molecular weight polymers are produced by reaction of alkyl anions
with disilabicyclooctadienes, as for example shown in Eq. 5-31. These compounds can be considered to contain masked disilenes. However, the
mechanism of polymerization does not involve disilenes. Instead, regiospecific

attack of the alkyl anion occurs in the initiation step. The reaction then proceeds as a living anionic polymerization with elimination of the disilene fragment as a new silyl anion that can continue the reaction chain. This route is
useful for making alternating polysilane copolymers or stereoregular polysilanes, and the first fully ordered polysilane of the ABAB-type was obtained in
this way. Ordered polysilanes may show unusual electronic behavior, and are
hence of theoretical and perhaps practical interest.

(5-31)

Ring-opening polymerization

Only *strained* cyclosilane rings undergo ROP under kinetically controlled conditions. The cyclotetrasilane $(Me_2Si)_4$ polymerizes to $[Me_2Si]_n$ on long-term
standing. Poly(methylphenylsilane) has been synthesized by ROP of the
strained cyclosilane $(PhMeSi)_4$ on reaction with elemental potassium, butyllithium, or silylcuprates. The ROP route allows for the preparation of well-
defined polysilanes with controlled molecular weights and microstructures, low
polydispersities, and defect-free structures.

Dehydrogenative coupling of diorganosilanes

Metal-catalyzed dehydrogenation of diorganosilanes is an attractive alternative
for polysilane synthesis (Eq. 5-32). This reaction might be more easily controlled than the usual alkali metal condensation of dichlorosilanes. Due to the
mild reaction conditions, a greater variety of organic groups could probably be
introduced directly. Although there are promising results in this area, a sufficiently efficient catalyst is still lacking.

$$RR'SiH_2 \xrightarrow{\text{catalyst}} \left[\begin{array}{c} R' \\ | \\ Si \\ | \\ R \end{array} \right]_n + H_2$$

(5-32)

Currently, the best catalysts appear to be dialkyltitanocenes and dialkyl-zirconocenes, Cp_2MR_2 (M = Ti, Zr). Truely polymeric compounds are only formed from primary silanes, $RSiH_3$. With diorganosilanes R_2SiH_2, the products are oligomers of different size, with a maximum chain length of about 20 silicon atoms. The polymerization is accelerated by the addition of olefins, which are hydrogenated and therefore act as "H_2 sponges". The main problem with the currently available catalysts, which operate by a σ-bond metathesis mechanism (▶ glossary), is that the metal center activates not only Si–H bonds, but also the Si–Si bonds of the oligomer; hence the metal complex also catalyzes depolymerization reactions. Another problem is the formation of cyclic products.

Redistribution reactions

Polysilanes can be prepared by the catalytic redistribution of Si–Si and Si–Cl bonds of disilanes at 250 °C as shown schematically in Eq. 5-33.

$$Cl\left[\begin{array}{c} Me \\ | \\ Si \\ | \\ Cl \end{array} \right]_n Cl + Cl-\begin{array}{c} Me \\ | \\ Si \\ | \\ Cl \end{array}-\begin{array}{c} Me \\ | \\ Si \\ | \\ Cl \end{array}-Cl \xrightarrow{\text{catalyst}} Cl\left[\begin{array}{c} Me \\ | \\ Si \\ | \\ Cl \end{array} \right]_{n+1} Cl + MeSiCl_3$$

(5-33)

The most effective catalyst seems to be hexamethylphosphoric acid triamide (HMPA). This synthetic route uses mixtures of disilane by-products from the "direct process" for the synthesis of methylchlorosilanes (Section 5.2). The approximate composition of this mixture is 55 % $MeCl_2Si$–$SiCl_2Me$, 35 % Me_2ClSi–$SiMeCl_2$, and 10 % Me_2ClSi–$SiMe_2Cl$. Monosilanes are distilled off as well as Me_2ClSi–$SiMe_2Cl$, which does not react. The resulting yellow pyrophoric polymer is highly branched and still contains some chlorine.

Chemical modification of polysilanes

The presence of functional groups in polysilanes is limited by the vigorous conditions of the usually employed sodium condensation reaction (Eq. 5-30). Any functions which react with molten sodium metal must therefore be introduced after the polymer is synthesized. Since the polysilane skeleton is rather sensitive towards acids or bases, the modification reactions must be well designed. Some examples are shown in Table 5-2.

Table 5-2. Examples for the modification of substituents in polysilanes.

$\equiv$Si–X–CH=CH$_2$	HBr/BBr$_3$	$\rightarrow$	$\equiv$Si–X–CH$_2$CH$_2$Br
$\equiv$Si–X–C$_6$H$_4$OSiMe$_3$	MeOH	$\rightarrow$	$\equiv$Si–X–C$_6$H$_4$OH
$\equiv$Si–Ph	HCl/AlCl$_3$ /C$_6$H$_6$	$\rightarrow$	$\equiv$Si–Cl
$\equiv$Si–Ph	CF$_3$SO$_3$H/CH$_2$Cl$_2$	$\rightarrow$	$\equiv$Si–OSO$_2$CF$_3$
$\equiv$Si–H	CH$_2$=CHR/AIBN	$\rightarrow$	$\equiv$Si–CH$_2$CH$_2$R

The thus introduced functional groups can be further converted to other groups by conventional reactions. The fraction of Si–Ph or other Si–aryl groups replaced by Si–Cl upon treatment with a HCl and a Lewis acid can be varied by the amount of HCl used and the reaction conditions. Up to 80 % of the phenyl groups of [PhMeSi]$_n$ can be replaced by reaction with trifluoromethane sulfonic (triflic) acid, CF$_3$SO$_3$H, in CH$_2$Cl$_2$. Some cleavage of Si–Si bonds and consequent degradation of the polymer takes place at high levels of substitution, but the molecular weight is little changed by replacement of up to 30 % of the phenyl groups. The chloro or triflate groups on silicon are highly reactive and can be replaced by a variety of nucleophiles without disturbing the Si–Si bond to give substituted polysilanes.

5.4.3 Crosslinking of Polysilanes

Crosslinking ($\blacktriangleright$ glossary) is essential if the polysilanes are used as precursors to SiC ceramics (Section 5.5). Without crosslinking, most of the polymer is volatilized before rearrangement to polycarbosilanes or thermolysis to SiC (see Section 5.5) can take place. Common methods for the crosslinking of polysilanes are as follows:

- When vinyl or other alkenyl groups are present as substituents, photolysis leads to crosslinking, which dominates over photochemical degradation of the polysilane chain. The crosslinking probably results by addition of initially formed silyl radicals to the C=C double bonds on neighboring chains.

- Related reactions can occur when organic molecules with two or more unsaturated groups are mixed with the polysilane. Typical examples are tetravinylsilane, $Si(CH=CH)_4$, or the hexavinyl-substituted compounds $(CH_2=CH)_3Si(CH_2)_nSi(CH=CH_2)_3$. The radicals necessary to cause crosslinking can be formed by photolysis of the polysilane or by heating the mixture with a free radical initiator such as AIBN.

- Polysilanes that contain Si–H bonds can be crosslinked by hydrosilylation reactions by, for example, addition of trivinylphenylsilane in the presence of a catalyst. This process allows "room temperature vulcanization" (▶ glossary) of polysilanes, analogous to the RTV silicone elastomers (Section 5.2).

5.5 Polycarbosilanes

Polycarbosilanes are a class of polymers with carbon and silicon atoms forming the backbone. The sequence of atoms in the industrially important carbosilanes is –Si–C–Si–C–. Polycarbosilanes are used exclusively as preceramic polymers for SiC; technical applications as polymeric materials are currently not at hand. However, new preparation methods and polycarbosilanes with other sequences of the Si and C atoms, such as $-Si-(C)_x-Si-$, $-Si-(C=C)_x-Si-$, –Si–aryl–Si–, etc. may change this situation in the future. Copolymers from polysilanes and organic polymers are not considered polycarbosilanes.

5.5.1 SiC Fibers from Polycarbosilanes (Yajima Process)

The preparation and properties of SiC, one of the most important high-performance non-oxide ceramic materials, have been discussed in Section 2.1.2. The traditional routes to SiC ceramic powders are not suitable for making ceramic films (at reasonable temperatures) or fibers.

SiC filaments are technically used in ceramic- (CMC), metal- (MMC), glass- or polymer–matrix (PMC) high-performance composites (▶ glossary). Continuous-fiber reinforced materials have the advantage that they do not fail catastrophically, because after matrix failure, the fiber can still support a load. CMCs provide fracture toughness values similar to those for metals. SiC filaments are particularly suitable for high-temperature use by incorporating them in BN, B_4C, AlN, SiC, Si_3N_4, TiC, TiO_2, TiB_2, carbon, etc. For example, SiC fiber-reinforced SiC composites are suitable for structural components of motors or turbines. SiC filaments have also been used to reinforce a variety of organic polymers. For example, epoxy/SiC composites possess compressive strength twice that of epoxy/graphite. SiC fiber–Al matrix composites may serve as examples for MMCs. They are lightweight, high-strength materials used for components in aircrafts, helicopters, missiles, bridge structures, etc.

SiC fibers are commercially available as Nicalon® and Tyranno® fibers. The latter contain 1.5–4 wt% titanium. The production is currently about 2×10^4 kg per year. Nicalon fibers have an overall composition of $Si_{0.58}C_{0.31}O_{0.11}$. About 40–50 % of the silicon atoms are surrounded by four carbon atoms (as in crystalline SiC), about 6 % by four oxygen atoms (as in crystalline SiO_2), and the remaining fraction is substituted by both oxygen and carbon atoms. Nicalon fibers are black due to the presence of graphite crystallites.

SiC fibers of 10–20 μm diameter exhibit tensile strengths of ~2.4–3.3 GPa (depending on the diameter), elastic moduli of 200–270 GPa and densities of 2.3–2.6 $g \cdot cm^{-3}$, which is distinctly less than for the bulk ceramic material. By comparison, SiC single crystals of 0.3–1.0 μm diameter (density ~3.2 $g \cdot cm^{-3}$ depending on the polymorph) exhibit tensile strengths of ~8 GPa and elastic moduli of ~580 GPa. Above 1200 °C, the mechanical strength of the SiC fibers is lost. Solid-state reactions between the oxygen introduced during the curing step (see below) and SiC generate gaseous CO and SiO. The gas evolution causes significant damage. Furthermore, rapid crystallization of β-SiC occurs above 1200 °C.

The properties of ceramic fibers are greatly influenced by the composition and structure of the preceramic polymer. The requirements for an ideal preceramic polymer have been discussed in Section 5.1. Polymers for fiber spinning must additionally be adjusted to the requirements of the fiber-making process (Figure 5-15).

Preparation of SiC fibers

This is carried out as a step-wise process:

- Step 1: *Spinning of the green fiber.* Spinning of SiC fibers is typically carried out by extruding the melted polymer through a spinneret ("melt-spinning"), but dry spinning is also possible.

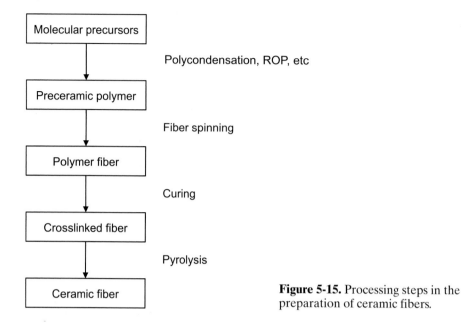

Figure 5-15. Processing steps in the preparation of ceramic fibers.

In the latter process, the solution of the polymer is extruded into a heated drying chamber where the solvent is volatilized, leaving the solid polymer fiber. The polymer viscosity should exhibit non-Newtonian behavior during spinning, i.e., it should be shear-dependent. This behavior is obtained when the particles in the liquid strongly interact with each other. A compromise must be found between the rheological properties of the polymer and the requirements for obtaining high ceramic yields (▶ glossary). The viscosity should be sufficiently high that the green fiber retains its shape and supports itself. If the polymer is too highly crosslinked, the ceramic yield would probably be high, but melt spinning will not be possible.

- Step 2: *Curing*. The green (▶ glossary) polymer fibers need to be cured (crosslinked) below their melting point to render them infusible, so that the fiber form is retained on pyrolysis. The polymer therefore must have some latent reactivity for crosslinking.

- Step 3: *Pyrolysis*. After spinning, the polymer is pyrolyzed to give the desired ceramic material. The pyrolysis chemistry and the reaction conditions must be carefully controlled to ensure removal of the gaseous products and uniform densification. The polycarbosilane fibers are converted to β-SiC by firing at 800–1400 °C in N$_2$ atmosphere. The weight loss during this process is 20–40 %, and the volume decrease is 60–80 %.

5.5.2 Chemical Issues of Polymer Preparation, Curing and Pyrolysis

Polymer preparation

Kumada rearrangement. When poly(dimethylsilane) is heated under argon, the rearrangement shown in Eq. 5-34 ("Kumada rearrangement") takes place.

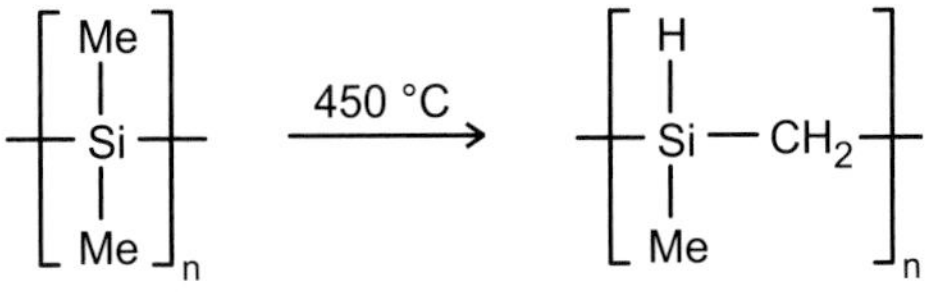

(5-34)

In reality, the structure of the obtained polycarbosilane is more complex. Particularly, there are other structural elements than the –Si(H)Me–CH$_2$– units in the idealized formula. A ladder-like structure with rings and chains was proposed from spectroscopic investigations, as shown schematically in Figure 5-16. There are three to four branchings per ten silicon atoms. However, the carbon and silicon atoms are always alternating. The polycarbosilanes isolated after removal of volatile compounds are glassy resins that are meltable and soluble in organic solvents.

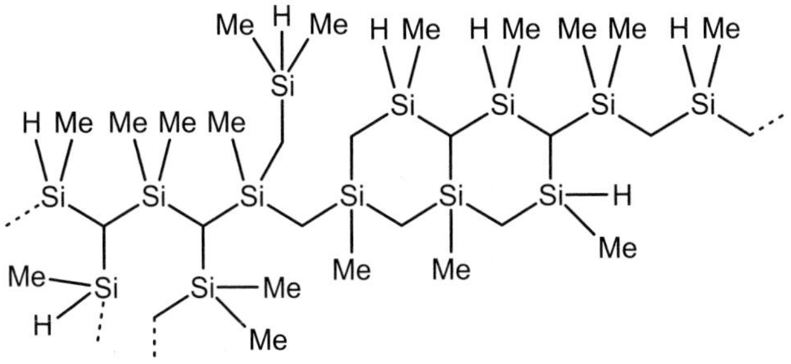

Figure 5-16. A structural model for the polycarbosilane obtained by Kumada rearrangement of poly(dimethylsilane).

Considerable effort has been invested in optimizing this step, because the Kumada rearrangement is the critical step in the generation of polycarbosilanes with the correct processing properties for fiber spinning. In the most common method, poly(dimethylsilane) is heated to 320 °C in flowing argon, melted, refluxed for 5 hours, and then heated to 470 °C to remove volatile materials. The resulting polycarbosilane is obtained in about 55 % yield and has an average molecular weight $M_n = 1500$. Total conversion should lead to poly(silapropylene), $[Si(H)MeCH_2]_n$. However, the typical chemical analysis for Yajima polycarbosilanes prepared from poly(dimethylsilane) by the Yajima process is around $SiC_{1.8}H_{3.7}$ instead of SiC_2H_6.

The primary reaction in the Kumada rearrangement probably involves pyrolytic cleavage of Si–Si bonds. The thus formed silyl radicals give $CH_2=SiMe-$ and Me_2HSi-terminated fragments which subsequently combine to form Si–CH_2–Si linkages. Alternative mechanisms involving the initial formation of silenes and silylenes are also consistent with the experimental results.

Lewis acid catalysts improve the conversion of polysilanes to polycarbosilanes. The commercially used catalyst is poly(borodiphenylsiloxane) $[B(OSiPh_2O-)_{3/2}]_n$, prepared by condensation of boric acid, H_3BO_3, and Ph_2SiCl_2. Other boron compounds such as $B(OR)_3$ or $B(NEt_2)_3$ were equally successful. Polymers are typically obtained in about 60 % yield at ambient pressure and ~350 °C by this modification of the process. Spectroscopic investigations indicate that the polymers also contain Si–Si bonds and are more branched than the polymers prepared without the use of the catalyst. A disadvantage is that oxygen is already introduced in this step. The role of the catalyst in the formation of the polycarbosilane probably is to enhance dehydrogenative coupling of Si–H bonds to form Si–Si crosslinks.

C–Cl/Si–Cl dehalogenation. There are several approaches to obtain the desired Si/C sequences in carbosilane chains by dehalogenative coupling of C–Cl and Si–Cl groups. The most obvious precursors are chloromethyl(chloro)silanes, $ClCH_2SiR_2Cl$, because they possess the desired CH_2–Si unit. Reaction of these precursors with elemental magnesium either gives the cyclic compounds $[SiR_2CH_2]_n$ (n = 2–4) (mostly the 1,3-disilacyclobutanes [n = 2] which can be ring-opened; see below) or polymers, depending on the electronic and steric properties of the substituents at silicon and the reaction conditions. For example, while $ClCH_2SiMeCl_2$ only gives oligomeric and polymeric products, a high portion of the disilacyclobutane is obtained when $ClCH_2Si(OR)Cl_2$ is similarly reacted with magnesium. A 53 % yield of the polymer $[SiMe_2CH_2]_n$ and only 7 % of cyclic $[SiMe_2CH_2]_2$ is obtained when $Me_2ClSiCH_2Cl$ is added >to a suspension of Mg in THF, while the inverse addition, i.e., addition of Mg to the silane solution, gives 13 % of [SiMe$_2$CH$_2$]$_n$ and 69 % of cyclic compounds.

Highly branched polymers having the correct Si:C ratio for the formation of SiC (see below) were prepared by coupling of ClCH$_2$SiCl$_3$ with magnesium, followed by reduction with LiAlH$_4$ (Eq. 5-35). The extensive branching is due to the SiCl$_3$ group which can couple with up to three carbon atoms. The formula [SiH$_2$CH$_2$]$_n$ (as in Eq. 5-35) is therefore only idealized.

$$ClCH_2\text{-}SiCl_3 \xrightarrow{Mg/Et_2O} +CH_2\text{-}SiCl_2+_n \xrightarrow{LiAlH_4} +CH_2\text{-}SiH_2+_n$$

$$(5\text{-}35)$$

The most direct route to carbosilanes is the coupling of CH$_2$X$_2$ and R$_2$SiX'$_2$. Several polycarbosilanes with moderate molecular weight were prepared by this route by condensation of dihalomethane derivatives, mostly CH$_2$Br$_2$, with dichlorosilanes (Eq. 5-36). The general approach can also be used to prepare polycarbosilanes with other Si/C sequences, as exemplarily shown in Eq. 5-37 and 5-38.

$$n\ RR'SiCl_2 + n\ CH_2Br_2 + 4n\ Na \xrightarrow[-NaBr]{-NaCl} \left[\overset{R}{\underset{R}{Si}}\text{-}CH_2\right]_n$$

$$(5\text{-}36)$$

$$Li+C\equiv C+_x Li\ +\ Cl\text{-}\overset{R'}{\underset{R}{Si}}\text{-}Cl \xrightarrow{-LiCl} \left[\overset{R'}{\underset{R}{Si}}(C\equiv C)_x\right]_n$$

x = 1,2

$$(5\text{-}37)$$

$$Li\text{-}C\equiv C\text{-}C\equiv C\text{-}Li\ +\ ClMe_2Si\text{-}SiMe_2Cl$$

$$\longrightarrow \left[\overset{Me\ Me}{\underset{Me\ Me}{Si\text{-}Si}}\text{-}C\equiv C\text{-}C\equiv C\right]_n$$

$$(5\text{-}38)$$

Ring-opening polymerization (ROP). Several types of polycarbosilanes can be prepared by metal-catalyzed ROP of 1,3-disilacyclobutane derivatives (Eq. 5-

39). Hexachloroplatinic acid, H_2PtCl_6, is a typical catalyst. This route provides strictly linear polycarbosilanes if there are no substituents at the silicon atoms that can undergo crosslinking reactions (▶ glossary; such as R = H or vinyl).

$$(5\text{-}39)$$

For example, ROP of 1,3-disilacyclobutane (R = R′ = H) gives a highly branched polymer. Unbranched poly(silaethylene), $[SiH_2CH_2]_n$, was prepared by ROP of 1,1,3,3-tetrachloro-1,3-disilacyclobutane (R = R′ = Cl in Eq. 5-39) followed by $LiAlH_4$ reduction of the chloro-substituted polymer, $[SiCl_2CH_2]_n$. Linear $[Si(H)MeCH_2]_n$, analogously prepared from the corresponding disilacyclobutane (R = Me, R′ = Cl) by this route, has different properties to the polycarbosilane obtained by the Kumada rearrangement. This shows again that the latter has a different structure, as discussed above.

Unbranched $[SiR_2CH_2]_n$ is chemically related to polyolefins $[CR_2CH_2]_n$ and polysiloxanes $[SiR_2O]_n$ which have similar glass transition temperatures. The asymmetrically substituted polymers $[Si(Me)(n\text{-}alkyl)CH_2]_n$ exhibit very low T_gs down to –91 °C.

Chlorine substituents in the polymer can be replaced by other groups such as OEt or saturated and unsaturated organic groups. Likewise, Si–H groups can be added to olefins, which is another approach to introduce a specific side chain into the polycarbosilane.

ROP of 1,3-disilacyclobutanes gives polycarbosilanes with alternating silicon and carbon atoms in the polymer chain. Polycarbosilanes with other Si/C sequences can be prepared by anionic ROP of silacyclobutanes (giving [Si–C–C–C]$_n$ sequences) or silacyclopent-3-enes (giving [Si–C–C=C–C]$_n$ sequences).

Hydrosilylation and metathesis of alkenylsilanes. Hydrosilylation allows another direct route to polycarbosilanes. For example, vinyldichlorosilane, $SiH(CH=CH_2)Cl_2$, contains both a Si–H and an olefinic group. Thus, the Si–H group of one molecule can add to the vinyl group of the next, etc. to give a polycarbosilane. The obtained chlorosilane polymer is then reduced by $LiAlH_4$ (Eq. 5-40). Hydrosilylation mainly affords linear –SiCH$_2$CH$_2$– units, and only a minor portion of branched –SiCH(Me) – units.

$$\underset{\substack{| \\ Cl}}{\overset{\substack{Cl \\ |}}{H-Si-CH=CH_2}} \xrightarrow{\text{cat.}} H\left[\underset{\substack{| \\ Cl}}{\overset{\substack{Cl \\ |}}{-Si-CH_2CH_2}}\right]_n \underset{\substack{| \\ Cl}}{\overset{\substack{Cl \\ |}}{Si-CH=CH_2}} \xrightarrow{\text{LiAlH}_4} H\left[\underset{\substack{| \\ H}}{\overset{\substack{H \\ |}}{-Si-CH_2CH_2}}\right]_n \underset{\substack{| \\ H}}{\overset{\substack{H \\ |}}{Si-CH=CH_2}}$$

$$(5\text{-}40)$$

Allyl- or ethynyl-substituted silanes, $R_2SiH(CH_2CH=CH_2)$ or $R_2SiH(C\equiv CR')$ can be similarly polymerized by hydrosilylation, giving polymers with other Si/C sequences.

Other organic polymerization reactions (▶ glossary) also allow the formation of polycarbosilanes. For example, high molecular unsaturated polycarbosilanes were prepared by metal complex-catalyzed metathesis (▶ glossary) polymerization (Eq. 5-41). Any other organosilicon compound with two unsaturated substituents can be similarly used.

$$\underset{\substack{| \\ Me}}{\overset{\substack{Me \\ |}}{H_2C=CH-Si-CH=CH_2}} \xrightarrow{\text{cat.}} \left[\underset{\substack{| \\ Me}}{\overset{\substack{Me \\ |}}{-CH_2-Si-CH_2-}}\right]_n + n\ C_2H_4$$

$$(5\text{-}41)$$

Crosslinking

The polycarbosilanes obtained by one of the above-mentioned methods must be cured to give reasonable ceramic yields (▶ glossary) of SiC upon pyrolysis. For example, the linear polycarbosilanes obtained by ROP of substituted disilacyclobutanes decompose with very low ceramic yields if they contain no latent functionality (Si–H or Si–CH=CH₂ groups) because cyclic or linear, low-boiling oligomers are formed which distill off. The polycarbosilanes obtained by the Kumada rearrangement, on the other hand, are of sufficiently low molecular weight ($M_n \sim 1\text{--}2\times10^3$) that they can be melt-spun. However, the green fiber (▶ glossary) must be cured. The commercially available SiC fibers are cured by heating the spun polycarbosilane fibers at ~200 °C in air for 30 minutes. This converts surface Si–H groups into Si–O–Si groups. Pyrolysis of the air-cured fibers gives ceramic yields (▶ glossary) of 80–85 %, compared to 60–70 % without curing. Since significant amounts of oxygen are introduced into the fiber material, several alternative approaches were investigated.

- High-molecular weight polycarbosilanes were prepared by thermal decomposition of poly(dimethylsilane) or poly(silastyrene) in an autoclave instead of the two-step process described before. They do not melt, but are soluble and can be dry spun. The thus obtained fibers offer much better high-temperature stability as a consequence of their low oxygen content. The effect of the thermal pretreatment ("thermosetting") was also demonstrated for the completely linear and atactic (▶ glossary) [Si(H)MeCH$_2$]$_n$ obtained by ROP of 1,3-dichloro-1,3-dimethyl-1,3-disilacyclobutane followed by LiAlH$_4$ reduction (Eq. 5-39). The ceramic yield of the untreated polymer is quite low (20 %). However, a pretreatment under nitrogen at 400 °C results in dehydrocoupling of Si–H bonds. Due to this crosslinking reaction (▶ glossary), the ceramic yield upon pyrolysis is improved to 66 %.

- A variation of the Nicalon process consists in heating the polycarbosilane with Ti(OBu)$_4$. The titanium compound provides Ti–O–Si crosslinks. The polymer therefore has a higher molecular weight and is more easily spun. Ceramic fibers with higher ceramic yield (70 %) are obtained after pyrolysis at 1200 °C in N$_2$ atmosphere. They are marketed under the trade name Tyranno.

- The mechanical properties of the fibers at high temperatures were improved by γ-ray or electron radiation instead of curing by oxidation due to the very low oxygen content. In contrast to the original Nicalon fibers, the so-called "HI Nicalon" fibers retain a tensile strength of >2 GPa after being heated in argon for 10 hours at 1500 °C. However, this process makes the fibers more expensive.

- The incorporation of vinyl and Si–H groups into the starting polysilane allows crosslinking (▶ glossary) of the polysilane (or polycarbosilane) chains by hydrosilylation reactions. Yajima polycarbosilane, which contains Si–H bonds, can be also rendered infusible by hydrosilylation with appropriate amounts of an unsaturated compounds prior to pyrolysis as discussed in Section 5.4.3. Low oxygen contents are obtained with this crosslinking (▶ glossary) method, but the carbon content is increased.

- Much effort has been made to modify the curing atmosphere. For example, essentially oxygen-free, dense SiC fibers (3.1 g·cm^{-3}) were made by curing the green polycarbosilane fiber in NO$_2$ instead of air. This reduces the oxygen level to about 5 %. If the NO$_2$-cured fiber is then treated with 5 % BCl$_3$ over a 25 to 140 °C range, the fibers then can be heated to 1800 °C to produce oxygen-free fibers. Boron serves as a sintering aid for SiC densification during CO/SiO evolution.

Pyrolysis

Pyrolysis of polycarbosilanes (obtained by the Kumada rearrangement) in an argon atmosphere proceeds in several steps. Low-molecular weight carbosilanes distill off below ~350 °C. Between ~350 °C and 500 °C, the molecular weight of the polycarbosilane increases by formation of additional Si–C–Si linkages. The thermolysis reactions progressively end as the pyrolysis proceeds. In the next stage, from ~500 °C to 900 °C, the main transformation into an inorganic material occurs, with the loss of hydrogen and methane. As Si–H bonds are more reactive than C–H bonds, the SiC_4 tetrahedra found in SiC are formed first. Tetrasubstitution of the carbon atoms, i.e., final conversion of $\equiv$Si–CH_3, $\equiv$Si–CH_2–Si$\equiv$ or $HC(Si\equiv)_3$ units in $C(Si\equiv)_4$ nodes, occurs at higher temperatures. At 900 °C the material is amorphous and can be described as a hydrogenated Si_xC_y (y >x). Crystallization of β-SiC starts at 1000–1200 °C. H_2 evolution continues in this temperature range, and represents the final stage in conversion of the polycarbosilane to the inorganic SiC network. At 1200 °C, all atoms have a local environment expected for crystalline β-SiC. The final materials contain significant portions of oxygen and free carbon. SiC fibers obtained from the titanium-modified polycarbosilane remain amorphous on heating to 1200 °C and crystallize at higher temperatures into a mixture of β-SiC and TiC.

If the pyrolytic degradation of $[Si(H)MeCH_2]_n$ would proceed optimally, then the overall reaction would produce $SiC + H_2 + CH_4$. However, the control of the Si:C ratio in the resulting fibers is still a crucial issue. They can contain an excess of elemental carbon or (less often) elemental silicon. The nature of the pyrolysis atmosphere, and the reaction conditions in general, plays a very important role in the loss of carbon. For example, a hydrogen-rich atmosphere favors the formation of methane and thus decreases the carbon content of the ceramic material.

5.6 Polysilazanes and Polycarbosilazanes

If the oxygen atom in polysiloxanes (silicones) is replaced by an NH or an NR group, one obtains polysilazanes (see Figure 5-1). However, polysilazanes are more difficult to handle, mainly due to their high reactivity towards water, particularly at low pH, and therefore have not achieved the same degree of importance. The main reason for developing polysilazanes is that they are preceramic polymers for silicon nitride (Si_3N_4).

Conventional routes for the preparation of Si_3N_4 ceramic powders have been discussed in Chapter 2, and the production of Si_3N_4 films by CVD in Section 3.2.5. Preceramic routes were mainly developed for the production of strong, high-modulus, oxidation-resistant continuous Si_3N_4 fibers that could be used in ceramic-matrix composites (CMCs). A Si_3N_4 fiber ("TNSN") commercially available from Toa Nenryo Kogyo K.K. (Tokyo) exhibits a tensile strength of 2.5 GPa, an elastic modulus of 250 GPa, and a density of about 2.5 $g \cdot cm^{-3}$.

Thermal stability and mechanical performance of amorphous Si_3N_4 is limited by crystallization which starts above 1000 °C (in the presence of elemental silicon the crystallization temperature is even lower). In carbon-containing systems the onset of crystallization is shifted to higher temperatures. Therefore, there are considerable efforts to prepare ternary Si/C/N (silicon carbonitride) fibers from polycarbosilazanes with both Si–N–Si and Si–C–Si linkages in the main chain. As in the case of SiC, the presence of oxygen in the fiber material impairs its thermal stability.

5.6.1 Preparation of Polysilazanes and Polycarbosilazanes

Reaction of chlorosilanes with amines. The most common way to form Si–N bonds is the reaction of ammonia or amines (ammonolysis or aminolysis) with halogenosilanes. Tetrachlorosilane, $SiCl_4$ reacts with ammonia under various conditions (in the liquid or gas phase) to give hydrolytically sensitive polymeric silicon diimide "$Si(NH)_2$" of unknown structure. This compound gives pure Si_3N_4 after pyrolysis at 1250 °C. Although the ammonolysis of $SiCl_4$ is useful for the production of Si_3N_4 powders or in the CVD process (Section 3.2.5), it is not suitable for obtaining fusible or soluble preceramic polymers.

Reaction of dihalogenosilanes R_2SiCl_2 with ammonia or primary amines gives oligomeric or polymeric silazanes (Eq. 5-42). As discussed in the previous sections, the proportion of cyclic products (mainly cyclotrisilazanes and cyclotetrasilazanes) is largely determined by the steric properties of the substituents, both at silicon and nitrogen. The higher the steric bulk, the larger is the proportion of cyclic products.

$$R_2SiCl_2 + 3\ R'NH_2 \longrightarrow \left[\begin{array}{c} R \\ | \\ -Si-N- \\ | \quad | \\ R \quad R' \end{array} \right]_n + 2\ [R'NH_3]Cl$$

(5-42)

Polysilazanes with various substituents (including vinyl groups for cross-linking; ▶ glossary) were prepared as preceramic polymers for silicon nitride or carbonitride ceramics. To obtain high ceramic yields (▶ glossary), methyl substituents are the second best choice after hydrogen. Thus, methylated poly-silazanes obtained from the ammonolysis reactions of $Me(H)SiCl_2$ (which is cheap and non-dangerous), Me_2SiCl_2, or $MeSiCl_3$, or from the reaction of H_2SiCl_2 with $MeNH_2$ were investigated for their use as preceramic polymers, especially for the production of ceramic fibers.

Simple polysilazanes (Si:N ratio = 1) are nitrogen-deficient with respect to the Si/N stoichiometry of Si_3N_4. Therefore, chlorosilanes were also reacted with hydrazine instead of ammonia or amines to increase the nitrogen content.

Ammonolysis or aminolysis of chlorosilanes results in large amounts of ammonium chlorides (Eq. 5-42) which are difficult to remove. This by-product is undesirable since it introduces chlorine in the preceramic polymers and acts as a catalyst in the splitting of Si–N bonds. This problem has been overcome by the Si–Cl/Si–N metathesis reaction (▶ glossary) using the readily available hexamethyldisilazane (Eq. 5-43). Volatile trimethylchlorosilane is the main by-product. This reaction can be applied for the synthesis of a wide range of polymers.

$$R_2SiCl_2 + Me_3Si-NH-SiMe_3 \longrightarrow \left[\begin{array}{c} R \\ | \\ Si-NH \\ | \\ R \end{array} \right]_n + 2\ Me_3SiCl$$

$$(5\text{-}43)$$

N–H/Si–H dehydrocondensation. The formation of Si–N bonds by hydrogen elimination from Si–H and N–H groups is promoted by several catalysts. Various monomeric hydrogenosilanes R_2SiH_2 or $RSiH_3$ were reacted catalytically with amines and ammonia. However, the hydrogenosilanes particularly interesting for the preparation of preceramic polymers, $MeSiH_3$ or SiH_4, are difficult to handle. Therefore, it is easier to prepare oligomers by ammonolysis and modify them catalytically in order to obtain a suitable ceramic precursor.

Dehydrocondensation reactions are also catalyzed by a strong base such as KH, although mainly four-membered rings are obtained. It probably proceeds by initial deprotonation of an NH group, followed by elimination of a hydride ion at a neighboring silicon atom. The resulting silylimine ($R_2Si=NR'_2$) entities then dimerize in a head-to-tail manner (Eq. 5-44).

$$2 \quad -\underset{\underset{H}{|}}{\overset{|}{Si}}-\underset{\underset{H}{|}}{N}- \quad \xrightarrow{\text{base}} \quad \overset{\overset{|}{N}}{\underset{\underset{N}{|}}{\underset{}{\searrow}Si\underset{}{\nwarrow}Si\underset{}{\swarrow}}} \quad + H_2$$

$$(5\text{-}44)$$

This method is very convenient for polymerizing oligomeric silazanes possessing –SiH–NH– groups. For example, ammonolysis of $Me(H)SiCl_2$ gives an oil that mainly contains six- and eight-membered silazane rings $[Si(H)Me–NH]_n$ (n = 3 or 4). KH-catalyzed crosslinking (▶ glossary) of the cyclosilazanes gives a polymer of the approximate composition $[Si(H)Me–NH]_{0.4}[SiMeN]_{0.6}$, which is one of the more important starting polymers for the production of Si_3N_4 or Si/C/N ceramics.

Poly(silylcarbodiimide). Reaction of $SiCl_4$ with bis(trimethylsilyl)carbodiimide, $Me_3Si–N=C=N–SiMe_3$ in organic solvents with pyridine as catalyst proceeds at room temperature by elimination of Me_3SiCl and formation of polymeric $[Si(N=C=N)_2]_n$ in which each silicon atom is linked to four neighboring silicon atoms via –N=C=N– units. The reaction is comparable with the reaction of $SiCl_4$ with water (elimination of HCl; formation of Si–O–Si links). Depending on the solvent, gels or powders can be obtained.

ROP of cyclosilazanes. High molecular weight linear polysilazanes can be prepared by anionic ROP of four- or six-membered cyclosilazanes, in some cases also by cationic or thermally induced ROP. The reactivity for the ROP process decreases with the steric bulk of the substituents. ROP of cyclosilazanes can also be promoted by transition metal complexes. Note the analogy of this reaction (Eq. 5-45) with that of Eq. 5-14.

$$n \quad \begin{array}{c} \overset{H \quad Me_2}{\overset{N-Si}{Me_2Si \diagup \quad \diagdown NH}} \\ \underset{H}{N} \quad \quad | \\ | \quad \quad SiMe_2 \\ \diagdown Si-N \diagup \\ Me_2 \quad H \end{array} \quad + \quad Me_3Si–NH–SiMe_3 \quad \longrightarrow \quad Me_3Si–NH \underset{}{\overset{}{-}} \left[\begin{array}{c} Me \\ | \\ Si \\ | \\ Me \end{array} -NH\right]_{4n} SiMe_3$$

$$(5\text{-}45)$$

Transamination. Transamination reactions play a very important role in the formation of the $N(Si\equiv)_3$ nodes during pyrolysis of the polysilazanes (see below) or in CVD processes (see Section 3.2.5). This reaction can also be employed for the preparation of preceramic polymers with the advantage that no chlorosilanes are required. Tris(dimethylamino)silane, $HSi(NMe_2)_3$, is prepared from elemental silicon and dimethylamine under Müller–Rochow conditions (see Section 5.2). In the presence of ammonia or primary amines and a catalyst, this molecular precursor readily undergoes transamination reactions (Eq. 5-46). The intermediate aminosilanes with Si–NHR or Si–NH$_2$ groups can undergo condensation reactions to give Si–NR–Si or Si–NH–Si linkages by elimination of RNH_2 or NH_3 (similar to the formation of Si–O–Si bonds by water elimination from Si–OH groups).

$$(Me_2N)_3SiH + 2\ RNH_2 \xrightarrow{\text{catalyst}} \left[\begin{array}{c} NHR \\ | \\ -Si-N- \\ |\quad | \\ H\quad R \end{array}\right]_n + 3\ Me_2NH$$

$$(5\text{-}46)$$

When tris(N-methylamino)methylsilanes, $CH_3Si(NHMe)_3$ are heated, transamination leads to the evolution of methylamine, $MeNH_2$, and the formation of highly crosslinked preceramic polysilazanes.

5.6.2 Curing and Pyrolysis Reactions

Latent reactivity in polysilazanes, leading to crosslinking (▶ glossary) upon thermal treatment, mainly results from Si–H and N–H groups (SiH/NH dehydrocondensation; see above) or from vinyl groups present in the polymer (olefinic polymerization or hydrosilylation). The efficiency for crosslinking follows the order:

hydrosilylation > SiH/NH dehydrocondensation > polymerization of vinyl groups ~SiH/SiH dehydrocondensation.

Similar results were obtained for polycarbosilazanes.

Pyrolysis of polysilazanes with organic substituents in an inert atmosphere at about 1000 °C results in single-phase, amorphous ceramic materials that contain substantial portions of carbon. For example, pyrolysis of $[Si(H)Me–NH]_{0.4}[SiMeN]_{0.6}$ at 1000 °C under N_2 gives a material of the chemical composition $Si_{2.0}N_{1.8}C_{1.0}$. The amorphous materials can be transformed in mono- or multiphased crystalline ceramics by an additional heat treatment. The metastable (▶ glossary) material is than transformed into the thermodynamically stable phases Si_3N_4, SiC and graphite.

As for the polycarbosilanes (see Section 5.5), the pyrolysis atmosphere plays a crucial role on the mechanisms and the final composition of the materials. Although nitrogen may act as a nitridating reagent above 1000 °C, ammonia is much more reactive and thus a more effective nitridation agent. If pyrolysis is carried out under ammonia atmosphere, carbon-free ceramics are obtained. It should be pointed out that the nitridation reaction with ammonia is so efficient that silicon nitride can even be prepared by pyrolysis of poly*carbo*silanes at 500–1000 °C in an ammonia atmosphere.

5.6.3 Multicomponent Systems

Multicomponent nitride and carbide ceramics are highly interesting with respect to the "tailoring" of materials properties. However, they are difficult to synthesize by traditional high-temperature routes. The preceramic polymer route offers great advantages for the synthesis of metastable, single-phase multicomponent systems because it operates at comparatively low temperatures. Multicomponent systems consisting of main group elements connected by covalent bonds (as in silicon and boron nitrides and carbides, for example) have high onset temperatures of crystallization and exhibit a high thermal stability. Below the crystallization temperature, these systems are amorphous with rigid, crosslinked networks. Although these solids are not in thermodynamic equilibrium, materials of high kinetic stability are accessible.

Addition of boron to SiC, Si_3N_4 or Si/C/N greatly enhances ceramic properties, including a reduced crystallinity and an improved thermal and oxidative stability.

There are two different strategies for the synthesis of preceramic polymers for the Si/B/(C)/N system:

- Chemical modification of polysilazanes or poly(silylcarbodiimides) by attaching a pendant boron-containing group to the silazane backbone. This approach is particularly useful for the preparation of Si/B/C/N materials with a low B:Si ratio. An advantage of this method is that one can start with a polysilazane with high molecular weight and tailored properties. An example is given in Eq. 5-47. When the polysilazane is heated with $B(NMe_2)_3$ in xylene for several days, the $B(NMe_2)_2$-doped polymer is formed by transamination. Prolonged heating results in a second transamination step which creates $B(NMe_2)$ crosslinks between the polysilazane chains.

$$\left[\begin{array}{c} Me \\ | \\ -Si-N- \\ | \quad | \\ Me \quad H \end{array}\right]_n \xrightarrow[- HNMe_2]{+ B(NMe_3)_3 \,/\, T\uparrow} \left[\begin{array}{c} Me \\ | \\ -Si-N- \\ | \quad | \\ Me \quad B(NMe_2)_2 \end{array}\right]_n \xrightarrow[- HNMe_2]{+ [SiMe_2\text{-}NH]_n \,/\, T\uparrow} \left[\begin{array}{c} Me \\ | \\ -Si-N- \\ | \quad | \\ Me \quad B-NMe_2 \\ \qquad | \\ \left[\begin{array}{c} Me \\ | \\ -N-Si- \\ | \\ Me \end{array}\right]_n \end{array}\right]_n$$

$$(5\text{-}47)$$

- Preparation of (co-)polymers from B- and Si-containing monomers. When boron is incorporated in the polymer backbone, retention of B in the pyrolysis products is much better. This allows the formation of ceramics with high B:Si ratios. For example, the molecular precursor Cl_2B–NH–$SiCl_3$ is accessible in two steps from hexamethyldisilazane (Eq. 5-48). Its reaction with methylamine results in the formation of a soluble, fusible and spinnable poly(N-methylborosilazane). Pyrolysis under an inert gas atmosphere leads to a black material of the composition $SiBN_3C$ which stays amorphous up to 1900 °C under nitrogen. This quaternary ceramic is reported to have a 10-fold oxidation resistance compared to SiC or Si_3N_4, an excellent thermal stability up to 2000 °C in He or N_2, and a remarkably high oxidation resistance.

$$(Me_3Si)_2NH \xrightarrow[- Me_3SiCl]{+ SiCl_4} Me_3Si-\underset{\underset{H}{|}}{N}-SiCl_3$$

$$\xrightarrow[- Me_3SiCl]{+ BCl_3} Cl_2B-\underset{\underset{H}{|}}{N}-SiCl_3 \xrightarrow{MeNH_2} \text{polymer}$$

polymer → (pyrolysis)

inert gas → $SiBN_3C$

ammonia → $Si_3B_3N_7$

$$(5\text{-}48)$$

When pyrolysis is performed under ammonia at 1000 °C, an amorphous ceramic with the composition $Si_3B_3N_7$ is formed, the temperature stability of which is about 150 °C higher than Si_3N_4 or a Si_3N_4/BN composite. It should be noted that the Si:B ratio in both $SiBN_3C$ and $Si_3B_3N_7$ is 1:1, from the molecular precursor through all process stages to the final ceramic.

5.7 Other Inorganic Polymers

Efforts to synthesize inorganic polymers with other atoms in the main chain than have been discussed in Sections 5.2 to 5.6 are motivated by the possibility of accessing new polymeric materials or preceramic polymers with interesting and useful properties. The intention of the non-comprehensive selection of examples in this section is to show the manifold possibilities, particularly when the polymer backbone consists of more than two elements or when transition metals are included. Most polymers are so new that much more research is needed before their technical potential can be assessed, or even before they can be utilized commercially. Therefore, this section merely scratches the surface of what has already been investigated on a fundamental level, and what is feasible in terms of possible applications.

5.7.1 Other Phosphorus-Containing Polymers

Polyphosphates, as well as silicates, are not considered polymers in the context of this section. However, if the bridging oxygen atom in $[PO_3^-]_n$ is formally replaced by a $O-C_6H_4-O$ unit, the structurally related "phoryl resins" are obtained (Figure 5-17). They are prepared by reaction of $RO(O)PCl_2$ with aromatic diols. Although they have some interesting properties, such as good transparency and hardness (▶ glossary), they lack long-term hydrolytic stability.

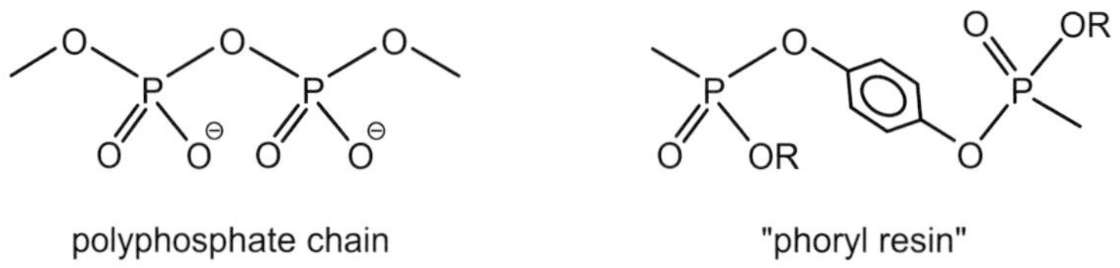

polyphosphate chain   "phoryl resin"

Figure 5-17. Structural relation between polyphosphates and the phoryl resins.

Recently, some new inorganic polymers have been developed in which part of the PR_2 units of polyphosphazenes is replaced by a CR, SR or S(O)R unit [poly(carbophosphazenes) (**2a**), poly(thiophosphazenes) (**2b**) and poly(thionyl-phosphazenes) (**2c**), respectively]. They are prepared by thermal ROP of the corresponding molten heterocycles **1** (Eq. 5-49). Typical reaction temperatures are between 90 °C (for **1b**) and 165 °C (for **1c**).

1

a: X = C
b: X = S
c: X = S(O)

$$(5\text{-}49)$$

The polymers are very sensitive to moisture as a consequence of the hydro-lytically sensitive P–Cl and/or S–Cl groups. As for the corresponding poly-phosphazenes (Section 5.3), hydrolytically stable derivatives are obtained if the perchlorinated polymers **2** are treated with nucleophiles, such as aryloxides or primary amines. The most promising derivatives are currently the poly(thionyl-phosphazenes) (**2c**), which exhibit a much higher stability than the poly(thio-phosphazenes) owing to the 4-coordinate sulfur atom. In **2a** and **2b**, the C–Cl or S–Cl groups are more reactive than the P–Cl groups, which allows a regio-selective substitution in every case. In the poly(thionylphosphazenes) **2c**, the regioselectivity is inverse, i.e., the P–Cl group is preferentially substituted. Amines readily substitute both the P–Cl and S–Cl groups at ambient temperature. The resulting moisture and air-stable poly[(amino)thionyl-phosphazenes] range from elastomers to glassy polymers. In the reaction with aryloxides, only substitution of the P–Cl bonds can be achieved, while alkoxides lead only to degradation of the polymer backbone. The poly[(aryloxy)thionylphosphazenes] (containing $P(OR)_2$ and S(O)*Cl* groups!) are also air-stable but readily decompose in solution in the presence of strong bases at elevated temperatures..

The glass transition temperatures of **2** is governed by two opposing effects compared with the parent polyphosphazenes. The C=N bond in **2a** and the S=O group in **2c,** as well as the smaller size of C and S compared to P, tend to increase T_g. On the other hand, the presence of only five large substituents per repeat unit compared to six substituents for polyphosphazenes lowers T_g.

When the substituents are large enough, this effect overrides the lower flexibility introduced by the replacement of a phosphorus atom by a S(O)R or C=N moiety.

5.7.2 Poly(oxothiazenes)

When gaseous S_4N_4 is passed over metallic silver at elevated temperatures, it is catalytically converted to the cyclic dimer S_2N_2. The dimer, in turn, slowly polymerizes in the solid state to $[SN]_n$ (Figure 5-18). The polymer is thermally stable to $140\,°C$ and highly crystalline with almost planar chains. It is composed of layers of fibers, which are soft and malleable, and have a golden metallic luster. The most important property of this compound is its highly isotropic electric conductivity along the chains down to low temperatures like a metal, which led to its use as electrode material. At $-273\,°C$ it becomes a superconductor (▶ glossary).

The processing of $[SN]_n$ is hampered by its insolubility and its failure to melt without decomposition. Furthermore, it is easily oxidized. The situation could possibly be improved by having the sulfur atom in a higher oxidation state, i.e., by having substituents at the sulfur atom. A series of polymers of this type, the poly(oxothiazenes) have been prepared (Figure 5-18).

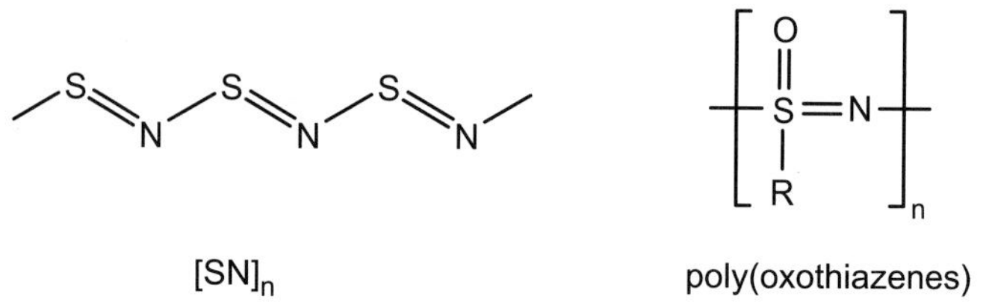

$$[SN]_n \qquad\qquad poly(oxothiazenes)$$

Figure 5-18. Structural formula of poly-SN and poly(oxothiazenes).

Polymers with aryl or alkyl substituents at the sulfur atom can be prepared by thermal condensation of sulfonimidates, $RO(O)S(R)=NH$ (elimination of ROH) or, at higher temperatures, of N-silylsulfonimidates, $RO(O)S(R)=N$-$SiMe_3$ (elimination of $ROSiMe_3$).

Poly(oxothiazenes) are very polar. For example, $[S(O)(Me)=N]_n$ dissolves in DMF, DMSO, or hot water. Its T_g ($\sim60\,°C$) is much higher than that of the corresponding phosphazene, $[NPMe_2]_n$ ($-46\,°C$). This shows that the backbone in poly(oxothiazenes) is less flexible, consistent with the trend already observed for poly(thionylphosphazenes) (see above).

5.7.3 Transition Metal-Containing Polymers

Transition metal compounds often have interesting optical, magnetic, or catalytic properties. Therefore, the incorporation of transition metal centers in the polymer backbone is an attractive approach to obtain new types of macromolecules with properties distinctly different from organic polymers or the inorganic polymers discussed so far.

In general, metal-containing polymers are obtained when bidentate (▶ glossary) molecules or ions of the type X–Y–X are used to bridge two metal centers.

$$\cdots\cdot\text{X--ML}_n\text{--X--Y--X--ML}_n\text{--X--Y--X--}\cdots\cdot$$

(ML_n is a metal complex fragment containing n coligands L not integrated in the polymer backbone; X is any coordinating group which provides a stable link to the ML_n fragment; and Y is an organic or inorganic spacer.)

Not any group X–Y–X can be used, because chelation (▶ glossary) is entropically favored. This is related to the ring/chain problem previously discussed for other inorganic polymers (Section 5.1). Therefore, rigid groups Y are more suitable to obtain real polymers.

An example of very rigid polymers is based on phthalocyanines as the metal complex moiety (Figure 5-19).

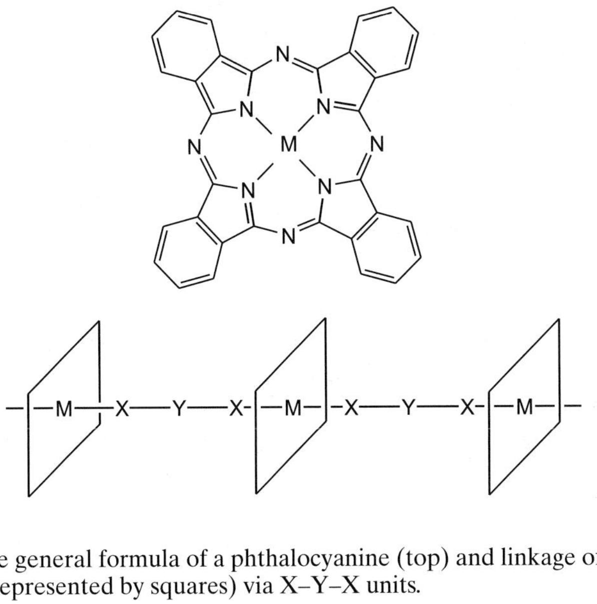

Figure 5-19. The general formula of a phthalocyanine (top) and linkage of the phthalocyanine units (represented by squares) via X–Y–X units.

There is a great number of phthalocyanines (and structurally related porphyrines) of various metals (M = Cr, Mn, Fe, Co, Rh, Ru) which can be stacked by appropriate groups X–Y–X. The coordination of the groups X must always be *trans*, owing to the very fixed geometry of the phthalocyanine ring system. Substituents can be introduced at the aromatic rings to modify, for example, the solubility of the polymer. Stacking can be achieved by a single oxygen atom (the resulting polymers have a –M–O–M–O– backbone). More extended X–Y–X entities have also been used, such as CN^- (giving M–CN–M–CN– linkages), 4,4'-bipyridine, pyrazine derivatives, 1,4-diisocyanobenzene, 4,4'-bipyridylacetylene, etc. The synthesis of the mostly insoluble polymers is accomplished by reacting the coordinatively unsaturated or dihalogenated metal phthalocyanine or porphyrine with the bidentate ligand (▶ glossary). The oxygen-bridged derivatives are obtained by condensation of $PcM(OH)_2$ (Pc = phthalocyanine).

Some of these polymers have an excellent thermal and chemical stability. They can become electrically conducting after chemical or electrochemical doping. One-dimensional conductivities up to 0.1 $S \cdot cm^{-1}$ were obtained. Such polymers may also have interesting magnetic or electro-optical properties.

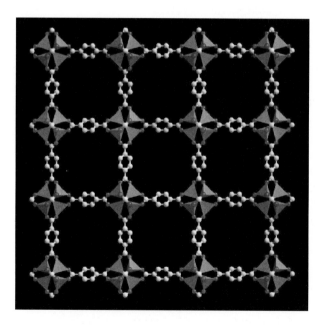

Figure 5-20. A two-dimensional section through the crystal structure of Zn_4O(terephthalate)$_6$.

Three-dimensional "polymers" can be obtained using the same approach. Figure 5-20 shows an example in which tetrahedral $Zn_4O(OOCR)_6$ clusters are crosslinked by phenylene groups in a cubic array. The compound was prepared from zinc nitrate and terephthalate. The crystals have a very low density (0.59 g·cm^{-3}); their framework is stable up to 300 °C. Their surface area and pore volume is higher than in most zeolites (see Section 6.4.1). Consequently, this compound exhibits also host–guest interactions.

Another class of polymers with potentially interesting properties are those in which ferrocenyl or related units are incorporated into the polymer backbone. The main reason for this interest is the high thermal stability and the interesting redox properties of the ferrocenyl unit. Interesting physical properties can particularly expected if the ferrocenyl units are separated only by short spacers. The most promising route is the ROP of ferrocenophanes (Eq. 5-50).

$$\text{(5-50)}$$

Bridging of the cyclopentadienyl rings by a single atom ([1]ferrocenophanes) induces tilting of the rings (the rings are parallel in non-bridged ferrocene). The thus generated strain is the driving force for ROP. For example, while high polymers are obtained from SiR_2-bridged ferrocenophanes (tilt angle ~21°), the corresponding disilanyl-bridged [2]ferrocenophanes (X = SiR_2SiR_2) are not strained enough (tilt angle 4–5°) for ROP. In contrast to that, the shorter C–C distance in ethylene-bridged [2]ferrocenophanes (X = CH_2CH_2, tilt angle ~21°) allows the preparation of poly(ferrocenylethylenes). Although most attention has been paid to the properties of poly(ferrocenylsilanes), the corresponding polymers with X = GeR_2, PR, P(S)R or S were also prepared by ROP of the corresponding [1]ferrocenophanes.

Most of the poly(ferrocenylsilanes) are soluble in organic solvents, despite their high molecular weights. While the $SiMe_2$ derivative is thermoplastic, the $SiHex_2$ derivative is an amorphous rubber with $T_g = -26$ °C. Electrochemical investigations showed two reversible oxidations in a 1:1 ratio. This proves an interaction between the ferrocenyl units along the chain. It was concluded that, initially, only every second iron atom along the chain is oxidized.

Poly(ferrocenyldisulfides) were prepared by atom abstraction polymerization from [3]trithiaferrocenophanes (Eq. 5-51).

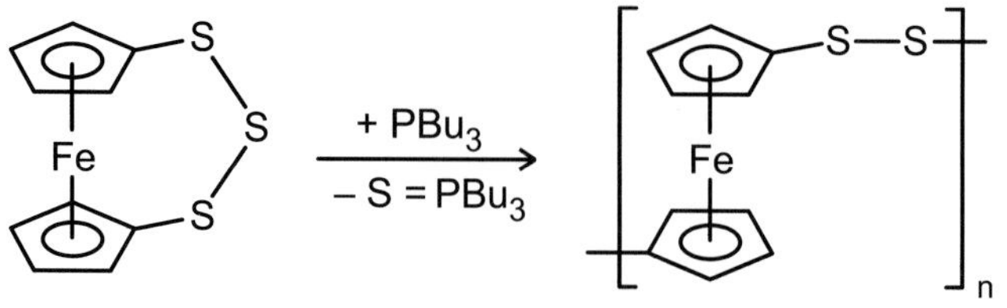

$$(5\text{-}51)$$

Among their interesting properties are their photosensitivity, the reversible cleavage of the S–S bonds (reduction by Li[BEt$_3$H], oxidation by I$_2$) and their electrochemical behavior.

5.7.4 Preceramic Polymers for BN

α-Boron nitride, BN, is a commercially important non-oxide high-temperature ceramic. It is hard, has a high mechanical strength, a low thermal expansion coefficient, and a high corrosion and oxidation resistance. As in the case of Si$_3$N$_4$ and SiC, fibers, coatings, and foams cannot be prepared from BN powders.

The ammonolysis of BCl$_3$ or B$_2$H$_6$ is not suitable for obtaining fusible or soluble preceramic polymers. The precursor family with the greatest potential for the formation of preceramic polymers for BN are borazene derivatives B$_3$N$_3$R$_3$R'$_3$ (Figure 5-21). They have the correct B/N ratio, and the six-membered ring is the basic structural motif of α-BN. Borazene itself, B$_3$N$_3$H$_6$, is a low-boiling liquid which is easily prepared. Controlled thermolysis of liquid borazene *in vacuo* at 70°C results in near-quantitative formation of soluble polyborazylene by H$_2$ elimination (Eq. 5-52). The molecular weight of the polymer depends on the time of heating. Analytical data indicate that the polymer is probably crosslinked with an average of three branchings per ring.

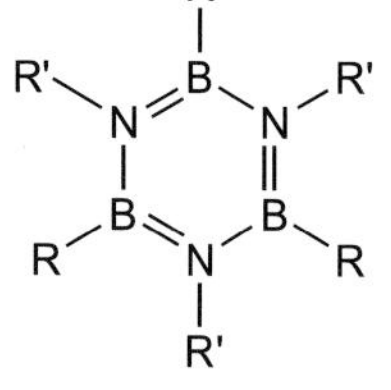

Figure 5-21. The structural formula of a borazene B$_3$N$_3$R$_3$R'$_3$.

$$\text{(5-52)}$$

X = H or link to another ring

Pyrolysis of the polymer under Ar or NH_3 to 1200 °C results in dehydrocoupling between adjacent borazene units and eventually in the formation of white, very pure BN powders with ceramic yields (▶ glossary) up to 92 %. High-quality BN coatings on alumina, Si_3N_4, SiC and carbon fibers have been achieved using polyborazylene precursors.

A variety of substituted borazene derivatives have been investigated. The most promising is *B*-triamino-*N*-tris(trimethylsilyl)borazene (R = NH_2, R′ = $SiMe_3$ in Figure 5-21). A fusible, soluble preceramic polymer is formed when this compound is heated at 200 °C. The polymer is melt-spun into fibers which are slowly calcined in NH_3 to give BN at 1000 °C. However, the synthesis of this borazene monomer is a tedious multi-step process.

A second class of borazene-based polymers are those in which the borazene units are not connected with each other but instead crosslinked by two ("two-point polymers") or three bridging groups per ring ("three-point polymers"). The most promising bridging groups are amino groups. The linkage of the borazene ring by amino groups is best achieved by reaction of B–Cl-substituted borazenes (R = Cl) with hexa- or heptamethydisilazane, $(Me_3Si)_2NR$ (R = H, Me). The driving force is the formation of stable, volatile Me_3SiCl.

For example, the reaction of *B*-trichloroborazene and hexamethyldisilazane (Eq. 5-53) can be performed in a variety of solvents. The solubility of the resulting three-point polymer in organic solvents appears to depend on the borazene/$(Me_3Si)_2NH$ ratio.

$$+ (Me_3Si)_2NH \longrightarrow \qquad \text{(5-53)}$$

However, if the polymers are dissolved in liquid NH_3 and then dried, they redissolve in organic solvents. Ammonia obviously breaks down some cross-links of the original polymer and thus makes it more soluble. These polymer solutions can be processed to form fibers, coatings, green bodies (▶ glossary), xerogels, or aerogels (see also Sections 4.5 and 6.3). They are converted in these forms in BN on pyrolysis to 1200 °C by elimination of ammonia. Labeling experiments show that the nitrogen in the final BN originates both from the rings and the bridges. Thus, opening of the borazene rings is a crucial reaction during pyrolysis.

5.8 Further Reading

H. R. Allcock, "Inorganic–organic polymers", *Adv. Mater.* **6** (1994) 106–115.

C. W. Allen, "Linear, cyclic and polymeric phosphazenes", *Coord. Chem. Rev.* **130** (1994) 137–173.

H.-P. Baldus, M. Jansen, "Novel high-performance ceramics – amorphous inorganic networks from molecular precursors", *Angew. Chem. Int. Ed. Engl.* **36** (1997) 328–343.

J. Bill, F. Aldinger, "Precursor-derived covalent ceramics", *Adv. Mater.* **7** (1995) 775–787.

M. Birot, J.-P. Pillot, J. Dunogues, "Comprehensive chemistry of polycarbosilanes, polysilazanes, and polycarbosilazanes as precursors for ceramics", *Chem. Rev.* **95** (1995) 1443–1477.

P. F. Bruins, *Silicone Technology*, Wiley, New York, 1970.

S. J. Clarson, J. A. Semlyen (Eds.), *Siloxane Polymers*, Prentice-Hall, London, 193.

R. Drake, A. I. MacKinnon, R. Taylor, "Recent advances in the chemistry of siloxane polymers and copolymers", in *The Chemistry of Organic Silicon Compounds* (Eds. Z. Rappoport, Y. Apeloig), Wiley, 1998, pp. 2217–2244.

W. Dressler, R. Riedel, "Progress in silicon-based non-oxide structural ceramics", *Int. J. Refractory Metals & Hard Mater.* **15** (1997) 13–47.

D. P. Gates, I. Manners, "Main-group-based rings and polymers", *J. Chem. Soc. Dalton* (1997) 2525–2532.

P. Greil, "Ceramic materials from organometallics", *Ceram. Trans.* **51** (1995) 171–178.

P. Greil, "Near net shape manufacturing of polymer derived ceramics", *J. Eur. Ceram. Soc.* **18** (1998) 1905–1914.

S. Hayase, "Polysilanes with functional groups", *Endeavour* **19** (1975) 125–131.

L. V. Interrante, Q. Liu, I. Rushkin, Q. Shen, "Poly(silylenemethylenes) – a novel class of organosilicon polymers", *J. Organomet. Chem.* **521** (1996) 1–10.

T. C. Kendrick, B. Parbhoo, J. W. White, "Siloxane polymers and copolymers" in *The Chemistry of Organic Silicon Compounds* (Eds. S. Patai, Z. Rappoport), Wiley, 1989, pp. 1289–1361.

H. R. Kricheldorf (Ed.), *Silicon in Polymer Synthesis*, Springer, Heidelberg, 1996.

B. Lacave-Goffin, L. Hevesi, S. Demoustier-Champagne, J. Devaux, "Synthesis and properties of polysilanes: versatile new organic materials", *ACS-Models in Chemistry*, **136** (1999) 215–236.

R. M. Laine, F. Babonneau, "Preceramic polymer routes to silicon carbide", *Chem. Mater.* **5** (1993) 260–279.

R. M. Laine, A. Sellinger, "Si-containing ceramic precursors", in *The Chemistry of Organic Silicon Compounds* (Eds. Z. Rappoport, Y. Apeloig), Wiley, 1998, pp. 2245–2316.

J. Livage, C. Sanchez, F. Babonneau, "Molecular precursor routes to inorganic solids", in *Chemistry of Advanced Materials – an Overview*, Wiley-VCH, New York, Eds. L. V. Interrante, M. Hampden-Smith, 1998, pp. 389–448.

I. Manners, "Ring-opening polymerization (ROP) of strained, ring-tilted silicon-bridged [1]ferrocenophanes: synthetic methods and mechanisms", *Polyhedron* **15** (1996) 4311–4329.

I. Manners, "Polymers and the periodic table: recent developments in inorganic polymer science", *Angew. Chem. Int. Ed. Engl.* **35** (1996) 1602–1621.

J. E. Mark, H. R. Allcock, R. West, *Inorganic Polymers*, Prentice Hall, Englewood Cliffs, 1992.

K. Matyjaszewski, M. Cypryk, H. Frey, J. Hrkach, H. K. Kim, M. Moeller, K. Ruehl, M. White, "Synthesis and characterization of polysilanes", *J. Macromol. Sci. Chem.* **A28** (1991) 1151–1176.

K. Matyjaszewski, D. Greszta, J. S. Hrkach, H. K. Kim, "Sonochemical synthesis of polysilanes by reductive coupling of disubstituted dichlorosilanes with alkali metals", *Macromolecules* **28** (1995) 59–72.

R. D. Miller, "Polysilanes – A new look at some old materials", *Angew. Chem. Adv. Mater.* **101** (1989) 1773–1780.

R. D. Miller, J. Michl, "Polysilane high polymers", *Chem. Rev.* **89** (1989) 1359–1410.

C. K. Narula, *Ceramic Precursor Technology*, Marcel Dekker, New York, 1995.

W. Noll, *Chemistry & Technology of Silicones*, Academic Press, 1968.

R. T. Paine, C. K. Narula, "Synthetic routes to boron nitride", *Chem. Rev.* **90** (1990) 73–91.

M. Rehahn, "Organic/inorganic hybrid polymers", *Acta Polym.* **49** (1998) 201–224.

R. Riedel, "Advanced ceramics from inorganic polymers" in *Materials Science and Technology* (Eds. R. W. Cahn, P. Haasen, E. J. Kramer), Vol. 17B, VCH, Weinheim, 1996, pp. 1–50.

R. Richter, G. Roewer, U. Böhme, K. Busch, F. Babonneau, H. P. Martin, E. Müller, "Organosilicon polymers – synthesis, architecture, reactivity and applications", *Appl. Organomet. Chem.* **11** (1997) 71–106.

E. G. Rochow, *Silicon and Silicones*, Springer Verlag, Berlin, 1987.

D. Seyferth, "Preceramic polymers: past, present, and future", *Adv. Chem. Ser.* **245** (1995) 131–160.

J. E. Sheats, C. E. Carraher, C. U. Pittman, M. Zeldin, B. Currell (Eds.), *Inorganic and Metal-Containing Polymeric Materials*, Plenum, New York, 1990.

F. O. Stark, J. R. Falender, A. P. Wright "Silicones" in *Comprehensive Organometallic Chemistry* (Eds. G. Wilkinson, F. G. A. Stone, E. W. Abel), Vol. 2, Pergamon Press, Oxford, 1982, pp. 305–363.

R. West, "Polysilanes", in *The Chemistry of Organo Silicon Compounds* (Eds. S. Patai, Z. Rappoport), Wiley, 1989, pp. 1207–1240.

R. West, R. Menescal, T. Asuke, J. Eveland, "Some recent developments in polysilane chemistry", *J. Inorg. Organomet. Polym.* **2** (1992) 29–45.

P. Wisian-Neilson, H. R. Allcock, K. J. Wynne (Eds.), *Inorganic and Organometallic Polymers II*, ACS Symp. Ser. 572, Am. Chem. Soc., Washington DC, 1994.

M. Zeldin, K. J. Wynne, H. R. Allcock (Eds.), *Inorganic and Organometallic Polymers I*, ACS Symp. Ser. 360, Am. Chem. Soc., Washington DC, 1988.

J. M. Zeigler, F. W. G. Fearon (Eds.), *Silicon-Based Polymer Science*, Adv. Chem. Ser. 224, Am. Chem. Soc. , Washington DC, 1990.

6 Porous Materials

In nature, many materials are porous, including wood, cork, sponge, bone, etc. or even the skeleton structure of the most simple organisms such as diatoms. Most synthetic materials are also – at least to some extent – porous. Although the evolution of natural porous structures is complicated and complex (see Section 4.3.1), the principle of minimal material and mass consumption for an optimum stability of the whole structure governs the formation process. In addition, the materials must be optimally adapted for their desired use.

Porous materials are also called cellular solids. The word "cell" is derived from the Latin *cella*: a small compartment, an enclosed space. Cellular solids are therefore clusters of cells, assembled by solid edges or faces. If the solid material is contained in the cell edges only, so that the cells are connected through open faces, the material is said to be open-celled (Figure 6-1). If the faces are solid, each cell is separated from its neighbors; the material is then called closed-celled. Everybody is familiar with organic porous materials such as polymeric foams used to make a vast range of items, from disposable coffee cups to the crash padding of an aircraft cockpit. Techniques now exist for foaming not only organic materials but also metals, ceramics and glasses. These materials are used increasingly, for example, for insulation or cushioning, to absorb the kinetic energy from impacts, and for catalysis.

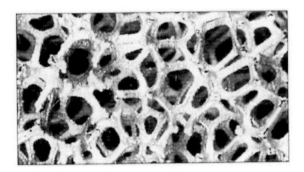

Figure 6-1. Structure of a high-porosity, open-cell metal.

Many physical properties such as density, thermal conductivity, strength, etc., depend on the porosity and the pore structure of a solid. Especially for industrial applications, a deliberate control of porosity is of great importance, for example in the design of catalysts, industrial adsorbents, membranes, structural materials, and ceramics. Furthermore, porosity is one of the factors that influence the chemical reactivity and the physical interactions of solids with gases and liquids.

In most textbooks inorganic porous materials are not even mentioned. We will give a short introduction into porosity, followed by a brief overview of different types of porous solids and their applications. The chosen examples are somewhat arbitrary, since it would be impossible to cover all types of inorganic porous materials. We try to introduce the reader to

- materials with different types of porosities regarding the size (from nanometer range up to millimeters) and arrangement of the pores (irregular or ordered);
- materials with different chemical compositions, e.g., from metals to oxides; and
- different preparative approaches to induce and control porosity in materials.

6.1 Introduction to Porosity

A solid is called porous, when it contains pores, i.e., cavities, channels, or interstices, which are deeper than they are wide. There are two ways to look at porous materials: They can be characterized by describing the pores, but also by describing the cell (pore) walls. Some porous materials are based on agglomerated or aggregated powders in which the pores are formed by inter-particle voids, while others are based on continuous solid networks around pores.

Not all pores are similar, and different types must be distinguished. A classification is usually done by describing their accessibility to an external fluid (Figure 6-2). From this viewpoint, pores totally isolated from their neighbors are called *closed pores* (a). They influence the macroscopic properties of the solid, but they are inactive in terms of chemical reactions. On the other hand, pores which are open to the external surface of the solid are called *open pores* such as (b), (c), (d), (e), and (f) in Figure 6-2. Some may be open only at one end, such as (b) and (f). They are described as *blind* (i.e., *dead-end* or *saccate*) pores. Others may be open at both ends (*through* pores) like (e). Pores may also be classified according to their shape: they may be *cylindrical* (either open (c) or blind (f)), *ink-bottle*-shaped (b), *funnel*-shaped (d) or *slit*-shaped. Close to, but different from porosity is the *roughness* of the external surface, represented around (g).

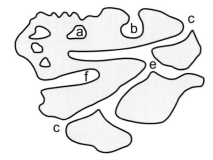

Figure 6-2. Different
types of pores.

    To describe qualitatively and quantitatively a porous solid, more information
is necessary, such as the porosity, the density, the specific surface area, or the
pore size and the pore size distribution of the porous solid. Table 6-1 shows the
definitions of some terms.

Table 6-1. Definition of terms used to characterize porous solids.

Density	true density	density of a material excluding pores and interparticle voids (density of the solid network)
	apparent density	density of a material including closed and inaccessible pores
	bulk density	density of the material including pores and interparticle voids (mass per total volume, with volume = solid phase + closed pores + open pores)
Pore volume V_p		volume of the pores
Pore size		also called pore width (diameter): the distance of two opposite walls of the pore
Porosity		ratio of the total pore volume V_p to the apparent volume V of the particle or powder
Surface area		the accessible (or detectable) area of solid surface per unit mass of material

For almost all these terms, the measured value depends strongly on the method used. Some methods detect only open pores, e.g., methods that use adsorption of molecules into the cavities, others may have also access to closed pores, e.g., spectroscopic, diffraction and scattering techniques. Moreover, for a given method the value can vary with the size of the molecular probe as shown in Figure 6-3 for a gas adsorption experiment.

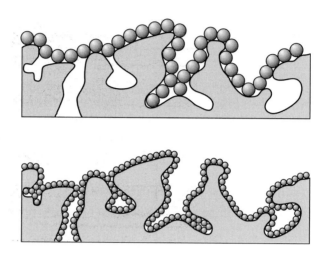

Figure 6-3. Schematic of gas adsorption on a porous material by gases of different molecular size. The surface area, when measured with large molecules, is smaller than when smaller molecules are used.

For most applications the pore size is of major importance. However, it is not very susceptible to a precise measurement, because the pore shape is usually highly irregular and variable, leading to a variety of pore sizes, or a broad distribution. Nevertheless, three different pore-size regions are defined by IUPAC:

- *Micropores*, which have diameters smaller than 2 nm
- *Mesopores*, which have diameters between 2 and 50 nm
- *Macropores*, which have diameters larger than 50 nm

This nomenclature is associated with the transport mechanisms occurring in the different types of pores (Figure 6-4).

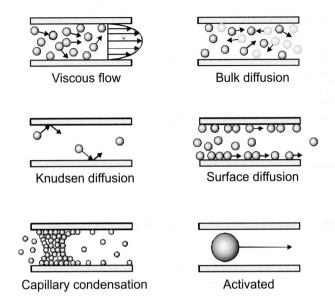

Figure 6-4. Transport mechanisms through pores.

Macropores. These are much larger than the mean free path length of a typical gaseous imbibing fluid. The dominant transport mechanisms are molecular (bulk) diffusion and, if a total pressure gradient exists, viscous flow. Overall permeabilities (▶ glossary) are highest in macroporous solids, but permselectivities are lowest.

Mesopores. These are on the same order or smaller than the mean free path and Knudsen diffusion becomes important. The classical inverse-square-root dependence of flux on molecular weight applies. In addition, surface diffusion of the molecules along the pore walls may contribute. This process is sensitive to the specific adsorption features of the molecule–solid combination. If a component has a sufficiently low volatility, then multilayer adsorption and capillary transport may occur, leading to an enhanced flux of the condensable species along the walls.

Micropores. These are encountered as the pore size is decreased to the size of the molecules. In this regime, much higher permselectivities are possible that depend on both molecular size and specific interactions with the solid. The so-called activated transport dominates in this class of porous solids.

6.2 Metallic Foams and Porous Metals

Porous metals and metallic foams are composite materials (▶ glossary) in which one phase is gaseous and the other a solid metal. The primary distinction between a porous metal and a metallic foam is the relative density. Porous metals have a high bulk density (high volume fraction of solid) and independent, distributed voids, while metallic foams have a low bulk density and interconnected voids. The porosity within these metallic materials varies from 30 to 98 vol%.

As foamed plastics (the organic counterparts), these metallic materials possess a unique combination of properties due to the porosity such as impact energy absorption capacity, air and water permeability (▶ glossary), unusual acoustic properties, low thermal conductivity and good electrical insulating properties. Therefore, foamed metals have a variety of different possible applications such as energy absorbing systems, porous electrodes, sound absorbers, filters, insulation materials, heat exchangers, construction materials or for electromagnetic shielding. Some devices made from foamed aluminum are shown in Figure 6-5.

Figure 6-5. Aluminum foam components for different applications.

A variety of different methods can be used for the production of metallic foams (Figure 6-6). Early attempts concentrated on casting techniques, similar to those used for plastics, with a gas serving as blowing agent. Casting techniques include all methods, in which the base metal or alloy (▶ glossary) is molten and then solidifies as a foam. Metallic foams can also be produced by metallurgy techniques. Metallurgical processes mainly consist of preparing the structural constituents, such as powder particles, chips, fibers, wire meshes, etc. and subsequently compacting and sintering them.

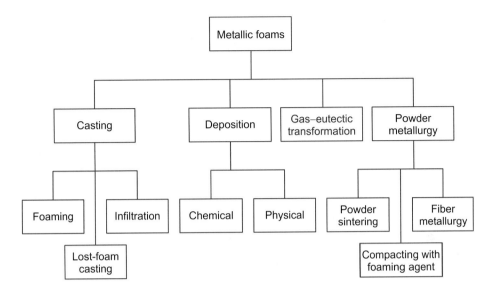

Figure 6-6. Summary of methods for the synthesis of porous metallic structures.

Not every process is applicable for every metal, and the materials obtained from the different techniques are quite different with respect to their structural characteristics. While casting techniques usually result in highly porous metal structures, processes based on powder metallurgy such as powder sintering result in materials with a medium porosity.

6.2.1 Casting Techniques

Foaming

In this method, a blowing agent, e.g., a metal hydride, is added to a molten metal. The mixture is heated to decompose the blowing agent by evolution of a gas. The gas expands, causing the molten metal to foam. After foaming, the resultant body is cooled to give a solid (Figure 6-7) that looks like a continuous sponge with bubbles encapsulated.

Controlling this foam generation process is rather difficult, and the foamed metal produced contains large bubbles that are distributed non-uniformly throughout the material. Different approaches have been used to overcome some of the problems, such as vigorous high-speed mixing to provide a good and uniform distribution of the foaming agent in the melt, or thickening of the melt, i.e., increasing the viscosity of the molten metal to prevent the escape of the gas bubbles. Problems, however, persist due to the relatively short time between the introduction of the foaming agent and the generation of the foam.

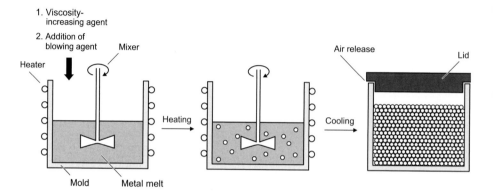

Figure 6-7. Foaming of a melt with a blowing agent.

Numerous variations of foaming exist; in some cases foam is generated by introducing gas directly into the melt during solidification.

The porosities typically reached by foaming techniques are very high (60–97 %), and closed-cell materials are obtained. The walls separating the cells (or pores) are usually poreless.

Lost-foam casting

Connected pores in plastic foam are filled with a castable inorganic material which is then hardened. Upon heating, combustion of the organic (plastic) foam leads to formation of a sponge-like solid. This solid is used as a mold for the metal which solidifies in its pores. The mold material is then removed and a metallic foam having the same sponge form as the original plastic is obtained. The process is used primarily for making cellular metals with low melting points such as copper, aluminum, lead, zinc, tin and their alloys (▶ glossary; Figure 6-8). The porosity of the final metallic material depends primarily on the porosity of the plastic foam used as template (▶ glossary); porosities up to 95 % can be reached.

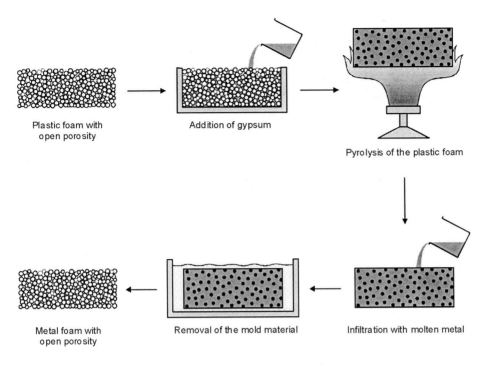

Figure 6-8. Principle of lost-foam casting.

Infiltration

This method produces an interconnected cellular structure of sponge metal with porosities of about 70 % by casting metal around densely packed granular templates (▶ glossary) introduced into the casting mold. These templates can be soluble (but heat-resistant) solid compounds, such as sodium chloride, which is later leached out to leave a porous metal. Other inorganic materials such as expanded clay, glass spheres, hollow corundum spheres, etc. can also be used as space-filling material which is removed in a second step (Figure 6-9). The advantage of using hollow spheres as the template is that it does not have to be removed.

   As an alternative to casting molten metal around granules, these templates (▶ glossary) can also be incorporated into metal melts. In this process, the metal is molten in a crucible, and to this vigorously stirred melt granules are introduced and thus uniformly dispersed. During the mixing, the metal is already allowed to cool until the mixture is sufficiently viscous to prevent segregation.

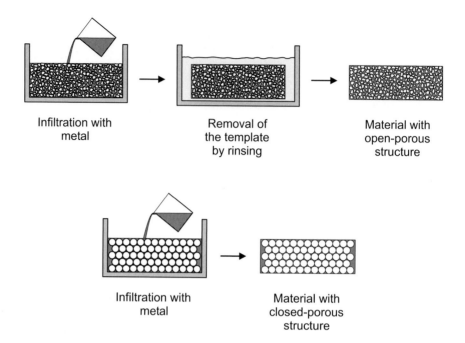

Figure 6-9. Process of casting metals around templates. Upper row: two-step process in which the template is removed after infiltration of the metal; lower row: one-step process using hollow spheres.

6.2.2 Gas–Eutectic Transformation

This process is usually applied for metal–hydrogen systems and is based on the saturation of the molten metal by the gas. Making the material consists of two steps:

1. Charging the melt with hydrogen to reach the eutectic composition (▶ glossary).
2. Solidification of the melt. During the solidifying process the saturated melt decomposes into a solid and a gas phase.

Foaming is not observed, because the gas is evolved as the melt freezes. The process is in many ways similar to conventional eutectic (▶ glossary) solidification, the distinction being that the liquid decomposes in a solid and a gas rather than into two solids. The main process variables that govern the porosity and the size, shape and orientation of the pores are the hydrogen level in the melt, gas pressure over the melt during solidification, direction and rate of heat removal, and chemical composition of the melt. Porosities of about 70 % and pore sizes in the range of 10 μm to 10 mm are typically obtained. Materials produced by this method are called "gasars", which is an abbreviation of the Russian term for "gas-reinforced".

6.2.3 Powder Metallurgy

Powder Sintering

Powder sintering is one of the simplest methods of making porous metals. In the easiest case, the metal powder is filled in a mold and then sintered. This process is called loose powder sintering. According to the sintering mechanisms (see Section 2.1.4), contacts between powder particles are established and grow during the time that the powder particles are heated in contact with each other. Application of pressure is not necessary. The materials produced are porous metals rather than foams, with porosities ranging from 40 to 60 vol%.

To increase the porosity, pore-forming or spacing agents are frequently added to the blend. Such agents decompose or evaporate during sintering, or are removed by sublimation or dissolution. The porosity can increase up to 90 % and a foam can be obtained. In general, the structure of a sintered porous metal depends on the shape of the particles or other constituents to be sintered, the degree of compacting before sintering, and the sintering conditions.

The maximum porosity is usually achieved by sintering of hollow spherical particles. First, the hollow spheres are close-packed, followed by a pre-sintering step of this structure. The metal powder is filled into the interstices of this material and the process is finalized by the actual sintering process (Figure 6-10).

Besides the above-discussed processes, alternative techniques can be used for the production of metallic foams such as sintering of a slurry saturated sponge, or foaming from a slurry.

Fiber metallurgy

Using metal fibers instead of metal powders in producing porous material, especially for filters, has some advantages:

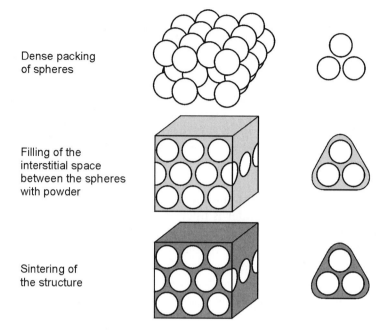

Dense packing
of spheres

Filling of the
interstitial space
between the spheres
with powder

Sintering of
the structure

Figure 6-10. Principle of sintering with hollow spheres.

- The porosity can be controlled within wide limits, from 0 to 95 vol%, with the material retaining its constructional properties even at the highest porosity.
- High strength and ductility (▶ glossary) can be obtained, surpassing several-fold the corresponding properties of materials produced from metal powders at any given porosity.

Fiber metallurgy involves the preparation of metal fibers by machining, drawing or other techniques. The fibers are then compacted to a preset degree followed by sintering in a non-oxidizing atmosphere to develop the required strength and porosity.

To improve strength, metallic fibers are sometimes coated with a low-melting temperature agent before compacting and sintering. Pre-coating provides better bonding between the constituents.

When tubular or solid fibers in parallel alignment or in grids are sintered, the result is an ordered, oriented structure. When the fibers are not oriented, the pores are arranged more or less randomly (Figure 6-11).

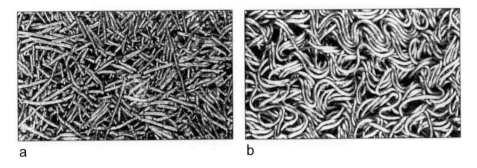

Figure 6-11. The structure of porous metals produced by a) sintering of short straight fibers; b) sintering of long crimped fibers.

Compacting with foaming agents

The metal powder is mixed with a foaming agent and subsequently compacted. As a result, a product is obtained in which the foaming agent is homogeneously distributed within a dense, virtually non-porous metallic matrix. This foamable material can be processed into sheets, rods, profiles, etc. using conventional techniques. The foamed metal parts are obtained by heating the material to temperatures above the melting point of the matrix metal. After expansion of the material to the desired degree, the foaming process is terminated by cooling below the melting temperature. The density of metal foams can be controlled by adjusting the amount of the foaming agent and several other foaming parameters. If metal hydrides are used as foaming agent, a content of less than 1 % is sufficient in most cases. The materials obtained by this process are usually foams with porosities in the range of 60 – 90 vol%.

6.2.4 Metal Deposition

Different deposition techniques can be used: CVD, electrochemical deposition and PVD (see Section 3.2.1). Deposition techniques are often used for making filters, catalysts, and structural parts. During electrochemical deposition, the metal is deposited on porous organic substrates, typically polyurethane; the substrate is pre-treated to make it more rigid and conductive. Typical porosities are around 90 %.

PVD methods are carried out in a vacuum chamber containing a cold porous substrate and a vapor source of the base metal or alloy (▶ glossary). Metal atoms condense on the substrate, forming a continuous three-dimensional grid of preset thickness. The substrate is then removed by a thermal, chemical or other method, leaving a porous metal whose macrostructure is a replica of the substrate. The process is suited for making metals with porosities up to 95 %.

Another very remarkable process is the production of porous silicon by anodization. A silicon wafer is immersed in a solution of hydrofluoric acid, ethanol and water, and subjected to an electric current for a brief time. The anodizing process tunnels, giving an interconnected network of pores with a cell size of 10 nm. The density of the obtained porous silicon is as low as one-tenth of that of the metallic silicon, yet the material remains crystalline.

6.3 Aerogels

Aerogels are another example of a material with an extraordinary high porosity. However, in this case the porous inorganic network is formed by chemical reactions instead of solidification of a molten compound around gas bubbles.

As defined in Section 4.5, a gel consists of a porous, three-dimensionally continuous solid network surrounding and supporting a continuous liquid or gaseous phase. In aerogels, the pores are filled with air. They are prepared from "wet" gels by replacing the original pore liquid by air. The highly porous three-dimensional network of metal oxide-based aerogels is usually formed by sol–gel processing. The chemistry of sol–gel processing was discussed in Section 4.5; this section therefore concentrates on the drying methods by which the wet gels can be transformed into aerogels. Drying has to be performed in a way, that the filigrane network structure is maintained during removal of the solvent. It is remarkable that a variety of inorganic, inorganic–organic or organic gels can be converted into aerogels despite a typical solid content of only 1–15 vol%.

The pore structure of aerogels is difficult to describe, but it is characterized by well-accessible, cylindrical, interconnected pores with pore sizes from the microporous to the macroporous regime. However, the majority of the pores fall in the mesoporous range with diameters between 2 and 50 nm and a broad pore size distribution. The structure of a silica aerogel is shown in Figure 6-12. Other mesoporous materials are fumed silica, Vycor glass, carbon soot, or M41S materials (see Section 6.4.2). Among them, aerogels are unique owing to their extremely high porosity (low density) and high specific surface area, and the possibility of making monoliths. Aerogels can also be obtained as granulates or powders. The bulk density of aerogels is in the range of 0.004–0.5 $g \cdot cm^{-3}$ owing to the high porosity (for comparison: the density of air is

0.00129 g·cm^{-3}). Aerogels belong definitely to the lightest inorganic solids available today.

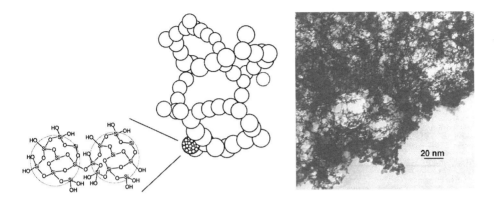

Figure 6-12. Silica aerogel; Left: Schematic of the structure; right: a transmission electron microscopy (TEM) image.

6.3.1 Drying Methods

Drying of a gel body is a complex process in which different stages during the evaporation of the pore liquid can be distinguished. Theses processes are described in Section 4.5.2. Drying typically results in cracking and collapse of the gel network, which leads to powders or strongly shrunken monoliths (Figure 6-13). For the production of aerogels, different ways have been developed in order to preserve the pore structure of the wet gels, i.e., to avoid irreversible shrinkage. The most important are the supercritical drying process and the so-called "ambient pressure" process.

Supercritical drying

In this procedure, the pore liquid is put into the supercritical state, in which there are no liquid-gas interfaces in the pores. A supercritical fluid (SCF) is the state of a compound, a mixture or an element above its critical pressure (p_c) and critical temperature (T_c) but below the pressure required to condense it into a solid (Figure 6-14).

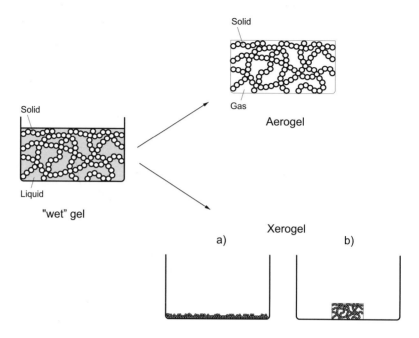

Figure 6-13. Top: Drying of a wet gel body to give an aerogel (the volume of the body remains approximately constant). Bottom: Conventional drying to give a xerogel powder (a) or monolith (b), associated with large shrinkage of the gel body.

The critical point marks the end of the liquid–vapor coexistence curve in the phase diagram. It is noteworthy that the melting curve extends over the supercritical region. For example, the pressure required to solidify CO_2 at its critical temperature is 5.7 kbar, but that for water is an enormous 140 kbar.

Technically, a SCF is a gas but not a vapor. The term "gas" refers to any phase which will conform in volume to the space available. A "vapor" is defined as a gas whose temperature is less than the critical temperature. Supercritical fluids exhibit simultaneously properties associated with both gases and liquids (see also Table 6-2). Thus, they are compressible like gases, but they also display solvencies similar to those of liquids.

A unique feature of supercritical fluids may be demonstrated by starting with a subcritical liquid at point 1 in Figure 6-14. If the liquid is depressurized isothermally from point 1 to point 5, the presence of a liquid–vapor interface meniscus is observed as the vapor pressure line is crossed. However, if the path 1–2–3–4–5 is taken, the fluid then passes from a liquid phase to a gas phase without a liquid–vapor interface at any point. This pathway is used in supercritical drying to avoid collapse of delicate microstructures by the strong capillary forces that arise at liquid–vapor interfaces.

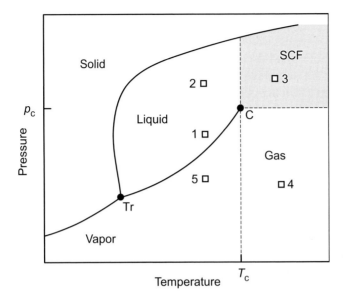

Figure 6-14. Schematic pressure–temperature diagram for a pure compound. The shaded area represents the supercritical fluid region (SCF), where C is the critical point. Tr represents the triple point, and 1 to 5 are random points in the phase diagram (see text).

Table 6-2. Comparison of some physical properties of gases, liquids and supercritical fluids.

Property	Gas	Liquid	Supercritical fluid
Density [$g \cdot cm^{-3}$]	10^{-4}–10^{-3}	0.6–1.4	0.1–1
Diffusion coefficient [$cm^2 \cdot s^{-1}$]	10^{-1}	10^{-5}	10^{-3}–10^{-4}
Solvency	No	Yes	Yes
Compressibility	Yes	No	Yes

Table 4-4 in Section 4.4.1 shows some typical critical pressure and temperature data for different solvents. For the drying of aerogels, the most convenient way is to put the pore liquid, which is usually an alcohol or acetone, into the supercritical state. However, problems may arise from the combination of high pressures with high temperatures as well as the flammability of these solvents. Alternatively, liquid carbon dioxide has been used. Supercritical CO_2 is of particular interest due to its low critical temperature (31 °C). Additionally, it is non-flammable, non-toxic and – especially when used to replace organic solvents – environmentally friendly. However, a time-consuming solvent exchange (organic solvent versus liquid CO_2) is required prior to supercritical drying.

How is supercritical drying of a wet gel performed in practice? The wet gel is placed into an autoclave and covered with additional solvent. Partial drying of the samples (which would lead to the formation of cracks) is thus avoided. After the autoclave is closed, the temperature is slowly raised, resulting in a pressure increase. Both the temperature and the pressure are adjusted to values above the critical point of the corresponding solvent (T_c, p_c) and kept there for a certain period of time. This ensures that the autoclave is completely filled with the supercritical fluid. The fluid is then slowly vented at constant temperature, which results in a pressure drop. When ambient pressure is reached, the vessel is cooled to room temperature and then opened. Drying is often performed in a way that the vessel is pre-pressurized with nitrogen to avoid evaporation of the solvent. The phase boundary between the liquid and the gas must not be crossed during drying at any time.

Ambient pressure drying

To make aerogels interesting for large-scale commercial applications, one should avoid supercritical drying as the most expensive and risky part of the preparation. Therefore, the interest in alternative ways for exchanging the pore liquid in the gels by air is very great.

The capillary forces exerted by the meniscus of the pore liquid and the pressure gradient upon the large shrinkage of the network (see Section 4.5.2) are the main reasons for the collapse of the filigrane structure. To obtain aerogels at ambient conditions, the network must be strengthened in order to avoid its collapse, i.e., its irreversible shrinkage. Additionally or alternatively, the contact angle (see also Section 4.5.4) between the pore liquid and the pore walls must be influenced by deliberate modification of the inner surface and by variation of the solvent properties to minimize the capillary forces. For silica gels, the following route was developed which allows drying under ambient conditions.

The water/alcohol mixture in the pores of the gel is first exchanged for a water-free solvent, and the Si–OH groups at the surface are silylated (e.g., by reaction with chlorotrimethylsilane). The reactivity of the gel surface is thus reduced, and the surface hydrophobized. The actual drying process is performed after removal of excess chlorotrimethylsilane by washing with a water-free solvent. As expected, the gel shrinks strongly during evaporation of the solvent from the pores. However, no irreversible narrowing of the pores with formation of Si–O–Si bonds is possible because no Si–O–Si bonds can be formed due to silylation. The gel therefore expands to nearly its original size after reaching the maximum value for the deformation. This is called "spring-back effect". However, the network of the gel must be stable enough to tolerate a reversible shrinkage to 28 % of its original volume (Figure 6-15).

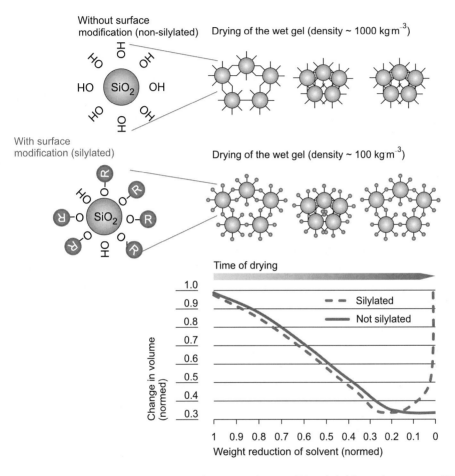

Figure 6-15. Ambient pressure drying. Top: irreversible shrinking of an unmodified silica gel. Middle: reversible shrinking of a silica gel modified by organic groups. Bottom: volume change of unmodified and modified gels upon ambient pressure drying.

6.3.2 Properties and Applications

As the metallic foams, aerogels possess a unique combination of properties due to their high porosity combined with low densities. Non-silicate aerogels are interesting materials for catalytic applications due to the high porosity combined with a high surface area. Silica aerogels are transparent, have excellent heat insulation properties, show a very good impact energy absorption capacity, and unusual acoustic properties. The sound velocities in SiO_2 aerogels are in the range of 100–300 m·s^{-1}, which are among the lowest for inorganic solids and worth mentioning, since sound velocities of 5000 m·s^{-1} were measured for quartz glass.

The thermal conductivity of silica aerogels is extraordinary low. The passage of thermal energy through a material occurs through three mechanisms: solid conductivity; gaseous conductivity; and radiative (infrared) transmission. The sum of these three components gives the total thermal conductivity of the material. Solid conductivity is an intrinsic materials property, which is very high for dense silica. However, the volume fraction of silica in aerogels is very low, and additionally the solid is composed of very small particles linked to a three-dimensional network with many dead ends. Therefore, thermal transport through the solid phase in not particularly effective. The open pores allow the passage of gases (air) which can transport thermal energy through the material. The final mode of thermal transport through silica aerogels involves infrared radiation. At low temperatures, radiative transport is very low, but at higher temperatures radiative transport becomes the dominant process of thermal conduction. Evacuation of the aerogel system and addition of an infrared opacifier such as carbon soot or TiO_2, greatly improves the high-temperature insulation properties. With these improvements there is no question that silica aerogels are among the best heat insulation materials currently available. Additionally, they are non-flammable and, in the absence of an opacifier, transparent. Aerogels can be utilized for the passive use of solar energy, for example for paneling house walls or for coating solar cells.

Silica aerogels are used by NASA for the collection of cosmic dust. This is a quite difficult task because these hypervelocity particles travel at nine times the speed of a bullet fired from a rifle. Although each of the captured particles is smaller than a grain of sand, high-speed capture could alter their shape and chemical composition – or vaporize them entirely.

To collect the particles without damaging them, a project called STARDUST was established, which used silica aerogels to perform the task. When a particle hits the aerogel, it will bury itself in the material, creating a carrot-shaped track up to 200 times its own length, as it slows down and comes to a stop. The situation is similar to that of an airplane setting down on a runway and braking to reduce its speed gradually. Since silica aerogels are transparent, scientists will use these tracks to find the tiny particles.

Figure 6-16.
Aerogel collector for cosmic dust.

The dust collector device consists of blocks of 1 and 3 cm thick silica aerogel tiles mounted in modular aluminum cells (Figure 6-16). For the STARDUST mission, cells are mounted on both sides of a two-sided, grid-shaped array that will deploy from the sample return capsule. After exposure, the cells assembly will fold up to a compact configuration for stowage into the earth return capsule.

6.4 Micro- and Mesoporous Solids with an Ordered Porosity

In this section we will introduce the reader to a templating (▶ glossary) approach towards ordered porous materials. Ordered porosity means that the pores are arranged in a regular fashion usually with a quite narrow pore size distribution. Figure 6-17 shows a schematic of representative materials in different pore sizes regimes with the corresponding pore size distribution.

As can be seen from Figure 6-17, materials in almost all pore size regimes exist. One of the major differences between the materials, besides the size of the pores, is the distribution of the pores. Pillared clays (see Section 2.4.3) and porous glasses have a broad distribution of the pores sizes.

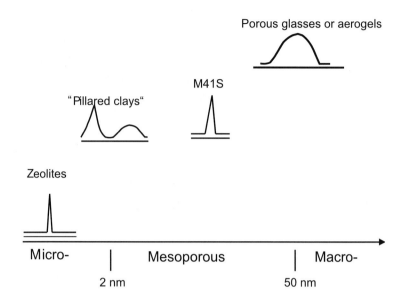

Figure 6-17. Simplified picture of representative porous materials with different pore sizes and the corresponding pore size distributions.

The probably best known *microporous* material with an ordered porosity synthesized by a templating approach are zeolites. Zeolites are characterized by an extremely narrow pore size distribution because the pores are intrinsic features of the crystal lattice. In the second part of this section the reader will be introduced to a new method by which structured *mesoporous* materials with a narrow pore size distribution, designated as M41S-materials, are prepared.

In general, three steps are involved in forming porous materials by templating: synthesis, drying, and template removal.

6.4.1 Microporous Crystalline Solids

Molecular or ionic compounds usually form crystal structures in which the molecules or ions are packed as closely as possible. In contrast, network-forming compounds (see also Section 4.1.1) can also crystallize in open framework structures in which the available space is not optimally occupied. In this section, the focus is on zeolites as an example of crystalline materials with network structures where the porosity is a result of the special three-dimensional, periodic arrangement of the network. Since the pores are part of the crystal structure, their narrow size distribution is not surpassed by any other material. Some other compounds, such as the aluminophosphates or the clathrasils (aluminum-free microporous silica), have similar structures but differ in the chemical composition of the framework.

Microporous inorganic solids are finding wide use as catalysts, molecular sieves and sorption media. The control of pore geometry and diameter is a key feature for many applications, especially those that rely on size and/or shape selectivity and a ready access to the pores. The porous nature of these crystal-line ceramic oxides also allows their use as molecular-sized reaction vessels, as specific hosts, and as shape-selective conversion media.

At present, about 40 naturally occurring zeolite minerals are known, and many more have been synthesized in the laboratory. Zeolites are crystalline aluminosilicates. The name zeolite, from the Greek *zeo* = to boil and *lithos* = stone, was coined to describe the behavior of the mineral stilbite which loses water on heating and thus seems to boil.

The framework forming the regular one-, two- or three-dimensional channel systems and cavities of molecular size, is constructed from linked tetrahedra. The most common zeolites are structurally based on corner-sharing silicon and aluminum TO_4 tetrahedra. If the network is built from $[SiO_4]$ tetrahedra only, the overall charge of the framework is zero, as already discussed for glass in Section 4.1.1. Substitution of Si(IV) by Al(III) results in a net negative charge of the framework, which is equal to the number of the constituent Al-atoms. This charge is balanced by exchangeable cations (M) located in the pores, in which also adsorbed water is located.

The formula for a zeolite thus is:

$$M_{x/n}(Al_xSi_yO_{2(x+y)}) \cdot z\,H_2O$$

The Si/Al ratio is adjustable over a wide range from 1 to infinity. The exchangeable cation M (n = cation charge) is generally a group I or II ion, although other metal, non-metal, and organic cations may also be used. Many elements which form tetrahedral TO_4 structural units, can take part in the con-struction of zeolites such as $[PO_4]$, $[BeO_4]$, $[GaO_4]$, $[GeO_4]$ and $[ZnO_4]$.

The three-dimensional structure of a typical zeolite is shown in Figure 6-18.

Before we turn to the question how the seemingly complicated structures of zeolites can be understood, we will focus on preparative aspects.

Synthesis

The diversity of structures in zeolites is enormous and, obviously, the synthe-sis conditions are crucial for which structure is obtained. The synthesis of zeolites is in principle based on two previously discussed processes: The sol–gel process (see Section 4.5) and hydrothermal processing (see Section 4.4). Zeolites are synthesized from aqueous solutions of sodium silicates and aluminates (see Section 4.4.2), which contain either alkali metal hydroxides or organic bases to achieve a high pH. A gel is formed from the silicate and aluminate ions by condensation reactions, as has been discussed in Section 4.5. If the silica content of the zeolite is low, the product can often just be crystallized at 70–100 °C. Silica-rich zeolites, on the other hand, are mostly obtained by hydrothermal treatment of the gels. In this case, the gel is put in an autoclave for several days; the zeolites are typically formed between 100 °C and 350 °C. Variable parameters that determine the type of zeolite obtained include the composition of the starting solution, the pH, the temperature and time program, and some history-related factors such as the aging conditions, the rate of stirring and even the order of mixing.

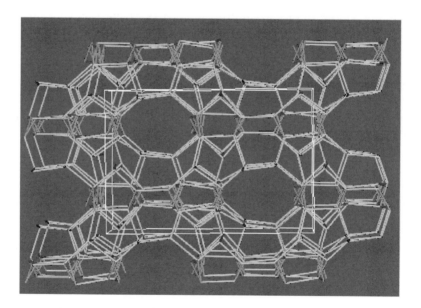

Figure 6-18. Three-dimensional view of zeolite ZSM-5.

A complicated interplay of all the different parameters governs the final zeolite structure. For example, if the Si/Al ratio of a given zeolite is to be changed, this is not possible by just changing the Si/Al ratio in the precursor solution. Instead, the whole set of parameters must be adjusted. Even today the synthesis of zeolites happens mostly by trial-and-error approaches.

Table 6-3 provides some "rules of thumb" on how different components of the reaction mixture influence the final zeolite structure.

Table 6-3. Influence of different components of the reaction mixture on the zeolite structure.

Reaction mixture composition	Primary influence
SiO_2/Al_2O_3 ratio	framework composition
H_2O/SiO_2 ratio	rate; crystallization mechanism
OH^-/SiO_2 ratio	silicate molecular weight
inorganic cations/SiO_2 ratio	structure; cation distribution
organic additives/SiO_2 ratio	structure; framework Al content

Besides acting as counter-ions to balance the zeolite framework charge, the metal or ammonium cations present in the reaction mixture additionally appear to have a structure-determining effect. The latter effect is also attributed to uncharged organic additives such as amines. With the use of amines and quaternary ammonium ions the number of zeolite structures has been greatly extended because these allow variation of the silica/alumina ratio over a wider range. Typical examples of organic additives are shown in Figure 6-19.

A templating theory has been postulated for the role of the cations and the organic additives in stabilizing the formation of structural sub-units in the reaction mixture. A chemical species is considered to be a template or structure-directing agent, if crystallization of a specific zeolite structure is induced that could not be formed in the absence of the reagent.

The notion that the zeolite structure growths around the template (▶ glossary) is probably too simple. For example, it has been shown repeatedly that, in the presence of a particular organic species, crystallization of various zeolite structures can occur. Conversely, one structure can be directed by a number of different amines.

The role of the "template" is more diverse and complex than initially perceived. It can:

- behave as a structure-directing or templating agent;
- act as a gel modifier, particularly influencing the Si/Al ratio;
- act as a void filler; and
- influence physically and chemically the formation and aging of the gels, and the crystallization process.

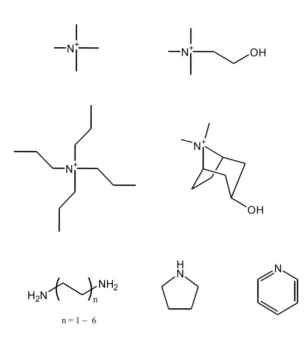

Figure 6-19. Some representative ammonium or amino compounds used as organic additives in zeolite syntheses.

Structure

There is a large number of complex zeolite structures, which are characterized by the building units, size and arrangement of the pores, and the interconnection of the pores.

The framework structure of zeolites can be constructed, by assembling structural units with increasing complexity starting from the tetrahedral building blocks; the pore structure is a result of the particular network structure. However, some general rules must be obeyed:

- The primary building units are tetrahedral.
- All tetrahedra share all their corners with neighboring tetrahedra.
- The arrangement of the building blocks must be periodic (crystallinity).

Another important feature is that there are no Al–O–Al linkages, and only Si–O–Al and Si–O–Si linkages are allowed. Therefore, the maximum substitution rate of silicon by aluminum is 50 %.

The three-dimensional arrangement of the TO_4 tetrahedra – also called primary building units – can occur in numerous ways. The wealth of zeolite structures is additionally due to the flexibility of the T–O–T bond angle, which can assume values in the extensive range of 120–180 °C. Nevertheless, only a small number of secondary building units (SBUs) are common to all zeolite structures.

For clarity it should be noted that in the graphic representation of zeolites each corner represents the center of a tetrahedron, i.e., the position of the silicon or aluminum atoms. This is shown for one of the SBUs – a six-membered ring structure – in Figure 6-20, as an example. All SBUs are shown in Figure 6-21.

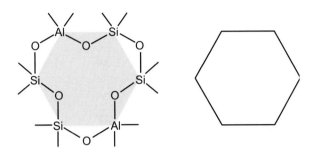

Figure 6-20. Six-membered ring structure of $[SiO_4]$ and $[AlO_4]$ tetrahedra (left) that can be found in zeolites, and its schematic representation as an SBU (right).

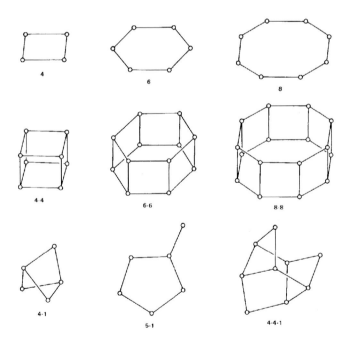

Figure 6-21. Secondary building units (SBUs) found in zeolite structures.

With these SBUs all known zeolites structures can be constructed. The SBUs provide a convenient method of topologically describing and relating different zeolites. Figure 6-22 shows an example in which the network structure (ZSM-5) can be imagined to be built only from "5-1" SBUs. The pores with openings formed from ten [TO$_4$] tetrahedra are a consequence of the particular arrangement of the SBUs.

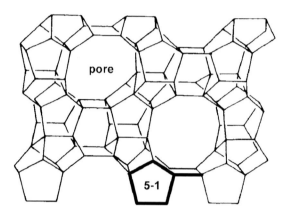

Figure 6-22. Network structure of ZSM-5.

For most zeolite applications the pore size is one of the key features. The pore or channel openings range from 0.3 to 1 nm, depending on the structure of the zeolite. Figure 6-23 shows cross-sections through pores of different zeolites. For ZSM-5 the pores are built from 10 tetrahedra, but for other zeolites larger or smaller sizes can be achieved by different arrangements of the SBUs. These pores can also differ in the shape of the opening.

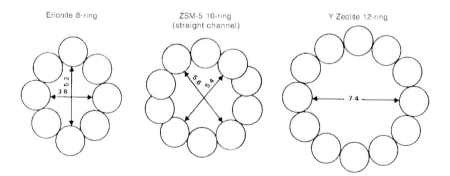

Figure 6-23. Examples of three types of pore openings in zeolites: Erionite (left) contains an 8-ring pore opening, ZSM-5 (center) a 10-ring system, and type Y zeolite (right) a 12-ring pore system (the diameters are given in Å).

To simplify the representation of zeolite structures, sheet (two-dimensional) projections are often used. The lower part of Figure 6-24 shows such a projection for ZSM-5 (compare this projection with the three-dimensional structure in Figure 6-22); the projection of mordenite is also shown in Figure 6-24 as an example for another zeolite structure.

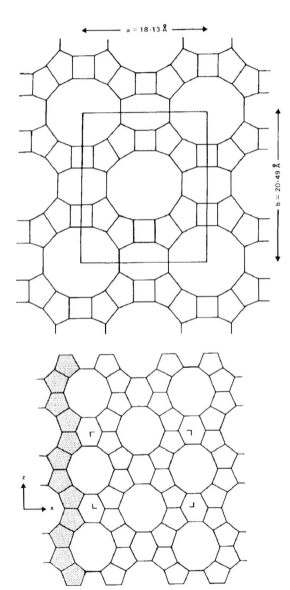

Figure 6-24. Sheet projection of mordenite with 8- and 12-membered channels (top) and ZMS-5 with 10-membered channels (bottom).

In passing, it should be pointed out that the combination of the SBUs is not as simple as playing Lego. In the representation shown in Figure 6-24, the primary units were reduced to one point (the corners of the SBUs). In reality, the three-dimensional shape must also be considered. For example, when we look at the tetrahedra that comprise the simplest SBU, the 4-membered ring, it is immediately obvious that there is more than one way of joining together the 4-ring building units (Figure 6-25).

Last, but not least, one must also consider the connectivity of the pores which can be viewed as one-, two-, or three-dimensional tubes or channels. For example, analcime consists of a system of non-intersecting, one-dimensional channels, while mordenite is a two-dimensional, intersecting channel system. Three-dimensional channel systems can be found in ZSM-5 (Figure 6-26).

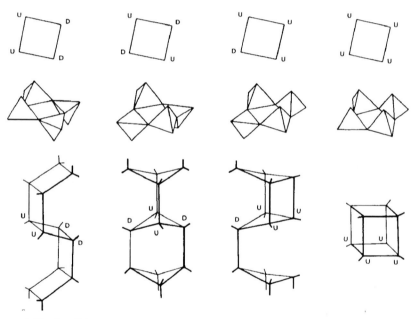

Figure 6-25. Possible linkages for the 4-membered ring SBU. Upper row: schematic representation of the SBU; U = up and D = down. Center row: the tetrahedral view. Bottom row: chain sequences formed from the differently oriented SBUs.

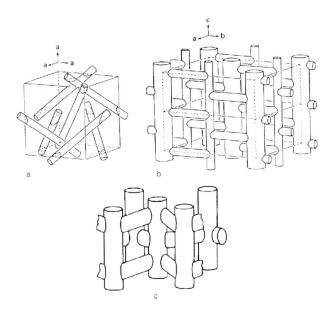

Figure 6-26. Channel representation for examples of one-, two-, and three-dimensional systems; a) analcime; b) mordenite; c) ZSM-5.

6.4.2 Mesoporous Solids with Ordered Porosity

In 1992, a new concept for the synthesis of porous materials was introduced by Mobil's scientists. In contrast to zeolites – which are synthesized with single, solvated organic molecules or metal ions as the template – supramolecular arrangements of molecules were used for the synthesis of porous materials. Larger pore sizes are accessible with this approach.

Typically, amphiphilic surfactant molecules (or lyotropic liquid crystals; ▶ glossary) are used as structure-directing agents.

The pores in the obtained materials are about an order of magnitude larger than those in zeolites, and can be tailored within the 2 to 10 nm range. The materials exhibit large specific surface areas, large pore volumes and, owing to the regularity of the template, a monomodal (▶ glossary), narrow distribution of pore sizes. In contrast to zeolites, the pore walls are not crystalline, but amorphous. Mesoporous silica has been prepared as transparent hard spheres, glass sheets, hollow spheres, fibers, thin films, or monoliths.

Owing to the larger pores and high surface areas, these new materials, denoted as M41S phases, are of great interest for applications in size- and shape-selective processes, e.g., as catalysts, catalyst supports, adsorbents, or as host structures for nanometer-sized guest compounds.

The M41S family of mesoporous materials has three important members which are distinguished by their different pore structure (Figure 6-27):

- MCM-41: one-dimensional hexagonally ordered cylindrical channel structure;
- MCM-48: three-dimensional bicontinuous cubic structure;
- MCM-50: two-dimensional system of lamellar silica sheets interleaved by surfactant bilayers. (MCM = Mobil composition of matter)

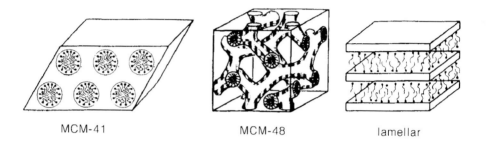

MCM-41   MCM-48   lamellar

Figure 6-27. Different structural types of the M41S family: hexagonal MCM-41; bicontinuous cubic MCM-48; lamellar MCM-50.

Similar to the zeolite synthesis, four reagents are required for the M41S synthesis: water, an amphiphilic molecule (▶ glossary, Figure 6-28), a soluble inorganic precursor, and a catalyst. Depending on the inorganic precursor, the synthesis conditions vary over a wide range: from seconds to days and from 180 °C to –14 °C. In the following paragraphs the different steps of the synthesis of these mesostructured materials are described, starting with the formation of a supramolecular arrangement of the template (▶ glossary) molecules, followed by the actual templating step, and finally the removal of the template.

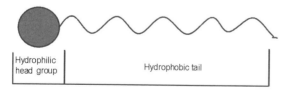

Hydrophilic head group

Hydrophobic tail

Figure 6-28. Amphiphilic surfactant molecule.

Supramolecular arrangement of the template molecule

As a result of their amphiphilic nature, surfactants can associate into supramolecular arrays. For example, cetyltrimethylammonium bromide (CTAB, $CH_3(CH_2)_{15}N(CH_3)_3{}^+Br^-$) will form spherical micelles (▶ glossary). This arrangement minimizes the unfavorable interaction of the hydrocarbon tails with water, but introduces a competing unfavorable interaction – the repulsion of the charged head groups. The balance between these competing factors determines the relative stability of the micelles.

The extent of micellation, the shape of the micelles, and the aggregation of micelles into liquid crystals (▶ glossary) depends on the surfactant concentration. A schematic phase diagram for a cationic surfactant is shown in Figure 6-29.

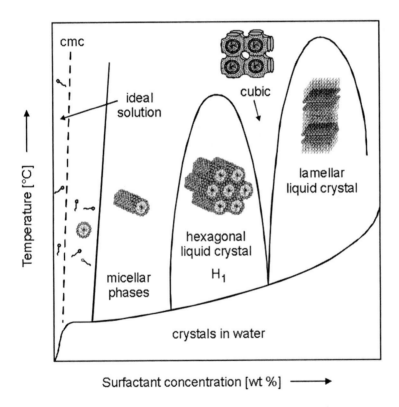

Figure 6-29. Schematic phase diagram for a surfactant in water.

At very low concentrations the surfactant is present as individual molecules dissolved in solution. When the surfactant concentration is gradually increased the following succeeding stages can be observed:

1. The surfactant molecules form small spherical aggregates (micelles, ▶ glossary) above the critical micelle concentration (cmc).
2. These spherical micelles can coalesce to form elongated cylindrical micelles when the amount of solvent available between the micelles decreases.
3. Formation of liquid crystalline (▶ glossary) phases (LC) occurs at slightly higher concentrations, starting with a hexagonal close-packed LC arrangement, followed by the bicontinuous cubic phase and finally resulting in lamellar structures (Figure 6-29).
4. At very high concentration, inverse phases can exist; here water is solubilized at the interior of the micelle (▶ glossary), and the head groups point inward.

Surfactants with a wide variety of sizes, shapes, functionalities, and charges exist, and can be used as templates (▶ glossary) for the synthesis of meso-porous solids.

Templating

The original MCM-41 synthesis was carried out under hydrothermal conditions by using CTAB as the structure-directing agent and tetraethoxysilane as the silica source.

Two mechanistic pathways leading to the formation of these mesoporous materials are discussed (Figure 6-30).

1. The surfactant molecules organize first, independent of the inorganic silicate polycondensation (▶ glossary). In this case, which is also called liquid crystal-initiated pathway, the siliceous framework polycondenses around the preformed surfactant aggregates.
2. The second mechanism assumes that anionic siliceous species associate first with the surfactant. This assembly than organizes in supramolecular arrays. This pathway is also called silicate anion-initiated pathway.

The two pathways discussed are probably only limiting cases. Experiments show that the formation of mesostructured materials is almost independent of which aggregation stage of the surfactant the inorganic species is added. This implies that the inorganic source does not simply petrify the LC array, but instead cooperatively coassembles at some stage with the surfactant to form LC phases.

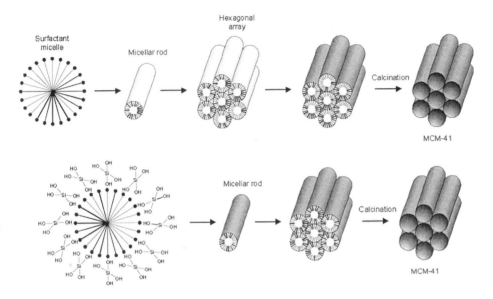

Figure 6-30. Possible mechanistic pathways for the formation of MCM-41; the gray shading represents the silica.

The M41S phases described so far are generally synthesized via an electrostatic templating procedure based on surfactants with cationic head groups, and anionic inorganic building units (S^+I^-; S represents the surfactant, I the inorganic species). A huge variety of this type of surfactants exist which can be utilized as templates (▶ glossary). Based on this charge interaction, other combinations of ions pairs, including charge-reversed (S^-I^+) and counter-ion mediated ($S^+X^-I^+$ or $S^-M^+I^-$, with X^- being halides and M^+ being alkali metal ions; Figure 6-31) synthesis pathways are possible. Besides these Coulombic forces, other interactions at the inorganic-organic interface can be used for templating such as covalent bonding of the template (▶ glossary) to the inorganic network-forming species (S-I) or neutral templating route based on hydrogen bonding (S^0I^0) (Figure 6-31).

By carefully choosing the template molecule, the pore diameter can be adjusted over a wide range. An increase in pore diameter can be achieved by the addition of an auxiliary organic molecule such as mesitylene, which is trapped in the micelle (▶ glossary) by increasing the volume of the hydrophobic core (Figure 6-32).

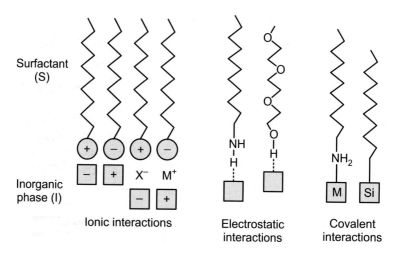

Figure 6-31. Schematic presentation of possible interactions at the inorganic–organic interface (S = surfactant, I = inorganic phase, M = metal, X = halide).

Removal of the template

The removal of the template (▶ glossary) is the last step in the synthesis of mesoporous structured materials. This can generally be performed in different ways:

- by solvent extraction;
- by calcination (▶ glossary);
- by oxygen plasma treatment; or
- by supercritical drying (see also Section 6.3).

The first two methods are most often applied. Extraction is performed by several washing steps. Often organic solvents with dissolved acids, i.e., ethanol with HCl, are used to remove the template from electrostatically templated materials. Covalently bonded templates cannot be removed by this procedure. Calcination is typically performed between 400 and 600 °C in various atmospheres such as nitrogen or air.

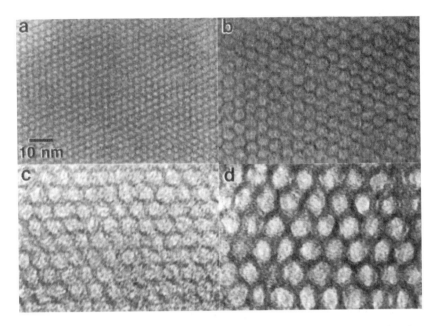

Figure 6-32. Transmission electron micrographs of MCM-41 materials having 2 nm, 4 nm, 6 nm, and 10 nm hexagonally arranged pore channels (increasing pore size from (a) to (d); the black areas represent the silica walls, the lighter areas the pores.

While framework structures such as the hexagonal or cubic phase are typically stable upon removal of the template with only a small degree of shrinkage, lamellar structures collapse during the removal of the template (▶ glossary).

The examples discussed show that a deliberate design of pores is possible in a wide pore size range, even with narrow pore size distributions. The application of more complex template systems could result in a huge extension of this approach also in the macroporous regime.

The underlying principle discussed above for silica materials may also be applied to the formation of other mesoporous materials with different chemical compositions.

6.5 Further Reading

M. Antonietti, C. Göltner, "Mesoporous Materials by Templating of Liquid Crystalline Phases", *Adv. Mater.* **9** (1997) 431–436.

J. Baumeister, "Verfahren zur Herstellung von Metallschäumen", *Techn. Mit.* **92** (1992) 94–99.

P. Behrens, "Mesoporous inorganic solids", *Adv. Mater.* **5** (1993) 127–132.

P. Behrens, G. D. Stucky, "Novel materials based on zeolites", *Compreh. Supramolecular Chemistry,* **7** (1996) 721–772.

J. S. Beck, J. C. Vartuli, "Recent advances in the synthesis, characterization and applications of mesoporous molecular sieves", *Curr. Opin. Solid State Mater. Sci.* **1** (1996) 76–87.

C. J. Brinker "Porous inorganic materials", *Curr. Opin. Solid State Mater. Sci.* **1** (1996) 798–805.

A. G. Evans, J. W. Hutchinson, M. F. Ashby, "Cellular metals", *Curr. Opin. Solid State Mater. Sci.* **3** (1998) 288–303.

L. J. Gibson, M. F. Ashby, *Cellular Solids – Structure and Properties,* Cambridge University Press, Cambridge, 1997.

G. J. Davies, S. Zhen "Metallic foams: their production, properties and applications", *J. Mater. Sci.* **18** (1983) 1899–1911.

C. G. Guizard, A. J. Julbe, A. Ayral, "Design of nanosized structures in sol–gel derived porous solids. Applications in catalyst and inorganic membrane preparation", *J. Mater. Chem.* **9** (1999) 55–65.

N. Hüsing, U. Schubert, "Aerogels – airy materials: chemistry, structure, and properties", *Angew. Chem. Int. Ed. Engl.* **37** (1998) 22–45.

M. L. Occelli, H. Kessler (Eds.), *Synthesis of Porous Materials – Zeolites, Clays, and Nanostructures,* Marcel Dekker, New York, 1997.

N. K. Raman, M. T. Anderson, C. J. Brinker, "Template-based approaches to the preparation of amorphous, nanoporous silicas", *Chem. Mater.* **8** (1996) 1682–1701.

E. Roland, P. Kleinschmit, "Zeolites", *Ullmanns Encyclopedia,* Vol. A28, 475–504.

J. Rouquerol, D. Avnir, C. W. Fairbridge, D. H. Everett, J. H. Haynes, N. Pernicone, J. D. F. Ramsay, K. S. W. Sing, K. K. Unger, "Recommendations for the characterization of porous solids", *Pure & Appl. Chem.* **66** (1994) 1739–1758.

D. W. Schaefer, "Engineered porous materials", *Mater. Res. Soc. Bull.* **14**(4) (1994) 14–17.

R. Szostak, *Molecular Sieves – Principles of Synthesis and Identification,* Van Nostrand Reinhold, 1989.

P. T. Tenev, J.-R. Butruille, T. J. Pinnavaia, "Nanoporous materials" in *Chemistry of Advanced Materials – an Overview,* Wiley-VCH, New York, 1998, Eds. L. V. Interrante, M. Hampden-Smith, pp. 329–388.

7 Nanostructured Materials

Many properties of solid matter are determined by the "infinite" three-dimensional arrangement of its building blocks rather than the properties of the building blocks themselves. How small can the dimensions of a distinct material in one to three dimensions become, yet still have the same properties as the bulk? For example, a macroscopic piece of metal is electrically conducting, but a single metal atom or a small metallic cluster is not. Thus, the question is, at which size will a piece of metal lose its electric conductivity, magnetism, etc. when it is cut in smaller and smaller pieces? Or, conversely, at which size will a metal atom cluster become electrically conducting when it gradually becomes larger?

It has been shown that the properties of matter can dramatically change at a particle size somewhere between 1 nm and 100 nm. In this regime, neither quantum chemistry nor solid state physics hold. The former describes the properties of atoms and molecules, and the latter the properties of solids.

The term "nanostructured materials" ("nanomaterials") thus refers to materials having a characteristic length scale in the lower nanometer range that influences their physical or chemical properties. Nanostructured materials may contain single or multiple crystalline, quasicrystalline or amorphous phases and can be metals, ceramics, semiconductors (▶ glossary) or polymers. They can be classified according to their dimensionality and/or the nature of the nanosized feature (crystallites, pores, etc.):

- Nanoparticles are essentially large atom clusters and can be considered zero-dimensional (0-D).
- Layered or lamellar structures are one-dimensional (1-D) nanostructures, in which only the thickness of the layer is in the nanometer range.
- Filamentary structures with diameters in the nanometer range and much larger lengths are 2-D nanostructures.
- The most common nanostructured materials are those in which all three dimensions are of nanometer size (3-D), such as nanostructured crystallites.

Nanomaterials have the potential of revolutionizing materials design for many applications. Their novel or unusual physical properties are mainly due to

- *Finite-size effects*: electronic bands (i.e., a quasi-continuous density of electronic states) are gradually converted to molecular orbitals as the size decreases. Confining electrons to small geometries gives rise to "particle in a box" energy levels owing to the coincidence of structural dimensions with the electron wavelength. This quantum confinement creates new energy states and results in the modification of (opto)electronic properties.

- *Surface and interface effects*: a high percentage of the atoms are surface or interface atoms. For example, about 50 % of the atoms in a spherical 3-nm particle are located at grain (▶ glossary) and interface boundaries (Figure 7-1). A nanocrystalline metal contains typically a high number of interfaces ($\sim 6 \times 10^{25}$ m^{-3} for 10 nm-sized grains). Nanoparticles can therefore be considered surface matter in macroscopic quantities.

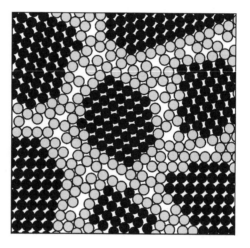

Figure 7-1. Two-dimensional model of a nanocrystalline material. The dark spheres represent the crystalline portion, and the gray circles the atoms in the interfacial region.

In *nanocrystalline materials* the dimension of the grains (▶ glossary) is in the nanometer range. Nanocrystalline materials can be considered to consist of two structural components: the small crystallites with different crystallographic orientations; and a network of intercrystalline regions (the "interfacial components"; see Figure 7-1), consisting of grain boundaries and triple junctions. Nanocrystalline materials are the best investigated type of nanomaterials, and maximum attention has been paid to their synthesis, sintering and characterization (see also Section 2.1.4). With nanocrystalline materials it appears to be possible to make metals hard and strong – like ceramics, and ceramics ductile – like metals (see below).

Nanocomposites are composite materials (▶ glossary) in which (0-D) nano-particles are embedded in a host phase. Supported nanoparticles of metals or metal oxides are widely used as heterogeneous catalysts. A high dispersion of the active component maximizes the contact area of the catalyst with reactant and support. Composites containing semiconductor (▶ glossary) nanoparticles (so-called quantum dots or Q particles) have very interesting optical and/or electrical properties (see below). The fracture toughness of ceramics can be considerably enhanced by dispersing nanoparticles or nanocrystalline fibers as a second phase.

Nanoporous materials contain pores with diameters in the nanometer range. These are discussed in Chapter 6.

Nanometer-scale (multi-)layers are very important in biological systems (see Sections 4.3 and 7.2). Artificially multilayered inorganic materials are composed of layers of different phases (heterostructures), for example alternating layers of a compound A and a compound B. Each layer consists of only a few monolayers of the corresponding compound. The principal characteristic of multilayers is a composition modulation, i.e., a periodic chemical variation. Some of the highly interesting properties of such multilayers are attributed to the tremendous interface area. Multilayers composed of single-crystal layers that possess the same crystal lattice (i.e., which are topotactic, ▶ glossary) are called superlattices. Semiconductor superlattices have important technological applications, due to novel electronic and photonic effects. The atomic-scale multilayering of different materials also leads to significant enhancements of the yield strength (▶ glossary), tensile strength, hardness (▶ glossary) and wear resistance.

Nanotubes carry some potential for interesting applications. Any layered compound can potentially form nanotubes by rolling up the layers or sections of the layers. The best investigated examples are carbon nanotubes (Figure 7-2), with a diameter of 1–3 nm. Nanotubes of BN, silica, and transition metal oxides, sulfides and halides have also been synthesized, the synthesis procedures depending on the composition of the tubes. The tubes may be closed (as in Figure 7-2), or open at one or both ends. Apart from the features characteristic to nanostructured materials, the form of a tube is particularly attractive, because the inner surface has different properties from the outer surface. The caps are the most reactive regions which allows the selective opening of closed tubes by etching.

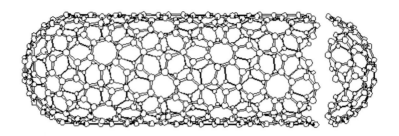

Figure 7-2. Schematic drawing of a carbon nanotube.

Nanotubes can be single-walled (as the carbon nanotube in Figure 7-2), or multi-walled. The latter can have a tube-in-a-tube morphology (Figure 7-3, left) or consist of a rolled-up sheet of a layered compound (Figure 7-3, right). Multi-walled nanotubes up to 1–2 mm length have been observed.

Figure 7-3. Multi-walled nanotubes. Left: tube-in-a-tube; right: rolled-up sheet.

Before we turn to the preparation of nanostructured materials, we will discuss some of the physical properties originating from the finite size and surface effects (Section 7.1). Thin layers have been discussed in several chapters of this book; some additional preparation methods for two-dimensional systems will be treated in Section 7.2. Filamentary (2-D) structures have only recently gained attention, and their preparations are to special to be treated in this book. Section 7.3 will then deal with methods for the preparation of nanocrystalline materials and nanosized particles.

7.1 Properties of Nanostructured Materials

Because of the finite-size and surface effects, physical and chemical properties of nanomaterials are normally very different both to those of the bulk materials and to those of isolated atoms, molecules or small (molecular) clusters of the same composition. The properties of nanostructured materials were mainly investigated for nanocrystalline materials (3-D) or nanoparticles (0-D); the investigation of 1-D and 2-D nanostructures lags behind due to the lack of general methods for the preparation of such materials.

Optical properties

The unique optical properties of very small particles were discovered in the 19th century by Michael Faraday (1791 – 1867) when he studied the various colors of gold colloids (▶ glossary). These colors can range from ruby red through purple to blue, depending on the colloid particle size. Mie developed the theory of plasmon resonance to explain the visible absorption bands exhibited by small particles. Visible light absorption takes place within the set of molecular orbitals with discrete energy levels which develop when a particle becomes small enough. With increasing size of the particle the spacing between adjacent energy levels becomes smaller, and the absorption band progressively red-shifts and eventually gives way to broad absorptions of the bulk compound. For example, a sharp band at about 360 nm is observed for CdS nanoparticles of 1 nm diameter, whereas 2 nm and larger particles show broad, red-shifted bands (see wavelength scale in Figure 3-2). The optical properties of semiconductor nanoparticles can be perturbed significantly by adsorption of solvent molecules or ligands. Since nanoparticles have a high percentage of surface atoms, perturbation of energy levels of surface atoms can affect the entire particle.

When semiconductor nanoparticles in the range of 1 to 15 nm are doped into polymers or glasses, the resultant composite materials (▶ glossary) exhibit third-order, non-linear optical properties (▶ glossary). Materials possess third-order, optical non-linearity if their refractive index depends on the intensity of the incident light. Under such conditions, the induced polarization of the material contains a term that has a 10^3 dependence on the electrical field of the incident light (hence the term "third-order").

Electric properties

Clusters in the range of a few nanometers show a transition from metallic to non-metallic behavior (i.e., quantum size effects). Figure 7-4 shows how the conductivity of metal particles changes with size ("size-induced metal-insulator transition", SIMIT effect). During this transition, the electronic band structure of bulk metals increasingly resolves into separate molecular orbitals. At the lower limit, i.e., when the particles become very small, they no longer exhibit metallic properties. In particular, there is no electrical conductivity within particles of that size.

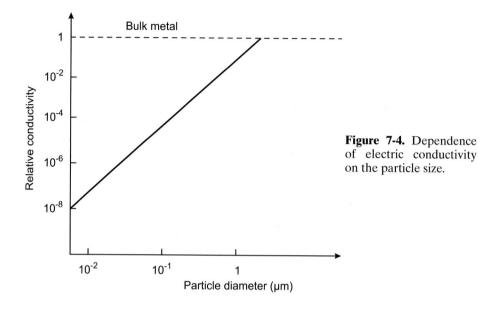

Figure 7-4. Dependence of electric conductivity on the particle size.

Magnetic properties

Ferromagnetic materials (▶ glossary) contain many magnetic domains. There is a critical size for each material below which the particles are single-domain. If a single-domain ferromagnetic nanoparticle is put in a non-magnetic matrix such that their magnetic interaction is negligible, a superparamagnetic material is obtained. The particles behave similar to paramagnets (▶ glossary), but rather than having a small moment of a few Bohr magnetons, the moment is a sum of the moment of all the atoms in the particle. Below a critical temperature T_B, the blocking temperature, superparamagnetic behavior disappears, and ferromagnetism sets in.

The phenomenon of giant magnetoresistance (GMR, also called "colossal magnetoresistance", CMR – decrease of electrical resistance of materials when exposed to a magnetic field) has been observed in nanocrystalline materials. It also occurs in certain metallic multilayers composed of alternating ferromagnetic (▶ glossary) and non-magnetic layers such as Fe/Cr and Co/Cu. Whereas the resistance drop is only 1–2 % in conventional coarse grained (▶ glossary) materials, the drop is as large as 50–80 % in nanocrystalline materials.

Materials with a high magnetoresistance are needed for high-density magnetic recording materials (heads in computer disk drives, for example).

Diffusion and sinterability

The small particle size and the many interfaces in nanocrystalline materials provide a high density of short-circuit diffusion paths. The increased diffusivity, together with the small grain size (▶ glossary), can have a significant effect on the sintering behavior, formability, mechanical properties (see below), ability to dope nanocrystalline materials efficiently at relatively low temperatures, and synthesis of alloy (▶ glossary) phases of immiscible metals and at temperatures much lower than those usually required in other systems. Nanopowders can sometimes be densified and sintered at temperatures substantially lower than those required for conventional ceramic materials (see Section 2.1.4.).

Mechanical properties

The reduction of grain size (▶ glossary) to the nanometer range results in many interesting mechanical properties.

The *hardness* (▶ glossary) and *strength* of sintered nanophase ceramics are considerably higher than those observed in conventional materials. For example, tensile and compressive strengths in nearly all nanoscale material systems studied have shown anomalously high values. One reason for this is that the mechanical strength of crystalline materials typically increases with decreasing grain size. Furthermore, the mechanisms of hardening/softening in nanocrystalline materials may be fundamentally different from that observed in coarser-grained materials.

Conventional coarse-grained ceramics are brittle, but nanocrystalline ceramics may be ductile (▶ glossary). *Superplasticity* is the capability of some polycrystalline materials to undergo large tensile deformation without necking or fracture. Elongations by a factor of ten are typical of this phenomenon. It would be of considerable industrial interest, as superplastic forming can be used to produce components having complex shapes from materials that are difficult to machine.

Chemical properties

The investigation of chemical phenomena related to nanosized structures such as colloids, micelles (▶ glossary), highly dispersed metal or metal oxide catalysts, or nucleation phenomena are not particularly new. What has changed is the perspective and the expectations. It can be anticipated that surface and size effects can influence chemical processes when they occur at nanostructures or within a nanostructured environment, as they affect physical properties. For example, as chemical reactions are governed by electron affinities/ionization potentials and orbital energies, there must be some relationship between the electronic structure of nanoparticles and their chemical properties. A number of chemical functions have been identified to be clearly size- and structure-dependent, e.g., electrochemical processes, sensing, adsorption and separation of molecules and in specific cases the catalysis of chemical reactions. In hetero-geneous catalysis with supported metals one finds size-sensitive ("demanding") reactions, where the reaction rates depend on the metal particle size, and size-independent ("facile") reactions. In "demanding" reactions, there appears to be a connection between the ionization potential of the metal particle and the reaction rates.

7.2 Mono- and Multilayers

Nanometer-scale (multi-)layers can be deposited by PVD and CVD methods. These gas-phase methods even allow the atomic-scale layering of materials (such as the ALE method) and have been discussed in Section 3.2.

The deposition of mono- or multilayers *from solution* is gaining much importance. Although these methods are currently mainly restricted to the preparation of layers of organic compounds, we will introduce them briefly because of their importance. These layers are formed by self-organization processes and are promising for a variety of technologies, including micro- and optoelectronics, optical information storage, corrosion resistance or chemical sensing. Monolayers of insoluble amphiphiles (▶ glossary) deposited at the water/vapor interface, known as *Langmuir films* or Langmuir monolayers, provide insight into fundamental issues of two-dimensional physics and are model systems for membranes. Densely packed molecular monolayers, absorbed from solution onto solid substrates are known as *self-assembled monolayers* (SAM).

It is common knowledge that oil spreads on water to form a thin film. Long-chain fatty acids, such as stearic acid, $CH_3(CH_2)_{16}COOH$, are typical amphiphiles (see Figure 6-28) with a hydrophilic –COOH head group and a long hydrophobic hydrocarbon tail. When a solution of such molecules in a water-immiscible solvent is placed on a water surface, the solution spreads rapidly over the available area. As the solvent evaporates, a monolayer is formed with the hydrophilic head groups immersed in the water (Figure 7-5). Reducing the area of surface available to the monolayer by a movable barrier system results in lifting the hydrophobic chains away from the surface (Figure 7-5b) until a close packing of the head groups is achieved with the long nonpolar chains directed almost vertically away from the surface (Figure 7-5c). In this stage, the surface area per molecule is similar to that occupied by the same molecule in a single crystal, e.g., 0.22 $nm^2 \cdot molecule^{-1}$ for stearic acid. The compact film can thus be considered as a two-dimensional solid.

Figure 7-5. Monolayer of an amphiphilic molecule on a water surface: a) expanded, b) partly compressed, c) close packed. From left to right, the surface available to the monolayer is reduced.

Langmuir films can be transferred from the water surface to a solid substrate by the so-called Langmuir–Blodgett (LB) technique (Figure 7-6). In the most commonly used method, the planar substrate (e.g., a glass slide) is first lowered through the monolayer into the water phase and then withdrawn. In order to maintain constant conditions during this process, the surface pressure is kept constant by a movable barrier. If the slide has a hydrophilic surface, deposition follows the sequence of events shown in Figure 7-6. As the slide is withdrawn (Figure 7-6a) the meniscus is wiped over the slide's surface and leaves behind a monolayer with the hydrophilic head groups turned towards the surface of the slide. Bonding of the monolayer to the slide is only complete after the initial liquid film between the slide and the monolayer has drained away or evaporated. After the first immersion, the surface of the slide is hydrophobic due to the deposited monolayer. Therefore, when the slide is dipped in the water phase for a second time, a tail-to-tail contact of the surface layer and the adsorbed monolayer on the slide is established (Figure 7-6b). The second withdrawal from the water phase resembles the first, except that the new monolayer is now deposited on the hydrophilic head groups of the layer already present (Figure 7-6c). Depending on the substrate, the monolayer, the subphase and the pressure, other types of deposition are possible.

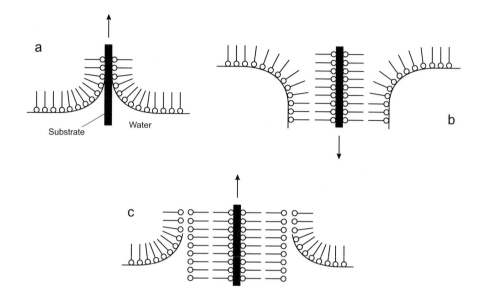

Figure 7-6. Layer deposition by the Langmuir–Blodgett technique. a) first withdrawal; b) second immersion; c) second withdrawal.

Figure 7-7 shows the structure of a biological lipid membrane (see also Section 4.3). Note the resemblance with the structure of LB films as schematically shown in Figure 7-6. LB films are therefore often used as model systems to study fundamental questions of biological membranes.

The synthesis of size-quantized microcrystals and ultrathin, particulate films of semiconductor sulfides (▶ glossary) by using surfactant monolayers has been discussed in Section 4.3.3.

An attractive alternative to transferred Langmuir monolayers are monomolecular films prepared by a self-organization process. Self-organization refers to the spontaneous assembly of molecules in a stable, well-defined structure by non-covalent interactions. SAMs form by spontaneous chemisorption and self-organization of long-chain, functionalized organic molecules when suitable substrates are immersed into a solution of the film-forming compound or treated with the vapor of such compounds. Formation of the SAM is a rather quick process. Among the best investigated SAM systems are (functionalized or non-functionalized) alkanethioles, $X(CH_2)_nSH$ (X = H or functional group), on gold surfaces. For n > 11, homogeneous 2- to 3-nm-thick layers are formed by dense packing of the parallel alkyl chains with the thiol groups interacting with the gold surface. The possibility for chemical manipulations at the functional group at the end of the chain opens a wide field for molecular engineering of surface properties.

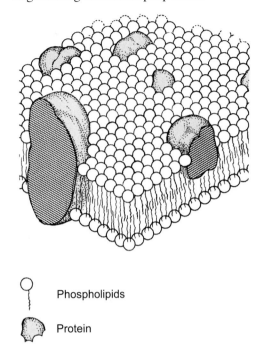

Phospholipids

Protein

Figure 7-7. The bilayer structure of a biological membrane.

An interesting new method of pattern formation with the potential of a high resolution is based on SAMs. Nanostructuring of electronic devices is becoming an increasingly important issue. Since the invention of the integrated circuit, the minimum feature size of individual structures in the circuits has halved about every five years – an effect often called the "Moore curve". Minimum feature sizes for DRAMs of 130 nm and 100 nm are predicted for the years 2004 and 2007, respectively, with the spatial definition of their interfaces of 1–10 nm.

There are two significant problems at these dimensions. The first is the different electrical, magnetic or optical properties inherent to nanosized materials. Most of the phenomena utilized in electronic structures have critical scale lengths in the nanometer domain owing to quantum effects. There will also be quantum effects in electron transport, and consequently, the classical device physics begins to break down.

The second problem concerns fabrication of the individual device and integrated circuit structures. The control of materials addition and removal to create patterns must be performed with near-atom-scale precision, and requires the accurate knowledge of the chemical behavior of and around small structures. Apart from pushing existing patterning technologies to its limits, this requires the development of new methods. Proximal probes such as a scanning tunneling microscope (STM) can provide scanned, low-energy beams which have the required spatial resolution.

As the patterns become smaller, higher-energy beams must be used for lithography (see Figure 5-11). It is probable that electron-beam, ion-beam, UV and X-ray lithography can be successfully refined for applications at least down to 10 nm. Below 100 nm, current technologies have serious resolution problems, and the high beam energy frequently causes unwanted ancillary damage.

The key requirements for a resist material in nanolithography applications are high resolution, high contrast, and compatibility with the transfer of the pattern to the device. Commercial resists are organic polymers, such as poly(methylmethacrylate) (PMMA). Polysilanes may have some potential in the future (see Section 5.4.1). The recently developed chemically amplified resists offer the advantage of a crosslinked material (▶ glossary) with a high contrast and a high resolution. In these resists, the irradiation activates an acid catalyst (negative resist) or an acid inhibitor (positive resist). Following exposure, the acid, in the acid-activated regions, catalyzes crosslinking reactions (▶ glossary) during the post-exposure annealing.

Two approaches showing how SAMs can be used for lithography ("soft lithography"), are illustrated in Figure 7-8.

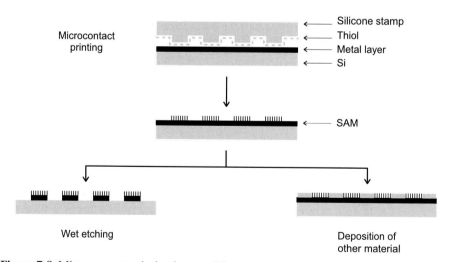

Figure 7-8. Microcontact printing in nanolithography.

In the first step, the pattern is stamped on the substrate surface with the self-assembling molecules as the "ink" (microcontact printing). SAMs are formed in the contact areas. The structured SAM can be used as etch mask for certain wet etching processes (Figure 7-8, left) or to deposit selectively other materials on the substrate area not covered by the SAM (for example, by selective CVD) (Figure 7-8, right).

7.3 Preparation of Nanoparticles and Nanocrystalline Materials

In this section, we will discuss methods specially optimized to yield nanoparticles or nanophase materials. Many of them are variations of the methods discussed in previous chapters.

The goal of all preparation methods is the production of nanoparticles that are uniform in size and shape (monodisperse systems). In principle, any method capable of producing very fine grain sized (▶ glossary) polycrystalline materials can be used. If a phase transformation is involved (e.g., gas to solid), then steps must be taken to increase the nucleation rate (▶ glossary) and decrease the growth rate during formation of the product phase to obtain nanocrystalline materials. Additionally, the agglomeration and coalescence of the particles must be prevented once they have been synthesized.

7.3.1 Formation of Nanoparticles from Vapors

The synthesis of nanoparticles via gas-phase reactions involves several stages, which are the same as discussed in Section 3.3 for the gas-to-particle conversion route of aerosol (▶ glossary) processes:

- evaporation (producing a supersaturated vapor);
- nucleation (▶ glossary);
- particle growth by coalescence and coagulation; and
- transport and collection of the particles.

Methods for the gas-phase growth of small metal clusters (up to some 100 atoms) will not be discussed here (such as supersonic beam expansion, laser vaporization, or laser photolysis of organic compounds). We will rather restrict ourselves to approaches leading to nanoparticles large enough for isolation and producing nanoparticles on a preparative scale.

Nanophase materials can be generated via aerosol processes (Section 3.3). One of the advantages of aerosol processes is that the product particle size can be controlled by adjusting the droplet size of the aerosol and the precursor concentration in the droplets.

The most extensively used technique for the synthesis of nanoparticles on a preparative scale uses thermal evaporation of solids. Therefore, the preparative aspects relevant to nanophase materials will be examplarily discussed for this method.

Gas condensation method

The most common apparatus is shown in Figure 7-9. This comprises an ultra-high vacuum chamber, equipped with a liquid nitrogen-cooled finger and scraper assembly, and an *in-situ* compaction unit to consolidate the powders collected in the chamber. The vacuum chamber is first pumped to a vacuum $<10^{-5}$ Pa, and then filled with a few hundred Pa of a high-purity inert gas, typically He.

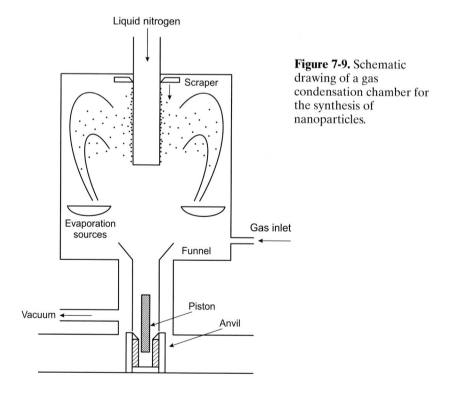

Figure 7-9. Schematic drawing of a gas condensation chamber for the synthesis of nanoparticles.

A solid compound with a high vapor pressure – mostly a metal – is evaporated by resistive heating from a high-temperature crucible (or by laser or electron beam evaporation, etc.). An evaporation temperature corresponding to a vapor pressure of about 10 Pa for the precursor is chosen. Formation of small clusters of fairly homogeneous size by homogeneous nucleation takes place near the vaporization source. Farther from the source, the clusters grow – mainly by cluster–cluster condensation – to give nanoparticles with a broader size distribution. A convective flow of the inert gas between the warm region near the vapor source and the cold surface carries the nanoparticles to the cooled finger, where they are collected. The convection may be combined with a forced flow of the inert gas.

The reason for using a relatively high pressure of an inert gas is that the frequent collisions with the gas atoms decrease the diffusion rate of the atoms away from the source region. The collisions also cool the atoms. If the diffusion is not limited sufficiently, then supersaturation is not achieved, and atoms or small clusters are deposited on the collecting surface. Thus, the main process parameters controlling the characteristics of the particles produced are the inert gas pressure, the evaporation rate, and the gas composition. The particle size can be decreased by decreasing either the gas pressure in the chamber or the rate of evaporation of the metal.

The particles are scraped from the cold finger and funneled into a piston and anvil device where they are compacted. This approach is a semi-batch process with the production of gram quantities per run, and is mainly used for preparing nanoparticles of the elements. Nanocrystalline alloys (▶ glossary) can be produced by using two or more evaporation sources. Oxide and other ceramic materials can be produced by mixing or replacing the inert gas with a reactive gas.

Metal oxide nanoparticles such as TiO_2 and Al_2O_3 have also been produced in a two-step process. First, metal nanoparticles were formed by evaporation of the metal, followed by controlled oxidation with pure oxygen or air. Magnesium oxide having ultrafine grain size (5-nm diameter) was produced by sublimation of MgO in a vacuum chamber. The method of post-reacting small metallic particles with a gaseous phase has also been applied to generate metal hydride or nitride nanoparticles, for example.

Solvated metal atom dispersion method

This is another method of growing clusters from atoms. The metal is vaporized and codeposited with a large excess of an organic solvent at liquid nitrogen temperature, $-196\,°C$ (Figure 7-10). Metal atoms or small metal clusters (dimers, trimers, etc.) are thus trapped in a solid (frozen) organic matrix and prevented from recombination to the bulk metal. Ultrafine, highly reactive particles are formed upon warming to room temperature.

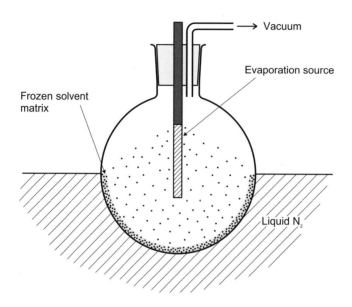

Figure 7-10. Schematic drawing of the reaction vessel for the solvated metal atom dispersion method.

This method has been used to prepare a wide array of nanomaterials, such as highly dispersed, active metals for organic syntheses or catalysis, bimetallic particles, or metal particles in polymers.

Generally, slow warming of the matrix using a large molar excess of solvent to metal or use of more polar (more strongly coordinating) solvents results in smaller particles.

The particle formation can be understood as follows:

- Upon slight warming and matrix softening, solvated atoms or small clusters become labile and begin to oligomerize.
- This aggregation is reversible within a narrow range of solvent viscosity and temperature. It competes with the reaction between the metal atoms and the solvent.
- As the clusters or particles grow, they become less mobile. Eventually, coordination of the solvent to the atoms at the cluster/particle surface becomes more favorable and stops further cluster growth.

7.3.2 Condensed-Phase Syntheses of Nanoparticles

The *sol–gel method* discussed in Section 4.5 can be performed in a way that nanosized spherical sol particles are initially formed, and these then aggregate to form the gel network ("particulate gels"). The nanoparticles may be covered by organic groups when organically substituted alkoxides are used as precursors. Unless extensive sintering occurs during post-treatment, the nanoparticulate structure is retained in the final material (see aerogels in Chapter 6.3).

Precipitation of a solid from a solution (Section 4.2) is a common technique for the synthesis of fine particles. Nanoparticles can be produced if reaction conditions and post-treatment conditions are carefully controlled. To form monodispersed particles, i.e., non-agglomerated particles with very narrow size distribution, all nuclei must form at almost the same time. Subsequent growth and processing must occur without further nucleation, agglomeration or aggregation of the particles (for definition of the terms, see Section 3.3).

Many materials containing fine particles, such as paints, inks or ferrofluids, are only useful if the suspended particles remain dispersed (non-agglomerated). For example, the desirable magnetic properties caused by single-magnetic-domain behavior cannot be realized if the ferromagnetic (▶ glossary) nanoparticles are not isolated from each other. Thus, it is very important to stabilize the particles against agglomeration.

The easy agglomeration of fine particles is caused by attractive van der Waals forces and/or forces that tend to minimize the total surface or interfacial energy of the system. Repulsive interparticle forces are therefore required to prevent agglomeration. The two commonly used methods, dispersion by electrostatic repulsion and steric stabilization were discussed in Section 4.5.1.

Metal nanoparticles

Fine metal powders have applications as electronic and magnetic materials, catalysts, explosives, and in powder metallurgy.

If a reducing agent is added to a solution of a metal salt, then small metallic particles will form. Dialysis can be used to remove remaining ions. A thickener, such as gelatin, can be added to prevent aggregation of the particles. Relatively narrow size distributions have been achieved with this technique, and the mean particle size can be controlled.

Many reducing agents can been used. One of the best investigated routes is the borohydride reduction of metal salts under an inert gas (Ar) atmosphere. It produces fine metal particles or metal boride particles, depending on the reaction conditions. For example, Co_2B is formed, when Co^{2+} is reduced by $NaBH_4$ in aqueous solution (Eq. 7-1), while elemental Co particles (Eq. 7-2) are formed in dry diglyme.

$$4\, Co^{2+} + 8\, BH_4^- + 18\, H_2O \xrightarrow{\text{water}} 2\, Co_2B + 25\, H_2 + 6\, B(OH)_3 \qquad (7\text{-}1)$$

$$Co^{2+} + 2\, BH_4^- \xrightarrow{\text{diglyme}} Co + H_2 + B_2H_6 \qquad (7\text{-}2)$$

Using this technique, nanoparticles of ferroelectric alloys (▶ glossary); such as Fe–Cu or Co–Cu, have been obtained at low temperatures.

Procedures for using the highly reducing alkali metals (mostly potassium) for preparing metal nanoparticles under mild conditions are available. Syntheses must be performed in absolutely moisture-free solvents and under anaerobic conditions. The alkali reduction method ("Rieke method") can be applied to almost any metal salt; an example is given in Eq. 7-3. The products are extremely reactive nanocrystalline metals.

$$AlCl_3 + 3\, K \xrightarrow{\text{xylene}} Al + 3\, KCl \qquad (7\text{-}3)$$

An important improvement is the addition of naphthalene to a diglyme slurry of potassium metal. Naphthalene is soluble in diglyme and reacts with potassium producing potassium naphthalide (Eq. 7-4). In this way the potassium is dissolved and a strong reducing agent is formed that reacts more rapidly with added metal salts. In a similar procedure, magnesium can be activated by adding anthracene.

$$(7\text{-}4)$$

Metal colloids (▶ glossary) have been made by thermolysis of transition metal carbonyls in an inert atmosphere. For example, thermal decomposition of $Fe(CO)_5$ (by which the carbonyl ligands are cleaved from the metal) in an organic surfactant solution produces "ferrofluids" of amorphous iron nanoparticles of about 8.5 nm diameter. The stability of the colloidal ferrofluid is achieved by the long-chain surfactant molecules.

The growth of metal particles from dispersed metal ions can also be achieved in the solid state. A solid support, usually silica, alumina, titania, etc., first has to be doped by metal ions. This can be achieved by several methods, such as:

- impregnation of the support with the solution of a metal salt;
- addition of metal salts to sol–gel systems (Section 4.5);
- coprecipitation (Section 4.2); or
- ion exchange.

When the dried, metal ion-doped materials are heated in a reducing atmosphere, metal particles are formed. They are stabilized by the surrounding solid matrix. The metal component can also be deposited on the carrier by CVD.

There is a fundamental problem if small particles are to be studied: the surface atoms are coordinatively unsaturated because they interact only with atoms inside the particle. Therefore, their electronic contribution to the behavior of the particle is different to that of the inner atoms which are fully coordinated. Surface atoms will be even more different if they are ligated, because the interaction of metal atoms with ligands leads to significant changes in the orbital energies. This is the case for any "stabilization" of the particles. The influence of the stabilizing ligand sphere is studied for well-defined giant clusters stabilized by ligands bonded to the surface atoms. The synthesis of these clusters is similar to that of colloids (▶ glossary): a metal salt is reduced in the presence of the protecting ligand. These clusters close the gaps between the "normal" metal clusters with lower nuclearity known from coordination chemistry and metal colloids.

Many of these clusters have a five-fold (icosahedral) symmetry rather than the three-fold symmetry characteristic of the close packing of atoms in bulk metals. In the latter, each metal is surrounded by 12 atoms – six neighbors in the plane of the atom, three above it and three below it. A metal atom in the center of an icosahedron is also 12-coordinate, but the arrangement of the neighbors are 1-5-5-1 (in planes) instead of 3-6-3.

An icosahedral 13-atom cluster core can rearrange to yield a close-packed arrangement. However, if rearrangement does not occur before further metal atoms condense to the core, the icosahedral core can grow by adding layers like an onion (Figure 7-11). Each face of the M_{13} ("one-shell") cluster is a M_3 triangle which can add additional atoms as in a close-packed arrangement.

Each additional shell consists of $10n^2 + 2$ atoms, i.e., there are 12, 42, 92, 162, and 252 atoms in the 1st to 5th shell. The icosahedral symmetry is retained, but the number of atoms along each edge is increasing (two in the first shell, three in the second, etc.). Adding up the atoms in the shells gives the "magic numbers" of 13, 55, 147, 309, and 561 metal atoms for one-shell to five-shell clusters. Note that each triangular face of the clusters is close-packed. However, the initial misfit around the central atom and the resulting strain remains. Therefore, the larger the core becomes, the more likely it is to rearrange to a close-packed form.

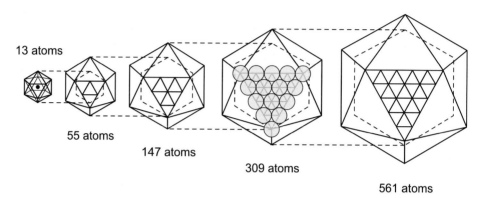

Figure 7-11. Giant cluster growth sequence.

Clusters of this type can be considered models for surface-protected metal colloid (▶ glossary) particles. The probably best investigated example is $Au_{55}(PPh_3)_{12}Cl_6$. It is a two-shell cluster with a diameter of 1.4 nm (without ligands). Other examples for giant metal cluster are $Ni_{147}(PPr_3)_{12}Cl_{24}$ and $Pd_{561}(phen)_{60}(OAc)_{180}$ (phen = o-phenanthroline). The Mössbauer spectra of $Au_{55}(PPh_3)_{12}Cl_6$ show contributions from four different types of gold atoms: 13 inner atoms forming the nucleus of the cluster, 24 uncoordinated surface atoms, 12 atoms coordinated to PPh_3 ligands, and six chlorine-bonded atoms. The signal of the Au_{13} core is close to that of bulk gold. These and other investigations show that the metallic properties of a cluster (or a particle, respectively) are mainly determined by the cluster nucleus and not by the surface atoms.

Semiconductor nanoparticles

Semiconductor nanoparticles (▶ glossary) of about 1 to 20 nm in diameter (quantum dots or Q particles) possess short-range structures that are essentially the same as the bulk semiconductors, but have optical and/or electrical properties which are dramatically different from the bulk. This has led to the development of methods for generating nanoparticles of binary compounds such as CdS, CdSe and other semiconducting compounds.

The simplest approach to stabilize colloidal semiconductor particles in the nanometer regime is to find an agent that can bind to the cluster surface and thereby prevents the uncontrolled growth into larger particles. In many cases this may be the solvent itself acting as the stabilizer. The more common approach is the use of a polymeric stabilizer that attaches to the particle surface, usually electrostatically. Methods have been developed by which protecting groups can be covalently bonded to the surface of the thus prepared semiconductor nanoparticles to stabilize them against agglomeration once the stabilizer is removed. For example, SPh groups were grafted on the surface of CdS nanoparticles by adding PhSH or PhS$^-$ as the capping agent to solutions of the growing clusters. The crystallographically characterized cluster with the formula $Cd_{32}S_{14}(SPh)_{36}(DMF)_4$ has an 82-atom tetrahedral core of the cubic-phase CdS capped by 36 phenyl groups and four solvent molecules (DMF). The diameter of the cluster core (i.e., without the organic shell) is 1.5 nm. Even bigger ligand-protected molecular clusters have been isolated in other metal chalcogenide systems, for example $Ag_{172}Se_{40}(SeBu)_{92}(dppp)_4$ (dppp = bis(di-phenylphosphino)propane).

Syntheses in confined spaces

The general idea behind this approach is to confine the growth of a particle by carrying out reactions in nanosized reactors ("ship-in-a-bottle" approach). Such reactors can be pores or channels in solids, or small liquid droplets.

Reversed micelles (water-in-oil microemulsions). When a small amount of water is added to a solution of surfactants in hydrocarbon solvents, the polar heads of the surfactant molecules gather together and thus disperse very small water droplets (Figure 7-12). These water droplets can act as nanosized reactors. For example, when Cd salts are dissolved in the water droplets and a sulfide source is added, CdS is precipitated in the water droplet. The particle size is governed by the size of the emulsion droplet, which is controlled by the water-to-surfactant ratio.

Sol–gel reactions (Section 4.5) were similarly carried out in microemulsions for the preparation of oxide nanopowders.

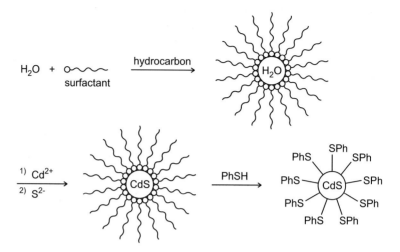

Figure 7-12. Microemulsion method of preparing surface-capped CdS nanoparticles.

Domains in ionomers. Small semiconductor particles were also prepared inside ionomers – copolymers containing ionic side-chain groups such as –COO⁻ or –SO₃⁻. The ionic groups tend to aggregate together forming domains, analogous to micelles (▶ glossary). Metal ions can be easily exchanged into these ionic domains where synthesis of the desired semiconductor (▶ glossary) clusters can be conducted. For example, stable PbS clusters of controllable size have been synthesized in a ethylene–methacrylic acid copolymer.

The cage framework of zeolites. This has been used to confine the growth of semiconductor nanoparticles. For example, Cd^{2+} ions were introduced into the zeolite cages by ion exchange. The dry Cd-exchanged zeolite was then treated with H_2S, and very small CdS clusters were formed in the zeolite framework.

Other porous solid materials can be similarly used, such as membranes, or even carbon nanotubes.

7.3.3 Mechanical Attrition

The grain (▶ glossary) diameter of powder particles can be reduced to the nanometer scale (2 – 20 nm) by high-energy ball milling. When mixtures of elemental powders are employed, the process results in the alloying of the powder particles.

The basic process of mechanical attrition is illustrated in Figure 7-13. Powders with typical particle diameters of about 50 μm are placed together with a number of hardened steel or tungsten carbide (WC)-coated balls in a sealed container which is shaken or vigorously agitated. The most effective ratio for the ball to powder masses is 5:10.

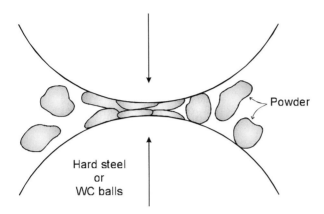

Figure 7-13. Basic process of mechanical attrition of powder particles.

During mechanical attrition the powder particles are subjected to severe mechanical deformation from collisions with the hard balls. The deformation is localized at the early stage in shear bands, with a thickness of about 1 μm extending throughout the entire particle and consisting of an array of dislocations with high density. Nanometer-sized grains (▶ glossary) are nucleated within these shear bands. For longer duration of ball milling this results in an extremely fine-grained microstructure with randomly oriented grains separated by high-angle grain boundaries.

7.4 Further Reading

A. S. Edelstein, R. C. Cammarata (Eds.) *Nanomaterials: Synthesis, Properties and Applications*, Institute of Physics Publisher, Bristol, 1996.

H. Gleiter, "Nanocrystalline materials", *Prog. Mater. Sci.* **33** (1989) 223–315.

A. Gurav, T. Kodas, T. Pluym, Y. Xiong, "Aerosol processing of materials", *Aerosol Sci. Technol.* **19** (1993) 411–452.

K. J. Klabunde, C. Mohs, "Nanoparticles and nanostructured materials", in *Chemistry of Advanced Materials – an Overview*, Wiley-VCH, New York, 1998, Eds. L. V. Interrante, M. Hampden-Smith, pp. 271–328.

L. N. Lewis, "Chemical catalysis by colloids and clusters", *Chem. Rev.* **93** (1993) 2693–2730.

H. S. Nalwa, *Handbook of Nanostructured Materials and Nanotechnology*, Academic Press, 1999.

G. Roberts (Ed.), *Langmuir–Blodgett Films*, Plenum Press, New York, 1990.

G. Schmid, "Large clusters and colloids. Metals in embryonic state", *Chem. Rev.* **92** (1992) 1709–1727.

G. Schmid (Ed.), *Clusters and Colloids – from Theory to Applications*, VCH, Weinheim, 1994.

C. Suryanarayana, "Nanocrystalline materials", *Int. Mater. Rev.* **40** (1995) 41–64.

C. Suryanarayana, C. C. Koch, "Nanostructured materials", in *Non-equilibrium Processing of Materials*, Pergamon, Amsterdam, 1999, Ed. C. Suryanarayana, pp. 313–344.

8 Glossary

Aerosol
The suspension of very fine particles of a solid or droplets of a liquid in a gaseous medium.

Alloy
A mixture of two or more metals.

Amphiphilicity
An amphiphilic molecule (amphi = either) has two distinct regions: a hydrophilic ("water-loving") head group which is easily soluble in water and a hydrophobic ("water-hating") or oleophilic ("oil-loving") tail.

Bidentate ligand
A molecule or ion with two Lewis basic (electron-rich) centers suitable for coordination to a metal center. Bidentate ligands can coordinate to only one metal (▶ chelating ligands) or bridge two metals (bridging ligands). Correspondingly, monodentate ligands have only one coordinating site, and polydendate ligands several sites.

Calcination
The heating of a solid to a high temperature, below its melting point, to create a condition of thermal decomposition or phase transition other than melting or fusing.

Ceramic yield
The portion of a material (mostly an inorganic polymer) that is transformed to ceramics upon pyrolysis. Ceramic yield = [weight of ceramic residue] × 100/ [weight of pyrolyzed material].

Chelating ligand
When a ▶ bidentate ligand is coordinated to one metal atom, a metallacycle (= metal-containing ring system) is formed. The ligand is then called a chelating ligand, the complex a chelate complex. Chelation (i.e., formation of a chelate system) is entropically favored.

Colloid
Colloids consist of a dispersed phase (mostly a solid) in a finely divided state that is uniformly distributed in a dispersion medium (or continuous phase; mostly a liquid). The size of the dispersed phase is so small that gravitational

forces are negligible and interactions are dominated by short-range forces, such as van der Waals attraction or surface charges.

Composite material
A materials system composed of a mixture or combination of two or more constituents that differ in form and (mostly) chemical composition. There are distinct phase boundaries between the two constituents.

Crosslinking
Formation of bonds between chains of macromolecules (polymers).

Dielectric
An electrical insulator material.

Dielectric loss
The energy that is converted into heat in a dielectric material when the material is subjected to a changing electric field.

Dielectric strength
The maximum electric field that a dielectric can withstand without electric breakdown.

Ductility
Ductility is the ability of a material to change shape without fracture.

Electromigration
Transport of atoms induced by flowing electrons under high current densities.

Enantioselectivity
Enantiomers are molecules of the same chemical composition that are mirror images of each other. A reaction that produces one of the two enantiomers in excess is called enantioselective.

Epitaxy
Two-dimensional structural similarity at the interface between two crystalline phases (a reactant and a product, the substrate and a crystalline film, etc.).

Eutectic mixture
An eutectic mixture is a solid solution consisting of two of more substances and having the lowest freezing point of any possible mixture of these components. The minimum freezing point for a set of components is called the eutectic point. At this point in the phase diagram, the solid phases and the liquid are in equilibrium.

Ferroelectricity
Ferroelectric materials can be polarized by applying an electric field. They are distinguished from ordinary ▶ dielectrics by their extremely large permittivities and the possibility of retaining some electric polarization after an applied voltage has been switched off. The crystal must belong to a non-centro-symmetric point group.

Ferromagnetism
Ferromagnetic materials become highly magnetized when a magnetic field is applied. After the applied field is removed, the ferromagnetic material retains much of its magnetization. Elemental iron, cobalt or nickel are ferromagnetic materials. Ferromagnetic behavior is only exhibited below a certain temperature ("Curie temperature") and is caused by the spontaneous parallel alignment of the electron spins of neighboring atoms in small domains.

Filler
A finely divided solid added to a liquid, semisolid or solid composition (e.g. paint, plastics) to modify the composition's properties.

Goldschmidt process
Johann Goldschmidt (1981–1923), a German chemist, discovered in 1898 the "thermite" reaction which produces molten iron from a mixture of iron oxide and aluminum which is ignited by a magnesium fuse.

Grain
A single crystal in a polycrystalline aggregate.

Green body
An unfired ceramic body. Green density is the density of a grren body.

Hardness
A measure of the resistance of a material to permanent deformation.

Heat capacity
The heat energy required to raise the temperature of a specific substance by 1 K at constant pressure and volume.

Liquid crystal
Certain organic materials exhibit states of intermediate order between the long-range order of solid crystalline compounds and the short-range order of ordinary liquids. These phases are called liquid crystals or mesophases since many of them can flow like a liquid but display birefringence and other properties reminiscent of the solid state.

Lyotropic liquid crystal
A liquid crystal phase which is dependent on the concentration of one component in another.

Mesogenic group
Compounds or substituents that form a mesophase (▶ liquid crystalline phase) in a certain temperature range.

Mesophase
See: liquid crystals.

Metastable solid
A solid phases that are off the thermodynamic minimum but are kinetically stable to quite high temperatures.

Metathesis
A double displacement reaction of the type
A-B + A′-B′ → A-B′ + A′-B
in which partners are interchanged.

Micelle
Micelles are spherical aggregates of ▶ amphiphilic molecules that assemble spontaneously. In water, the hydrophobic tails interact with each other and fill the interior of the micelle while the hydrophilic head groups stick in the water phase. In unpolar solvents, reversed micelles are formed, in which the hydrophilic head groups are directed towards the interior of the micelle and the hydrophobic tails into the solvent.

Monodispersity
In monodisperse materials all polymer chains or particles have the same size.

Monomodal
A distribution (e.g., size distribution) having one maximum. Correspondingly, a bimodal distribution has two maxima, and a polymodal distribution several maxima.

Non-linear optics
The optical properties of materials are normally considered to be independent of the intensity of light. In non-linear optical materials, the optical properties reversibly depend on the intensity of the incident light.

Nucleation
Formation of thermodynamically stable solid particles (nuclei) from a continuous phase (gas, liquid) by aggregation of molecular or cluster species. Heterogeneous nucleation is nucleation on foreign nuclei, dust, or surfaces. Homogeneous nucleation is nucleation in the absence of any foreign nuclei.

Ostwald ripening
A process in which larger particles grow at the expense of smaller ones. Larger, more stable particles are thus formed.

Paramagnet
A compound is paramagnetic if it contains unpaired electrons. Paramagnetic compounds or materials have a small positive magnetic susceptibility (= ratio of magnetization to applied magnetic field).

Permeability
The capability of a porous substance to allow a fluid to pass through it. A permselective material (e.g., a membrane) allows only one type of compound to pass through it.

Perovskite
A class of crystalline solids of the general formula $M^{2+}M^{4+}O_3$. The M^{2+} and oxide ions occupy the anion sites of the rock salt structure, while the M^{4+} ions are in one quarter of the octahedral sites (those only surrounded by oxide ions).

Photoresist
A light-sensitive polymer used in photolithography.

Piezoelectricity
Piezoelectric materials exhibit a linear relationship between electric and mechanical variables. They polarize under the action of an applied stress and develop electric charges on opposite crystal faces. By stretching or compressing a piezoelectric material, a voltage is generated. The reverse is also true: a voltage applied to the material causes it to become mechanically stressed.

Polycondensation
Chemical reaction in which compounds of high molecular mass are formed from monomers by cleavage of a small molecule, typically water, hydrogen, etc.

Polydispersity
The polydispersity index is a measure of the breadth of the molecular mass distribution of polymers (i.e. the size distribution of the polymer chains). It has a minimum value of unity for ▶ monodisperse systems.

Polymerization
Chemical reaction in which macromolecules (molecules of high molecular mass) are formed from monomers. In copolymerization reactions, two ore more kinds of monomers are employed.

Polymodal
A distribution (e.g., size distribution) having several maxima. Correspondingly, a bimodal distribution has two maxima, and a monomodal distribution one maximum.

Refractory material
Refractories are materials that resist the action of hot environments.

Resistivity
A measure of the difficulty of an electric current to pass through a unit volume of a material.

Rheology
The science of the flow and deformation of matter.

Semiconductor
A material whose electric conductivity is between that of a metal and an insulator. Conductivity increases with increasing temperature.

Solid solution
When one solid dissolves another solid without altering its crystal structure, these solids form a solid solution.

Spinel
A class of crystalline solids of the general formula MM'_2O_4. The M cations occupy tetrahedral, and the M' ions octahedral interstices in a cubic close packed array of oxide ions.

Step coverage
The ratio of the film thickness on the step sidewall to the film thickness on the top surface.

Strain
Change in length of a sample divided by its original length.

Stress
Force divided by the area over which the force acts. The force can be applied by tension, compression or shear.

Superconductor
A materials that conducts electricity with no resistance below a certain "critical" temperature, T_c.

Sustainable development
Sustainable development meets the needs of the present without compromising the ability of future generations to meet their own needs. Sustainability refers to the ability of a society, ecosystem, or any other ongoing system to continue functioning into the future without being forced into decline through exhaustion of key resources.

Tacticity
The regularity in which repeating units with a certain configuration are ordered in a polymer chain. In atactic polymers the side groups are randomly arranged on either side of the main chain, in isotactic polymers on only one side of the chain, and in syndiotactic polymers the groups alternate from one side of the chain to the other.

Template
A molecule or an assembly of molecules (e.g., a ▶ micelle) serving as a pattern or mould for the synthesis of another compound.

Thermochromic compounds
A thermochromic compound changes its color when a certain temperature is passed.

Thixotropy
The viscosity of the system is reversibly decreased upon action of mechanical forces (shaking, stirring, etc.). When the systems comes to rest, the original (higher) viscosity is restored.

Topotaxy
Three-dimensional structural similarity between two crystalline phases.

Triboelectricity
Charging of solids by friction.

Tribology
The study of mechanical phenomena in frictional processes, lubrication, and wear on surfaces.

Vesicle
Vesicles consists of a detergent (phospholipid) *bilayer* that encloses a small volume of a solvent (water, buffer solution, etc.). The hydrophilic heads of the ▶ amphiphilic molecules point towards the solvent both inside and outside the vesicle. The hydrophobic tails hold the vesicle together.

Vulcanization
A chemical reaction that causes▶ crosslinking of polymer chains.

Wave guide
An optical wave guide is a thin fiber along which light can propagate by total internal reflection and refraction.

Yield strength
The ▶ stress at which a specific amount of ▶ strain occurs in a tensile test.

Index